# DAIRY MICROBIOLOGY

## A Practical Approach

# DAIRY MICROBIOLOGY
## A Practical Approach

*Editor*
**Photis Papademas**
Department of Agricultural Sciences
Biotechnology and Food Science
Cyprus University of Technology
Limassol, Cyprus

CRC Press
Taylor & Francis Group
Boca Raton London New York

CRC Press is an imprint of the
Taylor & Francis Group, an **informa** business
A SCIENCE PUBLISHERS BOOK

CRC Press
Taylor & Francis Group
6000 Broken Sound Parkway NW, Suite 300
Boca Raton, FL 33487-2742



Printed on acid-free paper
Version Date: 20140916

International Standard Book Number-13: 978-1-4822-9867-3 (Hardback)



**Library of Congress Cataloging-in-Publication Data**

Dairy microbiology (Papademas)
Dairy microbiology : a practical approach / [edited by] Photis Papademas.
p. ; cm.
Includes bibliographical references and index.
ISBN 978-1-4822-9867-3 (hardcover : alk. paper)
I. Papademas, Photis, editor. II. Title.
[DNLM: 1. Dairy Products--microbiology. 2. Food Safety. 3. Probiotics. QW 85]

QR121
579.3'7--dc23 2014035401

**Visit the Taylor & Francis Web site at**
**http://www.taylorandfrancis.com**

**and the CRC Press Web site at**
**http://www.crcpress.com**

# Dedication

*The book is dedicated to*
*the late Dr. R.K. Robinson for leading the way*
*and to my son Nicholas for making it worthwhile.*

# Preface

Dairy products have been consumed in numerous forms over centuries by a vast majority of people (including the At-Risk Groups) from infancy. Besides nutrition, dairy products are responsible for pathogens and beneficial bacteria in humans; hence it is of paramount importance to have up-to-date knowledge on dairy microbiology.

The objective of this book is to provide a scientific background to dairy microbiology by re-examining the basic concepts of general food microbiology and the microbiology of raw milk while offering a practical approach to the following aspects: well-known and newfound pathogens that are of major concern to the dairy industry, e.g., *Cronobacter sakazakii* and its importance to infant formula milk or *Mycobacterium avium* subspecies *paratuberculosis* (MAP) that might be connected to chronic human diseases (Crohn's); the role of dairy starter cultures in manufacturing fermented dairy products; in developing novel functional dairy products through the incorporation of probiotic strains; insights in the field of molecular methods for microbial identification; controlling dairy pathogens owing to the compulsory application of food safety management systems (FSMS) to the dairy industry.

I hope that the book will provide dairy professionals and students alike, the latest information on this vast topic. Finally, I would like to sincerely thank the contributors both for their academic excellence and their "hands-on" experience that they openly offered for the successful completion of this project.

**Photis Papademas**
Department of Agricultural Sciences
Biotechnology and Food Science
Cyprus University of Technology
Limassol, Cyprus

# Contents

CHAPTER 1

# Basic Concepts of Food Microbiology

*Peter Raspor,*[1,*] *Sonja Smole Smožina*[2] and *Mateja Ambrožič*[2]

## INTRODUCTION

Food has never been safer than it is today, but at the same time it has also never been at greater risk. Concerns over food safety and quality as well as its production, processing and preservation have enhanced the importance of food microbiology as it encompasses the study of microorganisms which can have beneficial as well as harmful effects on the quality and safety of food. Food microbiology has begun to implement all necessary precautions to improve quality and safety of food products. The milk industry was one of the first to use food microbiology to select relevant starter cultures, to eliminate spoilage and reduce zoonoses, which were transmitted along food supply chain.

It is important to stress that milk production and processing was on the front lines of implementation of technical and regulatory novelties like novel process design, innovative equipment with most advanced pasteurization and sterilisation concepts, but also with variety of products that had more

[1] Vice Dean for Research and International Cooperation, Faculty of Health Sciences, University of Primorska, Head of the Institute for Food, Nutrition and Health, Polje 42 SI – 6310, Izola, Slovenia.
Email: peter.raspor@fvz.upr.si

[2] Chair of Microbiology, Biotechnology and Food Safety, Department of Food Science and Technology, Biotechnical Faculty, Jamnikarjeva 101, 1000 Ljubljana, Slovenia.
Emails: Sonja.Smole-Mozina@bf.uni-lj.si; mateja.ambrozic@bf.uni-lj.si

* Corresponding author

sophisticated functional properties like cheese with bacteriocins, cottage cheese with natural extracts, probiotics enhanced with prebiotics, and phytosterols enriched milk products, etc.

## Current Perspective Through Selected Past Events

Milk and milk products are indispensable components of our diet and are valuable sources of essential and nonessential nutrients. Milk is as ancient as mankind itself, because the natural role of milk is to nourish and provide immunological protection to the mammalian baby. Milk and dairy products have been recognised as a significant part of human diet as early as 8000–6000 BCE (Kervina 2006). In that period of history, ancient man learned to domesticate species of animals and to use by-products such as milk. The practice of preserving milk may be also as old as the domesticated animals and the use of fire, because milk is recognised as a highly perishable food item. Honey and milk are synonymous with welfare and there is even a dairy product mentioned in the Old Testament of the Holy Bible, which was translated by Luther as butter (Kervina 2006).

Although the collection and use of animal milk for human consumption can be traced back to the earliest societies of our history, the concept of milks sold in markets developed during the industrial revolution in the 19th century due to rapid growth of urban areas and improved and faster distribution chains and transportation facilities like railway and steamers (Eckles et al. 1936). These fast transport and cooling techniques prevented the undesirable changes in products and was the reason for the market expansion. Historically market milks have been defined as fluid milk products sold for direct human consumption. Until the end of 19th century, market milk was primary raw milk, collected from the farm, distributed fresh to the consumer and consumed fresh (Boor and Murphy 2002).

Milk and dairy products are important components of a healthy diet, but if consumed raw or unpasteurized, they can become a health hazard due to possible contamination with pathogenic bacteria. These bacteria can originate from clinically healthy animals from which milk is derived or from environmental contamination occurring during collection and storage of milk. In the first half of the 19th century scientists like William Dewees and Gail Borden (Holsinger et al. 1997) began to recommend that milk should be heated prior to consumption to increase its shelf life.

Pasteurization is a process based on the 1860s experiments of microbiologist Louis Pasteur (Westhoff 1978). He discovered that the use of temperatures (50–60°C), which did not alter the original characteristics of fluids, for a few minutes could prevent or delay microbial spoilage. Spoilage is a term used to describe the deterioration of a food's texture, colour, odour or flavour to the point where it is unappetising or unsuitable for human consumption (Arvanitoyannis et al. 2009). Microbial spoilage of food often involves the degradation of protein, carbohydrates and fats by the microorganisms and their enzymes. In milk the microorganisms that are involved in spoilage are psychrotrophic organisms that survive pasteurization or post-pasteurization contamination (Arvanitoyannis et al. 2009). With this discovery the ability to store and distribute milk far away from the farm has increased. Although milk adulteration was the primary concern at that time (Accum 1820, Sinclair 1906), the association of disease-causing organisms with raw milk was also becoming more evident. The epidemics like cholera, tuberculosis, typhoid fever, scarlet fever, anthrax, foot and mouth disease, and diphtheria (Straus 1913, Westhoff 1978), which killed thousands of people in the cities in the late 19th century, were among the human diseases that had been recognised as being transmitted through raw milk consumption, but data for thermal destruction of pathogenic microorganisms in milk were very limited, contrasting and confusing (Westhoff 1978). For example, North reported in 1925 that at least 26 reports appeared in the literature between 1883 and 1906 on the thermal death of *M. tuberculosis*, which was at that time considered to be the most heat resistant pathogen associated with milk. Reported times and temperatures ranged from 50°C to 100°C and 1 min to 6 h (North 1925).

The first application of pasteurizing heat treatments to milk may have been suggested and performed by Soxhlet, who pasteurized bottled milk for infants (Holsinger et al. 1997). The first commercial pasteurizer was made in Germany in 1882 (Sarg 1896) and pasteurization on a commercial scale quickly became common practice in Denmark and Sweden in the mid-1880s (Westhoff 1978). Initially, commercial pasteurization of milk was not readily accepted by consumers, but many companies had adopted the process in secret (Pegram 1991).

In the beginning of the 20th century, heat treatment of milk was slowly adopted. But, the extent and duration of commercial heat treatments lacked uniformity among processors' operation. To address these gaps, extensive research was conducted to establish both a standard for pasteurization and standards for pasteurization machinery (Westhoff 1978). The entire development required exchanging and sharing experiences and better control. In 1903, the 1st world dairy congress was organised in Brussels (Smith 1964). The International Dairy Federation was formed and numerous countries became members and contributed to the developments and legislations in dairy processes the world over.

Only after the World Wars in the 1920s and 1930s did milk consumption get promoted by politics as a normal and healthy food. The commercial success of the "holding" method and its acceptance as an adequate method of pasteurization resulted in the first pasteurized milk ordinance being published in the issue of Public Health Reports in 1924 in USA.

Pasteurization was defined as a heating process of not less than 61.1°C for 30 min in approved equipment, because experiments had pointed out defects in pasteurizing machinery (Westhoff 1978).

Although the holding method was the most widely used, new equipment designs like plate heat exchangers were being applied for use as high-temperature short-time (HTST) pasteurization methods. The first HTST pasteurization standards were included in the 1933 U.S. Public Health Service Milk Ordinance and Code (Westhoff 1978). Although pasteurization standards had now been established, even as late as 1938, milk-borne outbreaks were still responsible for 25% of all outbreaks due to contaminated food and water (United States Public Health Service/Food and Drugs Administration 2011).

The rickettsia responsible for Q-fever, *Coxiella burnetii* was first described by Derrick (1937). *Brucella* species are zoonotic bacteria, which are the principal causative agents of brucellosis in livestock and humans. Early investigations of Q-fever had already demonstrated that this organism was more heat resistant than *M. tuberculosis* and could be isolated from pasteurized milk that was processed according to minimum standards. In 1956 the pasteurization temperature was raised to 63°C to ensure destruction of *C. burnetii*, which was associated with Q-fever (Westhoff 1978).

Nowadays pasteurization is the principle method used to eliminate *C. burnetii* from milk, but throughout history many legal definitions of milk pasteurization appeared in the regulations with regards to time temperature combinations. The Codex Committee on Food Hygiene in 2004 defined pasteurization as a microbiocidal heat treatment aimed at reducing the number of any pathogenic microorganisms in milk and liquid milk products, if present, to a level at which they do not constitute a significant health hazard. Pasteurization conditions are designed to effectively destroy the organisms *Mycobacterium tuberculosis* and *C. burnetii*. According to validations carried out on whole milk, the minimum pasteurization conditions are those having bactericidal effects equivalent to heating every particle of the milk to 72°C for 15 seconds (continuous flow pasteurization) or 63°C for 30 minutes (batch pasteurization) (Codex Alimentarius Commission 2004).

## Starter Cultures

In the middle of the 19th century, Pasteur and others finally demonstrated that microorganisms cause food spoilage. Thousands of years before this sensational experimental proof, man had developed actions for preventing spoilage based on the use of experiences to identify suitable methods to prevent this spoilage. Raw milk in its natural state is highly perishable because it is susceptible to spoilage from naturally occurring enzymes and contaminating microorganisms. The preservation of food by fermentation is one of the oldest techniques used, because fermentation was simply the unavoidable outcome that resulted when raw food materials were left in an unpreserved state by endogenous microflora (Hutkins 2006). The oldest food processes include the baking of yeast leavened breads, brewing of beer, sake and wine and production of yoghurt and cheese.

Dairy starter cultures have a long history in the production of fermented milk products dating back several thousand years. The oldest method simply relies on the indigenous microorganisms present in the raw material. These cultures consist primarily of several members of the lactic acid bacteria. *Lactococcus lactis* as the workhorse of the dairy starter cultures or at that time called "*Bacterium lactis*" was the first bacterium to be isolated from a mixed population in pure culture and scientifically described by Joseph Lister in 1873 (Lister 1878).

The use of dairy starter cultures dates well before any knowledge of bacteria, and knowledge of which microbes are involved is quite recent and it is still not known for all dairy starter cultures used. In the beginning of 20th century the term "lactic acid bacteria" (LAB) started to be used to refer to milk souring organisms. The monograph by Orla-Jensen in 1919 formed the basis of the present classification of LAB (Orla-Jensen 1921). The criteria used by Orla-Jensen such as cellular morphology, mode of glucose fermentation, temperature ranges of growth and sugar utilization patterns are still very important for the classification of LAB. The advent of modern taxonomic tools like molecular and biological methods has increased the number of LAB genera from the four originally recognised (*Lactobacillus, Leuconostoc, Pediococcus* and *Streptococcus*) by Orla-Jensen.

Starter cultures have rapidly become an integral part of food industry. They consist of different microorganisms that are inoculated directly into food materials to overwhelm the existing flora and bring about desired changes in the finished products.

The essential roles of LAB starter cultures are (Tamime 2002):

- The production of lactic acid as a result of lactose fermentation.
- The production of volatile compounds that contribute toward the flavour.
- Possessing a proteolytic or lipolytic activity that may be desirable (maturation of cheese).
- Production of other compounds like alcohol (koumiss, kefir).
- The acidic conditions and production of bacteriocins prevents the growth of pathogens as well as many other spoilage organisms.

LAB are the backbone of the dairy starter culture industry. The world over there is a huge diversity of LAB and consequently also a diversity of fermented milk products. Tradition and experiences tell us that some fermented milk products can provide additional health benefits to the consumer. LAB are generally considered as beneficial microorganisms, some strains even as health promoting bacteria or probiotics to differentiate them from the technological cultures used to manufacture fermented milks (Fuller 1992). LAB constitute a group of gram positive bacteria united by morphological, metabolic and physiological characteristics, which produce lactic acid as one of the main fermentation products of carbohydrates (Von Wright and Axelsson 2011). The main LAB that compromise dairy starters are in the genera *Lactococcus, Streptococcus, Leuconostoc, Enterococcus*, and *Lactobacillus* (O'Sullivan 2006). Most of the time starter cultures are not made of a pure single strain but are often a mixture of many different strains and species. Dairy cultures also contain microorganisms outside the general LAB classification and these include certain bifidobacteria, brevibacteria, propionibacteria and fungi (O'Sullivan 2006). However, some genera like *Streptococcus, Lactococcus, Enterococcus* and *Carnobacterium* also contain strains that are recognised as human or animal pathogens (Von Wright and Axelsson 2011).

### *Challenges for Starter Cultures*

On one side finding the right culture for the right application often includes the simplification of a process or adding an attribute value to the consumer, who is the last link in food supply chain. Optimal speed of acidification, reaching target viscosity, developing a new flavour profile or safety of the final dairy product are just some issues concerning added values in production lines. Bacteriophages and their impact on starter cultures are still a challenge for milk starter developers. On the other side novel starter technology concept is eliminating the huge natural biodiversity of microorganisms and consequently also products, which were offered by traditional technologies. The future activities on the production side are challenged with the issue of how to preserve the natural heritage of many dairy products in this regard.

## Milk-borne Pathogens

Since ancient time milk and products derived from milk have been part of the human diet. Milk is a complex source of nutrients which includes protein, carbohydrate, lipid, vitamins and minerals. The components that make it nutritious for humans also provide ideal growth medium for producing both beneficial and detrimental microorganisms.

Milk-borne infections were relatively common before the advent of pasteurization in the late 19th century and even today, illness related to consumption of unpasteurized and also pasteurized dairy products remains a public health problem (Anaelom et al. 2010, European Food Safety Authority 2012). Thus, before the adoption of routine pasteurization, milk was an important vehicle for the transmission of a wide range of diseases. Pasteurization and improvements in veterinary medicine have seen a very large reduction in the incidence of milk-borne diseases. Historically the most serious human diseases disseminated by the consumption of contaminated raw milk are tuberculosis and brucellosis as representatives of zoonotic diseases. Zoonoses are infections and diseases that are naturally transmissible directly or indirectly, for example via contaminated foodstuffs, between animals and humans (European Food Safety Authority 2012). Tuberculosis is a serious disease of humans and animals caused by the bacterial species of the family Mycobacteriaceae, more specifically by the species in the *Mycobacterium tuberculosis* complex. Tuberculosis occurs in humans and many animal species including animals used for production of food like meat and raw milk for human consumption. *Mycobacterium bovis* is the causative agent of tuberculosis in animals used for production of food. The main transmission routes of *M. bovis* to humans are through contaminated raw milk and raw milk products or through direct contact with infected animals. Properly controlled heat treatment of milk inactivates *M. bovis* and this treatment has had a major impact on reducing the importance of milk as a vehicle of transmission of *M. bovis*, although consumption of unpasteurized milk and milk products continues to represent a hazard in relation to *M. bovis*. In the previous years, human infections due to *M. bovis* and human brucellosis cases were rare in the EU. In 2009 the EU notification rate for human tuberculosis due to *M. bovis* was 0.03 cases per 100,000 population and for brucellosis was 0.07 cases per 100,000 population (European Food Safety Authority 2012).

The concepts of "produce, sell and buy local" and also "back to nature" represent growing consumer demands for natural, unprocessed or at least minimal processed foods. These concepts have resulted also in an increased consumption of raw milk, even though numerous epidemiological studies have shown that raw milk is a food safety hazard (Oliver et al. 2009). Although pasteurization eliminates pathogens and consumption of unpasteurized dairy products is uncommon, dairy associated disease outbreaks continue to occur. During 1993–2006 in USA there were 121 dairy associated outbreaks. Of the 121 outbreaks, 54% involved cheese and 46% involved fluid milk. Of the 54% outbreaks involving cheese, 42% involved cheese made from unpasteurized milk. Of the 46% outbreaks involving fluid milk, 82% involved unpasteurized milk (Langer et al. 2012).

Raw milk contains numerous microorganisms that originate from the animal itself or from the environment, but at the same time raw milk is known to contain antimicrobial compounds like lactoperoxidase, lactoferrin, lysozyme and immunoglobulin. The role of these compounds is to confer a degree of protection on neonates of the species from which the milk was obtained and at the same time to protect the mammary gland itself from infection (Griffiths 2000).

The fermentation process is the result of the presence of microorganisms (bacteria, molds, yeast or combinations of these) and their enzymes in milk. Microorganisms present in milk can be classified into group of pathogenic and spoilage organisms, although some may play a dual role. The yeasts play a role in fermentation and maturation, as well as in spoilage of dairy products. The role of yeasts as spoilage organisms in dairy products is linked with their physiological characteristics and requirements like nutritional requirements, enzymatic activities, low water activities, etc. (Jakobsen and Narvhus 1996, Hansen and Jakobsen 2004). Nevertheless, yeasts have been used as starter cultures in specific applications in the dairy industry like kefir, koumiss and soft cheeses. Spoilage organisms are capable of hydrolysing milk components such as protein, fat and lactose in order to yield compounds suitable for their growth. Such reactions can lead to spoilage of milk, manifested as off-flavours and odours, and changes in texture and appearance. *Salmonella* spp., *E. coli, Campylobacter* spp., *Listeria monocytogenes, Staphylococcus aureus, Streptococcus* spp., *Yersinia* spp. and *C. Burnetii* are the most frequent pathogens and several publications have been published on the subject (Jayarao et al. 2006, LeJeune and Rajala-Schultz 2009, Oliver et al. 2009). Numerous other microorganisms such as lactic acid bacteria, micrococci, *Bacillus* spp., *Enterobacteriaceae, Pseudomonas* spp., etc. are also part of the initial biota of raw milk. Composition and levels found depend on the health status of the herds and the hygiene conditions under which the milk is collected (Chambers 2002). Levels and composition of the initial microflora are influenced by factors like health status of animals (e.g., contamination from the udder infections, udder diseases, faecal contamination of the udder, inhibitory substances or veterinary drugs used to treat diseased animals) and environmental sources of contamination (e.g., personnel; air borne and water borne contamination; contamination from milking and storage equipment at the farm level; contamination during transportation; and contamination at the dairy). Levels of initial microflora can be maintained if appropriate hygiene programs are implemented to control the initial contamination. Such programs include mastitis control programs; farm management and environmental provisions including feed; milking machine and milking procedure hygiene programs that include disinfectants and disinfectant rotation; and on farm cooling programs.

### *Challenges for Milk-borne Pathogens*

Milk-borne pathogens are traditionally considered as microorganisms, which can cause illness of the consumer if milk is not properly heat treated and maintained in the cold chain. Recently it has become obvious that viruses can be also transmitted by the milk supply route. Current life style is asking for natural and fresh products from all food sources and due to that habit some consumers are seeking out raw unpasteurised milk. Knowing the exact milk characteristics is a big challenge for food safety issues. Secondly, the food supply chain asks for longer and longer shelf life of food items, which represents an additional possibility for a number of psychrothrophic organisms that can survive and multiply in refrigeration supply chains. Unfortunately some psychrothrophic microorganisms are pathogenic like *Listeriamonocytogenes.* Keeping all these issues in mind, we have to pay attention to the milk supply chain for novel milk-borne pathogens, but also take care not to ignore traditional ones, which are coming back like *Mycobacterium.*

## Microbiology as Tool for Food Safety and Quality Issues

With important changes in lifestyles, demography, environmental conditions and shifts from the local to the global in the food trade, food supply is growing rapidly in size and diversity. Rising incomes and mobility have given rise to the demand to know how and where food is produced and to be assured of its safety. The increased demand for safer food has resulted in the development and introduction of new food safety standards and regulations to reach a higher level of food safety (Aruoma 2006).

Food quality and safety are important drivers for management of food processing and production systems in agribusiness as well as in the food industry. In the last decades consumers have become very critical about food quality and food safety due to several incidents of contaminated food like BSE, dioxin poisoning in feed and pathogenic strains of *E. coli* in sprouts. In order to build and maintain the trust of consumers in food quality and safety, quality management systems (QMS) and food safety management systems (FSMS) are used to control the quality and safety of foods (Raspor and Ambrožič 2012). Quality is divided into aspects of product safety, products quality and total quality, which embrace products safety and quality (Raspor and Jevšnik 2008). QMS refers to all activities that organizations use to direct, control and coordinate quality, including formulating a quality policy, setting quality objectives, quality planning, control and assurance and improvement (ISO 9000:2000 2000). A FSMS involves that part of the QMS that is specially focused on food safety. Nowadays in the food industry quality assurance systems include good manufacturing practice

(GMP), good hygiene practice (GHP), hazard analysis and critical control points (HACCP) and including private standards as well. Quality assurance is a term used for describing the control, evaluation and audit of a food processing system.

Food safety defined according to Codex Alimentarius as assurance that food will not cause harm to the consumer when it is prepared and/or eaten according to its intended use (Codex Alimentarius Commission/RCP 1 2003) is an international challenge requiring close cooperation between producers, processors, retailers and consumers on the one side and governmental and nongovernmental organisations on the other side. There are many factors affecting food safety such as global trade, socio-economic and technological development, urbanization and agricultural land use. In a food safety program we should be able to identify all hazards, analyse them, assess them, assess the likelihood of their occurrence and identify measures to their control. Hazard is a biological, chemical or physical agent in, or condition of, food with the potential to cause an adverse health effect (Codex Alimentarius Commission 2003). The aim of HACCP based systems is to ensure that food is produced safely. HACCP is a tool which identifies, evaluates, and controls hazards throughout the food chain from primary production stages to final consumption which are significant for food safety (Codex Alimentarius Commission 2003). It assesses hazards and establishes control systems that focus on prevention rather than relying mainly on end-product testing. A HACCP plan proves that the controls are in place and that the system is functioning effectively (Food and Agriculture Organization 1998).

In spite of major advances in science and technology, microbial food-borne illnesses are considered a significant public health issue (European Food Safety Authority 2012). To provide safety, stability and quality of food products detection of microbial contamination is therefore important. The spectrum of food-borne infections has been changing over time, as well established pathogens have been controlled or eliminated (Tauxe 2002) and new ones have emerged because of the changed consumer eating habits and patterns. In most cases food-borne diseases are associated with diarrhoea, vomiting, other gastrointestinal and/or extra intestinal manifestations, but secondary complications can occur. Managing the microbiological food safety risk with the goal of reducing the burden of microbial food-borne illnesses is still one of the important challenges today due to the fact that factors affecting the microbiological food safety are changing dramatically. Over time we have been witnessed to rapid and huge technological changes in food processing and production procedures and also due to the changes in methods of microbiological analysis.

Microbiological criteria are tools that can be used in assessing the safety and quality of foods. They are necessary to assist in setting critical limits in

HACCP systems, verification and validation of HACCP procedures, other hygiene control measures and in shelf life studies where storage trials and challenge tests are needed (Institute of Food and Science Technology, Professional Food Microbiology Group 1997). In microbiological risk assessments, knowledge of microbial population distribution and numbers in food forms an essential part of the information required, in conjunction with population exposure, infective dose and pathogenicity of organisms. When mathematical models are used to predict microbial growth, survival or decline, it is also necessary to appreciate the number that are inevitable and those that cause concern or signal the end of shelf life (Institute of Food and Science Technology, Professional Food Microbiology Group 1997). Microbiological criteria not only give guidance, but also set microbiological criteria defining the acceptability of the processes and for setting a limit above which a foodstuff should be considered unacceptably contaminated with the microorganisms for which the criteria are set (EC 2005).

Due to the reasons related to sampling, detection and unequal distribution of microorganisms in the food matrix, microbiological testing of finished food products done alone is not adequate to guarantee the safety of tested product. The safety of foodstuffs is mainly ensured by preventive approaches, such as good practices and application of procedures based on HACCP principles, but nevertheless microbiological criteria should form an integral part of the implementation of HACC based procedures and other hygiene measures (EC 2005).

Good practices or prerequisite programs (International Organization of Standardization 22000:2005 2005) are an essential element of food safety systems, but are often neglected as basics for successful food safety management systems. Good practices are basic conditions and activities, which are necessary to maintain a hygienic environment throughout the food chain suitable for the production, handling and provision of safe end-products and safe food for human consumption (International Organization of Standardization 22000:2005 2005).

### *Challenges for Milk Quality and Safety*

Consumers in developed countries ask for food products that are high and consistent in quality, possible to get in broad assortments throughout the year and also for competitive prices. All food producers and processors are responsible for quality products and to guarantee safety of foods. Collecting products from animals leads to transfer of microorganisms from animals to those products. Contamination of milk may also arise at any stage of milking and also during the later stages such as farm storage, transport and processing. That is why milk and milk products present a unique challenge for food safety, because numerous microorganisms including bacteria, yeasts and moulds constitute the complex ecosystem present in milk and fermented dairy products. Beside microbiological hazards chemical hazards also present hazards to the public health, because toxic chemicals present in animals' bodies can be shed into the milk. Chemical contaminants include industrially derived contaminants like dioxins, furans, PCBs and elemental heavy metals; biologically derived contaminants like mycotoxins and phytotoxins; pesticides and residues of plant agrichemicals like pesticides, and residues of animal remedies (O'Mahoney et al. 2009).

At this point climate changes should also be mentioned because it may have an impact on the occurrence of food safety hazards at various stages of any food chain, not just the milk supply chain. Climate change is widely recognized globally as the major environmental problem to be faced. There are multiple pathways through which climate related factors may impact food safety, such as changes in temperature, extreme weather events and others. Climate change may also affect aspects related to food safety systems such as agriculture, animal production and trade (Tirado et al. 2010). The experts selected feed related issues like raw materials, pasture, silage, storage and manufacturing feed and also animal health as the most critical factors that affect the occurrence of food safety hazards due to climate change (van der Spiegel et al. 2012).

An increasing number of people are consuming raw unpasteurized milk although numerous epidemiological studies have shown clearly that raw milk can be contaminated by a variety of detrimental microorganisms (Oliver et al. 2009). Reasons for increased interest in raw milk consumption is seen in enhanced nutritional qualities and consequently in health benefits. However, pasteurization reduces spoilage and eliminates pathogens. Spoilage microorganisms (sporoforms) in raw milk cannot be completely eliminated and growth can take place readily. For this reason shelf life of pasteurized milk, even when refrigerated, is limited. In many countries sale of raw milk for direct consumption is restricted or completely prohibited because of the potential risk to public health (LeJeune and Rajala-Schultz 2009) but a small amount of milk is still sold as raw or unpasteurized. In conventional distribution two main heat treatments, pasteurization and sterilisation are used for milk sold in the retail sector. The main aims of heat treatment developed for market milks are to eliminate or reduce the microbial population associated with pathogens, potential spoilage organisms and degradative enzymes (Lewis and Deeth 2009). A very good microbial quality of raw milk is also important to prevent production losses and to achieve an optimal shelf life of dairy products. To ensure a good microbial quality of milk, quality assurance systems of dairy farms are being developed and bacteriological schemes are being implemented. Many aspects of food safety and quality management are involved in the control of the microbial contamination, especially on farms as main areas of microbiological concerns and the maintenance of a cold chain along milk supply chain.

As mentioned above, dairy farmers as well as processors have to adopt practices of production and processing that satisfy the demands of consumers. Good dairy farming practice underpins the production of milk that satisfies the highest expectations of the food industry and consumers. On farm practices should also ensure that milk is produced by healthy animals under acceptable conditions for the animals and in balance with the local environment (Food and Agriculture Organization and International Dairy Federation 2004).

## Milk Supply Chain Issues

These days, production and processing are fragmented around the world. Food reaches consumers via supply chains that link many different organizations, companies and other partners, which could stretch across multiple borders. A food supply chain is a network of food related businesses involved in creation and consumption of food products, where food products move from farm to table (Selvan 2008). All actors within the food supply chain are linked by material, capital and as well as information flows, which are necessary due to the obligatory traceability system.

The global population is growing rapidly. Increasing population growth with economic development is resulting in increased demand for high quality food. Dairy products promote the good health and wellbeing of people, because they are an important part of a balanced diet contributing a majority of essential amino acids in our nutrition. The world dairy market is constantly growing and evolving (International Dairy Federation 2010). The amount of milk produced and consumed globally in 2009 reached a level of more than 703 million tonnes (International Dairy Federation 2010). By the IDF statistics in the total world milk production, cow milk represents around 84% and buffalo milk around 13%. Goat, sheep and camel milk represents a minor portion of the total world milk production. Milk production is a very important element of the whole dairy chain. The most important milk production regions are Europe and south Asia, which provide more than 50% of the global milk production. European countries alone provide 25.2% of the global milk production (International Dairy Federation 2010).

Dairy farmers' production systems are continuously being challenged to combine the responsibility of protecting human and animal health, animal welfare and the environment. This is increasingly becoming an important demand on economic husbandry. Recent studies by the United Nations Food and Agriculture Organization (Food and Agriculture Organization) have drawn attention to the considerable environmental footprint of the global livestock industry. Global climate change is also challenging the dairy industry to develop sustainable initiatives to reduce their environmental impact. Carbon footprint has become a widely used term and concept of global climate change. Carbon footprint by its definition is the total amount of greenhouse gas (GHG) emissions associated with a product, along its supply chain, and sometimes includes emissions from consumption, end of life recovery and disposal. It is usually expressed in kilograms or tonnes of carbon dioxide equivalent (Food and Agriculture Organization 2010). FAO reported (Food and Agriculture Organization 2010) that the dairy sector in 2007 emitted 1,969 million tonnes $CO_2$-eq of which 1,328 million tonnes are attributed directly to milk. Recent studies have estimated that cradle-to-farm gate emissions of milk globally contribute 4.0% of global greenhouse gas emissions such as methane, nitrous dioxide and carbon dioxide. The overall contribution of the global milk production, the processing and transportation to global GHG emissions is estimated at 2.7%. Or put in another way the global average of GHG emissions per kilogram of milk and related milk products is estimated at 2.4 kg $CO_2$ equivalent. This estimation includes emissions associated with milk production, processing and transportation of milk and milk products only. A significant source of emissions in the dairy supply chain is methane, produced from the natural digestive process of cows. Nitrous oxide and carbon dioxide are also by-products of dairy production.

Dairy industry worldwide has engaged to lower the GHG emissions in next key areas (Global Dairy Agenda for Action 2009):

- Emissions reduction (i.e., optimising animal feeding, optimising use of resources and optimising manure management).
- Energy efficiency (i.e., on farm energy use in milking and refrigeration, optimised processing, renewable energy).
- Transport efficiency (i.e., optimised milk collections, optimised transport and distribution).
- Reduction in loss of milk and milk products (i.e., shelf life improvement, reduce household waste).
- Resource efficiency (i.e., recycling of packaging, increase recovery of waste, use of packaging with lowest environmental impact).
- Management (i.e., development of a global standard for measuring monitoring and reducing GHG emissions).

Transport is highly related to almost all human activities. It stimulates economy, improves people's wellbeing and comfort, but at the same time contributes to negative environmental impacts such as pollution, depletion of ozone layers, depletion of resources, global warming, waste, noise, vibration, barriers and congestion in populated areas (Hesse and

Rodrigue 2004, Anderson et al. 2005, Russo and Comi 2010). Food reaches our plates nowadays via different routes and the present logistic challenge for food suppliers. Transportation and distribution are one of the most visible elements of our current food system. We can see trucks labelled with commercials of food giants on our roads every day. Transport of food and agricultural produce is a significant component of goods transport as a whole (Gebresenbet et al. 2011). Considering the importance of maintaining the food supply system and reducing the environmental impact of the transport system, increasing the efficiency of food distribution appears to be a major challenge not only in the dairy sector, but in the food sector in general. The transportation of foodstuffs is having a considerable environmental impact through fuel used for transportation and energy used for refrigeration. According to EUROSTAT statistics road plays a predominant role in European countries in both passengers and goods transportation. 46% of total goods required for transportation are done so by roads. Road transports consumed 26% of total final energy consumption of the 27 member countries of European Union. Statisatics showed that road transportations are responsible for emitting 93% of greenhouse gases of total European transport (Statistical Office of European Communities 2009). Road transportation of agricultural products and foodstuffs is largely a national operation, where majority of goods were transported less than 150 km, although certain foods are moved and delivered over considerable distances by road (Statistical Office of European Communities 2011).

Since the dairy industry concentrate their primary production and processing facilities we cannot ignore the fact that they substantially contribute to carbon footprint. In addition to this, food transport refrigeration is a critical link in the food chain, not only in terms of maintaining the temperature integrity, but also its impact on energy consumption and $CO_2$ emissions (Tassou et al. 2009). Refrigerated storage is one of the most widely practiced methods of preserving perishable foods like dairy products. It is important in maintaining the safety and quality of many perishable foods and it enables that food is supplied to their destination. Refrigeration stops or reduces the rate at which changes occur in food. These changes can be microbiological (i.e., growth of microorganisms), physiological (i.e., ripening, respiration), biochemical (i.e., browning reactions, lipid oxidation and pigment degradation) and/or physical (i.e., moisture loss). An efficient and effective cold chain is designed to prove the best conditions for slowing and preventing these changes for as long as it is practical (James and James 2010), but the data suggest that currently the cold-chain accounts for approximately 1% of $CO_2$ production in the world. The cold chain is vital part of modern global trade as it impacts on all food commodities.

## Current Microbial Issues

The analysis of foods for the presence of both pathogenic and spoilage microorganisms is a standard practice for ensuring food quality and safety. Modern quality management and control systems such as GMP and HACCP systems require methods and techniques that can be used on-line and give results in real-time. Especially in the food industry there is a need for more rapid methods to provide adequate information on the possible presence of pathogens in raw materials and finished products, for manufacturing process control and for the monitoring of cleaning and hygiene practices. Large scale production lines become even more sensitive for processing mistakes, which could result in unsafe milk products.

Microbiological assessment of quality and safety of foods generally and also milk traditionally relies upon the enumeration and specific detection of pathogenic and spoilage microorganisms. Conventional culturing methods are slow, material and labour intensive (de Boer and Beumer 1999). Modified versions facilitate obtaining results rapidly. Over the past two decades many improvements have been seen in both conventional and modern methods for the detection of microorganisms in food. Non-traditional testing methods relying on physical, chemical, immunological or molecular principles have been introduced to supplement or replace conventional testing methods (Deak 2009). Rapid techniques are particularly useful in modern procedures of quality management and control systems such as HACCP to ensure the microbiological quality and safety of foods in a preventive way that cannot be attained by end-product testing.

In the past decades the control of the safety of foods has been mainly carried out by product testing rather than process control. The main problem with doing end-product testing is the high number of samples to be eliminated before one can decide on the safety of the product batch. Microbiological methods are needed within a HACCP approach for risk assessment, the control of raw materials, the control of the process line and the production line environment, and for validation and verification of the HACCP program. Further development of on-line microbiology is important for rapid monitoring on HACCP plans.

### *Challenges for Fast and Reliable Microbiological Analytics*

The industry decision-makers, facing a scenario of increasingly more demanding milk hygiene standards feel the necessity of taking reliable and quick decisions about the milk and milk products (i.e., before the unloading of the tank lorries at the dairy). Microbiological control of raw milk quality at the dairy plant is therefore crucially important. This step is considered a critical control point (CCP) in the HACCP plans. Under the European

Union (EU) law, $10^5$ bacteria $ml^{-1}$ at this point of the process are the critical limits enforced (EC 2004).

The availability and application of culture-independent tools that enable a detailed investigation of the microflora and microbial biodiversity of food systems has had a major impact. However, it has become apparent that approaches that include a culturing step can lead to inaccuracies due to species present in low numbers being outcompeted in laboratory media by numerically more abundant microbial species (Hugenholtz et al. 1998) or the fact that others may simply not be amenable to cultivation in the laboratory (Head et al. 1998). For these reasons approaches to access the microbial composition of food have had to change dramatically. To address this, there has been an increased focus in recent year on the use of culture independent investigations through the direct analysis of DNA or RNA from food without a culturing step (Quigley et al. 2011). The shift from culture dependent assessment to culture independent analysis has led to a revolution in food microbiology.

Advent of biotechnology has greatly altered food testing methods. Improvements in the field of immunology, molecular biology, and automation and computer technology continue to have a positive effect on the development of faster, more sensitive and more convenient and reliable methods in food microbiology. Further, development of on-line microbiology, including ATP bioluminescence and cell counting methods, is important for rapid monitoring of cleanliness in HACCP programs. However, the important challenge is still sample preparation. More research is needed on techniques for separating microorganisms from the food matrices. The possibilities of combining different rapid methods, including immunological and DNA based methods are occurring more and more regularly in milk microbiology. Analytical technology is improving constantly and the current generation of assays potentially has the capability for near real time and online monitoring of multiple pathogens. Modern methods are based on molecular biology techniques like PCR, RFLP, DNA microarray assay, immunological techniques like ELISA, biophysical and biochemical principles with the application of biosensors like bioluminescence sensor, bio-analytical sensors utilizing enzymes, electrical impedimentry and flow cytometry (Mandal et al. 2011).

Concerns over milk safety and quality as well as production, processing and preservation have increased the importance of food microbiology. Raw milk is known to comprise a diverse microbial community. The high nutritional value of raw milk, its high water content and near neutral pH allows the growth of many, maybe even all microorganisms. These microorganisms include microflora of technological relevance like starter cultures as well. On the other hand the presence of spoilage bacteria can have significant negative effects on the quality of milk and milk products, while one of the most difficult and fundamental issues in food safety is the detection of pathogens that can have severe effects on human health. Current transformation of methods to fast and reliable microbiological analytics is a real challenge since it also needs rapid improvement of the skills and knowledge of diary/food microbiologist to be able to serve the system in a fast and reliable way.

## Conclusions

The food/milk supply chain is emerging as the most integrated system on the globe. Milk production, processing and consumption are connected to human nutrition requirements and the ability of current systems to keep quality and safety of milk and milk products during assigned shelf life periods. Novel technologies are offering efficient primary production and secondary processing with less and less impact on quality reduction, and at the same time promoting safety of final product to be consumed.

Substantial development was realised by environment engineering and process design, which was efficiently underpinned with fast and reliable analytical methods to trace physical, chemical and microbial hazards. The development in last fifteen years is slowly fusing together novelties on processing, technical and regulatory level. It is offering us new solutions in safe and quality management practices which can deliver numerous traditional and manmade novel products with the intention to serve needs of contemporary consumer who have less knowledge about food and less time for learning and handling basics nutritional needs.

## References Cited

Accum, F. 1820. A Treatise on the Adulteration of Food and Culinary Poisons. Exhibiting the Fraudulent Sophistications of Bread, Beer, Wine, Spirituous Liquors, Tea, Coffee, Cream, Confectionary, Vinegar, Mustard, Pepper, Cheese, Olive Oil, Pickles and Other Articles Employed in Domestic Economy. And Methods of Detecting Them. 2nd ed. Longman, Hurst, Rees, Orme and Brown, London. UK.

Anaelom, N.J., O.J. Ikechukwu, E.W. Sunday and U.C. Nnaemeka. 2010. Zoonotic tuberculosis: A review of epidemiology, clinical presentation, prevention and control. J. Public Health Epidemiol. 2: 118–124.

Anderson, S., J. Allen and M. Brown. 2005. Urban logistics—how can it meat policy markers' sustainability objectives? J. Transp. Geogr. 13: 71–81.

Aruoma, I.O. 2006. The impact of food regulation on the food supply chain. Toxicology 221: 119–127.

Arvanitoyannis, I.S., T.H. Varzakas and M. Koukaliaroglou-van Houwelingen. 2009. Dairy Foods. pp. 91–180. *In*: I.S. Arvanitoyannis [ed.]. HACCP and ISO 22000: Application to Foods of Animal Origin. Wiley-Blackwell. Oxford. UK.

Boor, K.J. and S.C. Murphy. 2002. Microbiology of market milks. pp. 91–122. *In*: R.K. Robinson [ed.]. Dairy Microbiology Handbook. Wiley-Interscience. New York. USA.

Chambers, J.V. 2002. The microbiology of raw milk. pp. 39–90. *In*: R.K. Robinson [ed.]. Dairy Microbiology Handbook. Wiley-Interscience, New York. USA.
Codex Alimentarius Commission. 2003. General principles of food hygiene CAC/RCP 1-1969. FAO/WHO, Rome, 31 pp.
Codex Alimentarius Commission. 2004. Code of hygienic practice for milk and milk products CAC/RCP 57-2004. FAO/WHO, Rome, 49 pp.
De Boer, E. and R.R. Beumer. 1999. Methodology for detection and typing of foodborne microorganisms. Int. J. Food Microbiol. 50: 119–130.
Deak, T. 2009. Testing methods in food microbiology. pp. 99–118. *In*: R. Laztity [ed.]. Food Quality and Standards Vol. III. Encyclopedia of Life Support Systems (EOLSS), Developed under the Auspices of the UNESCO, Eolss Publishers, Oxford. UK.
Derrick, E.H. 1937. Q fever, a new fever entity: chemical features, diagnosis and laboratory investigation. M.J. Australia 2: 281–299.
EC No 853/2004 of the European Parliament and of the Council of 29 April 2004 laying down specific hygiene rules for food of animal origin. Off. J. Eur. Communities L 226: 22–82.
EC No 2073/2005 of the European Parliament and of the Council of 15 November 2005 on microbiological criteria for foodstuffs. Off. J. Eur. Communities L 338: 1–26.
Eckles, C.H., W.B. Combs and H. Macy. 1936. Milk and Milk Products. McGraw-Hill, New York. USA.
European Food Safety Authority. 2012. The European Union Summary Report on Trends and Sources of Zoonoses, Zoonotic Agents and Food-borne Outbreaks in 2010. EFSA Journal 10: 2597: 442 pp. Available online: www.efsa.europa.eu/efsajournal.
Food and Agriculture Organization of the United Nations and International Dairy Federation. 2004. Guide to good dairy farming practice. FAO, Rome, 33 pp.
Food and Agriculture Organisation of the United Nations. 1998. Food quality and safety systems—a training manual on food hygiene and the HACCP system. Food Quality and Standards Service, Rome, pp. v–viii.
Food and Agriculture Organization of the United Nations. 2010. Greenhouse Gas Emissions from the Dairy Sector. A Life Cycle Assessment. FAO, Rome, 98 pp.
Fuller, R. 1992. Probiotics. The scientific basis. Chapman & Hall, London. UK.
Gebresenbet, G., I. Nordmark, T. Bosona and D. Ljungberg. 2011. Potential for optimised food deliveries in and around Uppsala city, Sweden. J. Transp. Geogr. 19: 1456–1464.
Griffiths, M.W. 2000. Milk and unfermented milk products. pp. 507–534. *In*: B.M. Lund, T.C. Baird-Parker and G.W. Gould [eds.]. The Microbiological Safety and Quality of Food. Aspen Publishers, Gaithersburg. USA.
Hansen, T.K. and M. Jakobsen. 2004. Yeast in the dairy industry. pp. 441–460. *In*: D.K. Arora, P.D. Bridge and D. Bhatnagar [eds.]. Fungal Biotechnology in Agricultural, Food and Environmental Applications. Marcel Dekker, New York. USA.
Head, I.M., J.R. Saunders and R.W. Pickup. 1998. Microbial evolution, diversity, and ecology: a decade of ribosomal RNA analysis of uncultivated microorganisms. Microb. Ecol. 35: 1–21.
Hesse, M. and J.P. Rodrigue. 2004. The transport geography of logistics and freight distribution. J. Transp. Geogr. 12: 171–184.
Holsinger, V.H., K.T. Rajkowski and J.R. Stabel. 1997. Milk pasteurization and safety: a brief history and update. Rev. Sci. Tech. Off. Int. Epiz. 16: 441–451.
Hugenholtz, P., B.M. Goebbel and N.R. Pace. 1998. Impact of culture independent studies on the emerging phylogenetic view of bacterial diversity. J. Bacteriol. 180: 4765–4774.
Hutkins, R.V. 2006. Microbiology and Technology of Fermented Foods. Blackwell Publishing, Ames. USA.
Institute of Food and Science Technology. Professional food microbiology group. 1997. Development and use of microbiological criteria for foods. Food Sci. Technol. Today 11: 137–176.
International Dairy Federation. 2010. The world dairy situation 2010. Bulletin of the IDF 446/2010. Brussels, 212 pp.

International Organization for Standardization. ISO 9000:2000. 2000. Quality management systems—Fundamentals and vocabulary, 2nd ed. ISO central secretariat, Geneva.

International Organization for Standardization. ISO 22000:2005. 2005. Food Safety Management Systems—Requirements for any Organization in the Food Chain. ISO Central Secretariat, Geneva, 61 pp.

Jakobsen, M. and J. Narvhus. 1996. Yeasts and their possible beneficial and negative effect on the quality of dairy products. Int. Dairy Journal 6: 755–768.

James, S.J. and C. James. 2010. The food cold-chain and climate change. Food Res. Int. 43: 1944–1956.

Jayarao, B.M., S.C. Donaldson, B.A. Straley, A.A. Sawant, N.V. Hedge and J.L. Brown. 2006. A survey of food-borne pathogens in bulk tank milk and raw milk consumption among farm families in Pennsylvania. J. Dairy Sci. 89: 2451–2458.

Kervina, F. 2006. History of milk. Litterapicta, Ljubljana. Slovenia.

Langer, A.J., T. Ayers, J. Grass, M. Lynch, F.J. Angulo and B.E Mahon. 2012. Nonpasteurized dairy products, disease outbreaks, and state laws-United States, 1993–2006. Emerg. Infect. Dis. 18: 385–391.

LeJeune, J.T. and P.J. Rajala-Schultz. 2009. Unpasterized milk: a continued public health threat. Clin. Infec. Dis. 48: 93–100.

Lewis, M.J. and H.C. Deeth. 2009. Heat treatment of milk. pp. 168–204. *In*: A.Y. Tamime [ed.]. Milk Processing and Quality Management. Blackwell Publishing, Oxford. UK.

Lister, J. 1878. On lactic fermentation and its bearing on pathology. Transactions of the Pathological Society of London 29: 425–467.

Mandal, P.K., A.K. Biswas, K. Choi and U.K. Pal. 2011. Methods for rapid detection of foodborne pathogens: an overview. Am. J. Food Technol. 6: 87–102.

North, C.E. 1925. Development of pasteurization. In commercial pasteurization, Part I. United States Public Health Bulletin 147: 20–39.

O'Mahoney, M., S. Fanning and P. Whyte. 2009. The safety of raw liquid milk. pp. 139–167. *In*: A.Y. Tamime [ed.]. Milk Processing and Quality Management. Blackwell Publishing, Oxford. UK.

O'Sullivan, D.J. 2006. Genetics of dairy starter cultures. pp. 221–244. *In*: K. Shetty, G. Paliyath, A. Pometto and R.E. Levin [eds.]. Food Biotechnology. CRC Press, Boca Raton. USA.

Oliver, S.P., K.J. Boor, S.C. Murphy and S.E. Murinda. 2009. Food safety hazards associated with consumption of raw milk. Foodborne Pathog. Dis. 6: 793–806.

Orla-Jensen, S. 1921. Dairy Bacteriology. P. Blakiston's Son & Co., Philadelphia. USA.

Pegram, T.R. 1991. Public health and progressive dairying in Illinois. Agric. Hist. 65: 36–50.

Quigley, L., O. O'Sullivan, T.P. Beresford, R.P. Ross, G.F. Fitzgerald and P.D. Cotter. 2011. Molecular approaches to analysing the microbial composition of raw milk and raw milk cheese. Int. J. Food Microbial. 150: 81–94.

Raspor, P. and M. Ambrožič. 2012. ISO 22000 Food Safety. pp. 786–816. *In*: D.W. Sun [ed.]. Handbook of Food Safety Engineering. Wiley-Blackwell. Oxford. UK.

Raspor, P. and M. Jevšnik. 2008. Good nutritional practice from producer to consumer. Crit. Rev. Food Sci. Nutr. 48: 276–292.

Russo, F. and A. Comi. 2010. A classification of city logistics measures and connected impacts. Procedia Soc. Behav. Sci. 2: 6355–6365.

Sarg, J. 1896. A New Dairy Industry. W.T. Barion and Co., Norfolk, UK.

Selvan, N.K. 2008. Food supply chain: emerging perspective. pp. 3–12. *In*: N.K. Selvan [ed.]. Supply Chain Management in Food Industry. The ICFAI University Press, Punjagutta. India.

Sinclair, U. 1906. The Jungle. Grosset and Dunlap Publishers, New York. USA.

Smith, J.A.B. 1964. The international dairy federation: what it is, and what it does. J. Soc. Dairy Technol. 17: 151–155.

Statistical Office of European Communities. 2009. Panorama of transport. Publications Office of the European Union: Luxembourg, 194 pp.

Statistical Office of European Communities. 2011. Farm to fork—a statistical journey along the EU´s food chain. Statistics in focus 27/2011. Publications Office of the European Union: Luxembourg, 12 pp.

Straus, G.L. 1913. Diseases in Milk. The Remedy Pasteurization. E.P. Dutton and Co., New York. USA.

Tamime, A.Y. 2002. Microbiology of starter cultures. pp. 261–366. *In*: R.K. Robinson [ed.]. Dairy Microbiology Handbook. Wiley-Interscience, New York. USA.

Tauxe, R.V. 2002. Emerging foodborne pathogens. Int. J. Food Microbiol. 78: 31–41.

Tassou, S.A., G. De-Lille and Y.T. Ge. 2009. Food transport refrigeration—Approaches to reduce energy consumption and environmental impacts of road transport. Appl. Therm. Eng. 29: 1467–1477.

The Global Dairy Agenda for Action—Climate Change. 2009. Available at: http://www.dairy-sustainability-initiative.org/Files/media/Declarations/Declaration_Final-Text_English-18-June-2009.pdf.

Tirado, M.C., R. Clarke, L.A. Jaykus, A. McQuatters-Gallop and J.M. Frank. 2010. Climate change and food safety: A review. Food Res. Int. 43: 1745–1765.

United States Public Health Service and Food and Drug Administration. 2011. Grade "A" pasteurized milk ordinance. Available at: http://www.fda.gov/downloads/Food/FoodSafety/ProductSpecificInformation/MilkSafety/NationalConferenceonInterstateMilkShipmentsNCIMSModelDocuments/UCM291757.pdf.

van der Spiegel, M., H.J. van der Fels-Klerx and H.J.P. Marvin. 2012. Effects of climate change on food safety hazards in the dairy production chain. Food Res. Int. 46: 201–208.

Von Wright, A. and L. Axelsson. 2011. Lactic acid bacteria: an introduction. pp. 1–16. *In*: S. Lahtinen, A.C. Ouwehand, S. Salminen and A. von Wright [eds.]. Lactic Acid Bacteria: Microbiological and Functional Aspects. CRC Press, Boca Raton. USA.

Westhoff, D.C. 1978. Heating milk for microbial destruction: A historical outline and update. J. food Prot. 41: 122–130.

CHAPTER 2

# The Microbiology of Raw Milk

*Apostolos S. Angelidis*

## INTRODUCTION

Raw milk quality is of utmost importance for the dairy industry because the principle requirement for the production of high quality dairy products is the production and use of high quality raw milk. This chapter aims to cover the main aspects of the microbiology of raw milk and is confined to the study of the milk of the most common domestic lactating animal species, i.e., cows, goats and sheep. The chapter discusses aspects pertaining mostly to bacteria (eubacteria) and to a lesser extent to fungi and viruses. Other groups of (micro) organisms that might be occasionally present in raw milk such as protozoa (e.g., *Toxoplasma*, *Cryptosporidium*), helminths or other parasites are not covered.

### Legislative Requirements for Raw Milk in the EU

Specific rules governing the production and handling of raw milk are laid down in Regulation (EC) No. 853/2004 of the European Parliament and of the Council. The same regulation contains legislative requirements pertaining to the composition [(Microbiological, Somatic Cell Count (SCC)] of raw milk. As of July 1, 2012, Regulation 853/2004 has been corrected once and amended by 12 more Commission or Council Regulations. As is the case for other European Regulations, the reader is advised to consult

Assistant Professor, Laboratory of Milk Hygiene and Technology, School of Veterinary Medicine, Faculty of Health Sciences, Aristotle University of Thessaloniki, Thessaloniki 54124, Greece.
Email: asangel@vet.auth.gr

the most recent version of Regulation 853. A way to do this is by performing a search for the Regulation in its consolidated form (i.e., including amendments and corrections) through an official web site. For instance, the reader can freely access the most recent consolidated form of Regulation 853 (and other European Union Law) by accessing the Eur Lex website (http://eur-lex.europa.eu/advanced-search-form.html?qid=1404982845826&action=update, accessed July 10, 2014) and clicking on the "Consolidated Legislation" tab.

Annex I of Regulation 853 contains the definitions of "raw milk", "milk production holding" and "processed dairy products", whereas Annex III lists specific requirements for different categories of animal-origin foods (e.g., different categories of meat and meat-products, fishery products, eggs, etc.). Section IX of Annex III is devoted to raw milk, colostrum, dairy products and colostrum-based products. According to the requirements laid down, raw milk should come from animals in good health and not displaying symptoms of infectious diseases that could be transmitted to humans through milk—including diseases of the genital tract, diarrhea, fever, or recognizable mastitis. Also the lactating animals must not have undergone illegal treatment as defined in Directive 96/23/EC. Special provisions are listed for two highly severe infectious zoonoses, brucellosis and tuberculosis and the interested reader should consult the Regulation for more detailed legislative requirements such as control plans and exceptions with competent authority authorization. The same section outlines general construction and cleaning requirements for milk production and storage premises, and milking and milk-storage equipment. The hygiene of milking is emphasized and so is the identification of animals undergoing medical treatment, for which proper milk withdrawal periods must be met. Emphasis is also given to milk cooling requirements. Hence, unless milk is going to be processed within two hr after milking or a higher temperature is deemed necessary for technological reasons for the manufacture of certain dairy products, upon milking "milk must be cooled immediately to not more than 8°C in the case of daily collection, or not more than 6°C if collection is not daily". The chill chain must be maintained throughout all steps of milk storage and transport and the temperature of milk should not exceed 10°C upon arrival at the destination establishment (e.g., milk processing plant).

Regulation 853 states that raw milk should not contain antibiotic residues above the allowable limits specified in Regulation (EEC) No. 2377/90. Regulation 853 also lists the criteria for raw milk with respect to its total mesophilic microbial content (plate count at 30°C) and SCC. Raw milk SCC is recognized as a reliable indicator of cow udder health. In addition, raw milk SCC is inversely related to cheese yield and quality. Hence, for cows' milk the SCC rolling geometric average over a two-month period, with a minimum of two samples per month, should not exceed 400,000. For microbiological counts, the rolling geometric average over a three-month period (with at least one sample per month) should not exceed 100,000 cfu/mL. The production of milk from lactating animals other than bovine (e.g., sheep, goats or buffalos) often presents practical difficulties. For instance, in south-western Europe, a considerable number of sheep and/or goat farms are located in mountainous regions, distant from milk processing facilities. The farms' herd size is often small and therefore does not produce sufficient revenues for the producers to invest on mechanization. Therefore frequent pick-up and collection of milk is not cost-effective or even plausible. In some cases access to electrical power is not for granted and automated milking systems and/or milk cooling equipment are not easy to install and/or operate. These limitations may have led European legislators to (understandably) adopt a more lenient limit with regards to the microbial content of milk from "other species", which is currently set at 1,500,000 cfu/mL. However, much better milk microbiological quality is required (i.e., a maximum of 500,000 cfu/mL) for raw milk from species other than cows, if such milk is intended to be used for the manufacture of dairy products in a fashion that does not involve a heat treatment (sanitation) step. An example where the "500,000 cfu/mL" criterion applies is the use of raw sheep milk for the manufacture of cheeses that undergo a minimum ripening time of two or more months. No SCC limits are listed in Regulation 853 for the milk of species other than bovine. This is likely the result of the fact that the SCC in small ruminants' milk (particularly in goats' milk) varies considerably, even among healthy animals and several non-infectious factors (i.e., factors other than inflammation of the mammary gland) can influence SCC. Milk secretion in small ruminants, particularly in goats, is apocrine in nature, and cytoplasmic particles, often similar in size to milk somatic cells, are normally present in their milk (Souza et al. 2012). A great body of research worldwide has been allocated on the investigation of (non-pathological) factors affecting milk SCC levels in sheep and goats, milk and such factors include, amongst others, parity, stage of lactation, or even sample handling, sample storage practices and measuring method (Paape et al. 2001, Contreras et al. 2007, Raynal-Ljutovac et al. 2007, Souza et al. 2012). Currently, there does not seem to be a consensus on establishing a unique and scientifically justifiable discriminatory SCC standard (limit) for sheep and goat milk quality because research studies on the SCC of small ruminant milk have sometimes resulted in conflicting outcomes and conclusions. Hence, at present, no legislative upper limits regarding small ruminant raw milk SCC values are in place in the EU.

As mentioned previously, the microbiological and SCC (where applicable) upper allowable limits for raw milk are expressed as geometric, rather than arithmetic means (averages). When evaluating bulk tank SCCs or microbial counts the use of geometric averages helps to decrease the variation or impact of high or low counts (often the result of sampling or measurement error) on the mean. An example illustrating the "smoothing" effect of geometric means is illustrated using the hypothetical data presented in Table 1.

**Table 1.** Arithmetic and geometric average of total mesophilic counts obtained from the analysis of ovine bulk tank milk samples (n = 5) collected randomly over a three-month period.

| Sampling no. | Microbial counts (cfu/mL) | Log microbial counts (log cfu/mL) |
|---|---|---|
| 1 | 820,000 | 5.91 |
| 2 | 950,000 | 5.67 |
| 3 | 3,900,000 | 6.54 |
| 4 | 1,200,000 | 5.85 |
| 5 | 720,000 | 5.71 |
| Arithmetic mean | 1,518,000 | 6.084 |
| Geometric mean | 1,212,895 | $10^{6.084}$ |

The geometric mean of a range of n values can be easily calculated. Probably the easiest way of calculation is with the use of spreadsheet computer packages. In Microsoft Excel, for instance, one can use the GEOMEAN function. If the user has no access to a software package, the geometric mean can be calculated with a simple calculator by multiplying all the n values and then taking the 1/n root of the product. In the above example, the "manual" calculation of the geometric mean would be as follows: $(820{,}000 \times 950{,}000 \times 3{,}900{,}000 \times 1{,}200{,}000 \times 720{,}000)^{0.2}$. Alternatively, one can rely on the definition of the geometric mean (i.e., the antilog of the mean of the log counts). Hence the user can calculate the log (to the base 10) cfu/mL value for each observation and then calculate the mean of the log counts (here equal to 6.084). The anti-log of this value ($10^{6.084}$) is equal to the geometric mean. In the hypothetical scenario provided above, it can be seen that the use of the arithmetic mean would result in an average microbial count that would exceed the 1,500,000 EU limit for raw ovine milk. The use of the geometric mean, however, helps to reduce the influence of extreme values (e.g., of the measurement taken during the 4th sampling point (i.e., 3,900,000)) resulting in a final average value that is within the EU limit.

As the dairy industry in developed countries is moving towards the production of dairy products with extended shelf-life, it is anticipated that greater demands will be placed on raw milk in terms of its quality.

## Pre-Harvest Food Safety—Raw Milk Production

Livestock species including ruminants may serve as reservoirs of critical pathogenic microorganisms and several food-borne hazards (both microbiological and chemical) can potentially originate during animal production. Many food-borne pathogens (e.g., *Salmonella* spp., *Campylobacter* spp., and Shiga-toxin producing *Escherichia coli*) can have habitats in lactating mammals (gastrointestinal tract and hide) and can therefore contaminate raw milk during milking. The contamination of animals is through ingestion of food or water contaminated with fecal material. Such pathogens further spread in the (farm) environment via animal feces and farm practices such as the use of recycled wastewater and non-treated manure as fertilizer.

The need for application of preventive food safety programs at the farm level arose as a result of the risks and consequences associated with the presence of food-borne hazards (pathogens and chemical residues) in raw milk. Food safety is an imperative issue for consumers, food manufacturers and government officials (Hoffmann et al. 2012) and it is now well recognized that food safety measures and approaches must be applied throughout the entire food chain, i.e., "from farm to table". Preventive measures and approaches applied during the pre-harvest phase of milk production aiming at minimizing food-safety risks include the prerequisite programs such as Good Agricultural Practices (GAP) and Good Hygienic Practices (GHP). In addition, the widespread and indubitable success of HACCP at the food industry (processing) level (van Schothorst and Kleiss 1994) prompted its application on the farm (on-farm HACCP-type programs). On-farm, HACCP-type programs for chemical hazards are feasible to develop and implement. A successful example is the 10-point Milk and Dairy Beef Residue Prevention Program that was originally developed by the American Veterinary Medical Association and the National Milk Producers Federation in 1993. This protocol originally included 10 "critical control points" and was designed to minimize veterinary drug residues in milk and meat. Since its first publication, the protocol has undergone some revisions and the latest revised version (2014 edition) is freely available at: http://www.nationaldairyfarm.com/sites/default/files/2014%20Residue%20Manual_WEB.pdf. In the US, the Grade "A" Pasteurized Milk Ordinance (Anonymous 2011) specifies the standards for production, handling, transportation, processing, testing and sale of milk.

It has been argued that the application of on-farm, HACCP-type programs (Cullor 1997) during the primary production of milk may not be effective or adequate to significantly improve food safety. This is mainly due to the lack of definitive critical control points that could eliminate or control identified microbiological hazards (Sperber 2005). Nonetheless, quantitative molecular methods, such as real time PCR, have been developed in the last decade which can target and quantify specific pathogens in environmental samples as well as in raw milk (Amagliani et al. 2012). Such methods are considerably faster and often more sensitive than conventional culture-dependent methods. In addition, recent applications make use of protocols in "multiplex" format, enabling the identification of more than one pathogen at the same time (Omiccioli et al. 2009). The significant shortening in the required time needed for analysis offers an advantage for highly perishable foods requiring continuous monitoring, such as raw milk. Hence, such methods may be used for more effective prevention at the pre-harvest level. Newer attempts to reduce pathogen contamination in the farm environment include the use of bacteriophages, or the addition of probiotics in the animal feed. These and additional current strategies (e.g., vaccination) to reduce food-borne pathogens in food animals have been reviewed by Oliver et al. (2009b) and practical food safety interventions during dairy production are reviewed by Ruegg (2003). A prototype application of a HACCP-type program in two parts of the production process (i.e., milk harvest and treatment of cows) has been described by Lievaart et al. 2005. In addition, the application of preventive, food-safety measures and approaches during the pre-harvest phase, as well as during milk collection, storage and transport has led to the improvement of milk quality in terms of its microbiological count (Nada et al. 2012).

## Sources of Microbial Contamination of Raw Milk

Milk is considered nature's perfect food for mammals and many benefits to human health have been associated with the consumption of high-quality milk and dairy products (Kliem and Givens 2012). Freshly milked raw milk contains active antimicrobial components (briefly discussed later in this chapter) which, depending mainly upon raw milk storage temperature, inhibit the growth of contaminating microorganisms for a short period of time. However, raw milk has a high $a_w$ value (*ca.* 0.99), is only slightly acidic (pH typically equal to 6.6–6.7) and is a medium rich in energy sources (lactose, proteins and lipids) as well as a source of a plethora of preformed building blocks for microbial metabolism and biosynthesis. Even under the currently available most optimal conditions of routine milk collection, some level of contamination of raw milk is practically inevitable. Hence, unless raw milk is properly handled and stored, contaminating microorganisms will eventually proliferate to high populations and lead to milk spoilage.

Milk is synthesized in specialized epithelial cells in the mammary gland of mammals. In cows, the udder is divided into two halves by the median suspensory ligament and each half further divided into two individual quarters, resulting in a total of four anatomically "independent" mammary glands ("quarters"). In the udder of goats and sheep there are only two mammary glands ("halves") (Nickerson and Akers 2001).

The parenchyma of the mammary gland is composed of glandular tissue (alveoli, which are the milk producing structures), ducts and connective tissue. Milk biosynthesis takes place in the alveolar epithelial cells, using precursors supplied by the animal's bloodstream. The (biosynthesized) milk components are released into the lumen of the alveolus. Alveoli form clusters (lobules) and the synthesized milk is drained via small ducts, which eventually converge into larger intralobular ducts. The parenchymal tissue of the mammary gland essentially consists of lobes formed by a cluster of lobules and drained by yet larger ducts. The large duct emerging from the lobe is called interlobar duct and milk is eventually collected (drained) into the mammary gland cistern (Nickerson and Akers 2001). During milking, the milk passes through the teats (i.e., through the teat cistern) and exits the udder through the teat canal. The smooth muscles surrounding the teat canal, aided by the presence of keratin in the canal lumen, help to maintain the canal closed between milkings and therefore prevent bacterial invasion from the external environment (teat skin, floor, etc.) into the teat cistern.

In healthy animals the synthesized and secreted milk is sterile. In lactating animals affected by intra-mammary infections (IMI), i.e., mastitis, the causative microorganisms will be present, in the milk usually in high numbers, depending on the severity of the infection. It should be noted, however, that occasionally in mild or less severe cases of mastitis the enumeration (or rarely even the detection) of pathogens in milk is not guaranteed. Occasionally, intermittent secretion can occur for some IMI pathogens. Even in healthy lactating animals, unless milk is drawn aseptically from the udder following a proper aseptic collection protocol, freshly milked raw milk will contain contaminating microorganisms at low to moderate levels. These initial microbial concentrations can vary considerably among lactating animals or among animals of different farms, but they are typically in the range of several hundreds to few thousands of cfu/mL (Morse et al. 1968). There are many possible sources of microbiological contamination of raw milk. Some of the obvious sources of milk contamination at the immediate environment of the milking parlor include the animal hide and the teat/udder skin surfaces that are contaminated with organisms from the bedding material or dirt from the ground, the air (dust), the milker's hands, the milking equipment, the animal feed and the water supply (Rendos et al. 1975, Hogan et al. 1989, Desmasures et al. 1997b, Murphy and Boor 2000, Zdanowicz et al. 2004). A recent study evaluated the diversity of the microbiota on cow teat skin using culture-dependent methods and direct molecular approaches (Verdier-Metz et al. 2012). The study pointed out a great diversity in the bacterial communities that may reside on teat skin. In fact, the authors reported that the microbial clones corresponded to 66 identified species, mainly belonging to bacterial genera commonly found in raw milk (*Enterococcus*, *Pediococcus*, *Enterobacter*, *Pantoea*, *Aerococcus* and *Staphylococcus*), as well as to several unidentified species.

The teat canal, i.e., the opening through which milk is secreted, can be also colonized by microorganisms. For instance, the microbiological investigation of different anatomical sites of nulliparous heifers revealed the presence of different coagulase-negative *Staphylococcus* spp. in samples taken aseptically from the teat canals (White et al. 1989). Further studies have also confirmed the presence of *Staphylococcus* spp. as well as the presence of additional species of bacteria such as *Streptococcus uberis* and *Escherichia coli* in the teat canals of lactating cows that were free of clinical mastitis (Paduch et al. 2012). A detailed study using a culture-independent approach (PCR-amplified 16S rRNA gene sequencing) revealed a broader diversity in the bacterial species residing in the teat canals of lactating dairy cattle. Among the species identified were bacteria commonly associated with soil, water and the gastrointestinal tract of cows, as well as uncultured bacteria (Gill et al. 2006). The teat canal is the usual port of entry for infectious pathogens responsible for IMIs in lactating ruminants. A study in dairy cattle revealed a positive association between teat end hyperkeratosis (typically induced by machine milking) and teat canal microbial load of environmental pathogenic bacteria. *E. coli* bacterial counts and the presence of this pathogen in teat canals were associated with the hyperkeratosis scores of teat ends (Paduch et al. 2012). Less obvious reservoirs of microbial contamination of raw milk are biofilms formed in the interior of rubber or stainless steel equipment used for milk collection, storage and transport. Biofilms are complex microbial communities residing in a matrix or organic matter (proteins, polysaccharides and DNA) adhered to solid surfaces (Marchand et al. 2012).

Thorough investigations have been conducted involving the microbiological examination of multiple environmental samples from sites within dairy farms including air samples, samples from animal sites and raw milk. These studies have shown that the diversity and numbers of different microbial groups in raw milks can be influenced or at least correlated with several factors including herd management practices, sampling stage (farm vs. dairy bulk tank) farm hygiene, season and milking practices. Most of the bacterial and fungal species found in milk are also recovered from the stable and milking parlor environments. However, some bacterial species were isolated only from raw milk and their origin remains unknown (Vacheyrou et al. 2011). Less intense hygienic milking practices seem to preserve a wider bacterial diversity in raw milk (Verdier-Metz et al. 2009), whereas differences in the microbiological composition can be observed between milk sampled from farm tanks (more Gram-positive isolates) and milk sampled from the dairy bulk tanks after being stored at low temperatures (more Gram-negative isolates) (Fricker et al. 2011). These and other extensive studies (Callon et al. 2007, Fricker et al. 2011, Tormo et al. 2011, Mallet et al. 2012) have also highlighted the great microbial diversity in raw milk and have led to the identification of bacterial species that had not been previously associated with raw milk. Finally, these and other studies have highlighted the necessity of combining culture-dependent and culture independent methods for a more accurate representation of the raw milk microbiota.

## Enumeration of the Raw Milk Microbial Flora

Accurate and efficient methods for quantifying the microbial load of incoming milk are of great importance to the milk industry and milk producers. Total viable counts are used as indicators of on-farm hygiene practices, milk quality and udder health (Hutchison et al. 2005). The microbial flora in raw milk can be measured, or at least closely estimated by the standard plate count (SPC) method. Despite some limitations (mentioned below), the SPC method is accepted as a standard method by international organizations such as the American Public Health Association (APHA), the International Dairy Federation (IDF) and the International Standardization Organization (ISO). The sampling, storage, transport and mixing of the milk sample(s) to be used for SPC analyses as well as the conduct of the SPC method are very critical in obtaining meaningful counts and internationally accepted protocols should be strictly followed (International Standardization Organization 2003, 2008, 2010). The SPC method has certain advantages. It does not require sophisticated equipment, costly consumables or highly educated personnel. However, the personnel should be careful, meticulous and well-trained. Raw milk is subjected to successive 10-fold serial dilutions with the use of appropriate sterile diluents (e.g., quarter-strength Ringer's solution, 0.1% peptone water). The diluents used for the dilution process must provide adequate osmotic strength to prevent bacterial lysis. The ten-fold serial dilutions of milk are then pour-plated into standard methods agar (SMA) and incubated aerobically at 30°C for 72 hr. SMA is non selective. Plates containing 10–300 (IDF) or 25–250 (APHA) colonies are used to calculate the SPC, i.e., the total number of colony-forming units of bacteria, yeasts and molds per mL of milk. The cfu/mL value is thus based on the number of microorganisms that can grow and produce visible colonies in the agar plate after the fore mentioned time, temperature and atmospheric conditions of incubation. The SPC method is also used routinely in dairy- and other food microbiology teaching laboratories, as it is very practical for conveying the necessary scientific principles and rationale to the undergraduate dairy science/ microbiology student. On the other hand, the SPC technique is very tedious both in terms of conduct as well as with respect to colony counting. This method is also prone to certain limitations. Bacterial cells in raw milk (and other foods) often form clumps; each such clump will give rise to a single colony after incubation of inoculated plates. In addition, in cases where two or more bacterial cells lie in close proximity in the agar upon solidification, they will give rise to a single colony after incubation. Injured cells may not form (visible) colonies during the time frame of incubation whereas, depending on their spatial distribution in the agar milieu (e.g., towards the surface as opposed to towards the bottom of the petri dish), some of the strictly anaerobic bacteria present in raw milk will most likely not grow. Hence, viable but non-culturable microorganisms are not counted. For all the aforementioned reasons, the SPC method might underestimate the total number of viable microorganisms present in raw milk. Most importantly, the method is impractical for the dairy industry because of the prohibitive length of time necessary to obtain results. Additional colony count methods that are more suitable for routine testing (i.e., routine methods) have been developed and used, such as the plate loop method (Thompson 1960) and the spiral plate method (Gilchrist et al. 1973).

An alternative approach to determine the bacterial counts of raw milk is the total microscopic count method. This method relies on counting stained bacteria on a special glass slide on which a known volume of milk is applied (typically 0.01 mL), following sample treatments to fix the sample. Several optical fields are counted using bright field illumination (Hill 1991). The method is relatively fast, yet tedious to the operator, particularly when a high number of samples must be examined. Another limitation is the inability to discriminate between living and dead bacteria, i.e., the method estimates total bacterial counts as opposed to total viable counts.

Additional available direct or indirect methods for assessing the bacteriological content of raw milk are available and their advantages and disadvantages are described in comprehensive reviews (O'Toole 1983, International Dairy Federation 1991a, Vasavada 1993, Suhren and Reichmuth 2000). Today, however, most milk processors rely on direct, automated, rapid (*ca.* 10 min), and high-throughput (up to 200 milk samples per hr) enumeration methods for determining total bacterial counts in incoming raw milk. These automated bacterial cell counters operate by degradation and separation (via centrifugation) of milk somatic cells and milk particles and constituents that are similar in size to bacterial cells, followed by fluorescent staining and counting of individual bacterial cells using flow cytometry. The results of automated analyzers are expressed as IBC/mL (Individual Bacterial Counts per mL), rather than CFU/mL. IBC estimates can be converted to CFU values via specific algorithms obtained through correlation analyses between the two parameters (Cassoli et al. 2007). Several parameters of automated flow-cytometry based bacterial counters have been evaluated (Lachowsky et al. 1997, Bolzoni et al. 2000, 2001). Manufacturers of automated, high-throughput milk analyzers continuously come up with more advanced and more reliable models. Thus, the interested user should consult the most updated relevant technical/

performance sheets. In recent years, automatic flow-cytometry based milk analyzers have been evaluated for their reliability to enumerate bacteria in ewes' (Tomáška et al. 2006) and goats' milk (Sierra et al. 2009, Ramsahoi et al. 2011). Finally, it should be emphasized that although the determination of the total viable counts in raw milk provides an estimate of overall milk quality for regulatory purposes, it does not provide any indication regarding the presence, types or levels of pathogenic microorganisms in raw milk.

## Microbial Groups Associated with Raw Milk

A great variety of bacterial and fungal species have been isolated from raw milk worldwide. As mentioned in the previous section, these contaminating organisms can originate from the lactating animal, the housing/milking environment, the animal feed or even humans. Both non-pathogenic and pathogenic microorganisms can therefore be present in freshly milked raw milk (Desmasures et al. 1997a, Uraz and Çitak 1998, Hassan and Frank 2011, Vacheyrou et al. 2011, Amagliani et al. 2012, Delavenne et al. 2012, Hill et al. 2012, Jackson et al. 2012).

Although total viable counts and SCC in raw milk are useful indicators of milk quality (udder health and milking hygiene), some milk processors perform supplementary microbiological tests on incoming raw milk as indicators of milk production conditions. Such tests include the laboratory pasteurization count (LPC) also called the thermoduric count, the coliform count (CC) and the preliminary incubation count (PIC). The output of these testing procedures can help identify and eliminate, or at least minimize sources of raw milk contamination. The LPC estimates the number of thermoduric bacteria left in milk after simulating batch pasteurization (62.8°C for 30 min). LPC is used mainly as an indicator of milking equipment sanitation and proper maintenance. The CC is obtained by plating milk on selective media that allow the growth of coliform bacteria. In raw milk from farm animals free from coliform mastitis, coliform counts have been used as indicators of the degree of fecal contamination of raw milk. However, the identification of coliform bacteria of non-fecal origin (environmental coliforms) has questioned the usefulness of the CC as an indicator of direct fecal contamination of raw milk. The PIC is obtained after holding the milk at 12.8°C for 18 h prior to plating. This incubation period at a relatively low temperature allows the multiplication of bacterial groups that can proliferate at low temperatures. The comparison of the PIC with the SPC of the unincubated sample indicates the level of contamination with psychrotrophic bacteria. Elevated psychrotrophic counts in raw milk are associated with improper cleaning or sanitizing procedures and/or poorly cleaned refrigerated bulk tanks. Although no regulatory standards exist for these supplementary microbiological tests, the procedures used for the analyses must be standardized to ensure accuracy of the results (Murphy and Boor 2000, Davidson et al. 2004). Some studies have further looked into possible associations or correlations both among different milk quality indicators, as well as among milk quality indicators and farm management practices (Boor et al. 1998, Pantoja et al. 2009, Elmoslemany et al. 2009, 2010). In general, these different milk quality indicators are not strongly correlated. A detailed study by Martin et al. (2011) aimed at evaluating the ability of currently applied raw-milk microbiological tests to predict the quality of commercially pasteurized fluid milk (2% fat). Raw milk samples taken from silo tanks just prior to pasteurization were examined using a variety of tests commonly applied to raw milk, including SCC, SPC, psychrotrophic bacterial count, ropy milk test, CC, PIC, LPC, and spore pasteurization count. Raw milk was pasteurized using four different time-temperature combinations, ranging from 76.7°C for 25 s up to 80.3°C for 33s. Pasteurized milk samples were subjected to microbiological (SCC, CC) and sensory evaluation analyses at several time points post-pasteurization up to the end of shelf-life (21 days). The authors reported poor correlations (typically $R^2 < 0.45$) between the results from the raw-milk tests and the results from tests used to evaluate pasteurized milk quality, suggesting the need for new tests that measure the specific biological barriers that limit the shelf-life and quality of pasteurized milk.

In scientific publications pertaining to dairy science and technology, the microorganisms present in raw milk are usually classified in different groups/types bearing distinct implications to different aspects of milk and dairy products' hygiene, processing or quality. These groups of microorganisms are discussed in the following sections.

### *Bacteria*

There are hundreds of different bacterial genera known to date and likely thousands of different species. Different bacterial genera can differ in their structural or physiological properties. For instance, most bacterial genera can be classified into Gram-positive or Gram-negative based on structural differences in their bacterial cell envelopes. In addition, differences in their physiological responses to oxygen, osmotic pressure, acidity or temperature as well as in their biosynthetic and metabolic activities are further used to classify bacteria into groups of special interest in food microbiology. Research in bacteriology over the last three decades or so has revealed that bacteria are quite versatile and adaptive organisms towards various forms of environmental stress (cold, osmotic, acid or heat stress). These adaptation mechanisms can be quite elaborate and include both physiological and genetic responses.

### *Gram-negative bacteria*

Gram-negative bacteria of concern in raw milk include genera, species or serotypes that are pathogenic to humans, as well as genera or species that negatively affect the quality of milk and dairy products. Some of the genera of Gram-negative bacteria that have been isolated with various frequencies from raw milk include *Acinetobacter, Aeromonas, Campylobacter, Citrobacter, Enterobacter, Escherichia, Flavobacterium, Klebsiella, Moraxella, Pseudomonas, Salmonella, Serratia, Yersinia* and *Xanthobacter* (Jayarao and Wang 1999, Jackson et al. 2012). Pathogenic Gram-negative bacteria are discussed in a later section. Gram-negative bacteria associated with milk spoilage can either belong to the group of "coliforms" or consist of non-coliform bacteria. Coliforms are defined as Gram-negative bacteria that can ferment lactose with the production of gas within 48 h of incubation at 32 or 35°C (Davidson et al. 2004). The four principal genera of coliform bacteria are *Escherichia, Enterobacter, Citrobacter* and *Klebsiella*. The non-coliform, Gram-negative group consists of a large, heterogeneous group of bacterial genera that include *Acinetobacter, Aeromonas, Flavobacterium, Moraxella,* and *Pseudomonas*. Among these, *Pseudomonas* spp. have been studied extensively because they are notorious spoilage agents of milk and dairy products. Jayarao and Wang (1999) tested raw bulk tank milk from 131 dairy producers in the US for the presence and levels of coliforms and other non-coliform Gram-negative bacteria. The results of their study suggested that the populations of both coliforms and non-coliforms vary considerably and include a wide variety of Gram-negative species, even species that had not been associated with raw milk at the time. Coliforms were detected in *ca.* 62% of the samples with populations ranging from 0 to 4.7 $\log_{10}$ cfu/mL, with an average of 3.4 $\log_{10}$ cfu/mL. Non-coliforms were present in *ca.* 76% of the samples with populations ranging between 0 and 6.2 $\log_{10}$ cfu/mL (average 4.8 $\log_{10}$ cfu/mL). *Pseudomonas* was the most prevalent genus, as it accounted for *ca.* 50% of the total isolates and for *ca.* 74% of the non-coliforms.

### *Spore-forming bacteria*

Spore-forming bacteria are a group of microorganisms that exhibit variable phenotypic characteristics with respect to oxygen requirements and temperature growth range and optima. Two main genera of spore-forming bacteria are of interest to dairy microbiologists, i.e., the Gram-positive bacteria belonging to the genera *Bacillus* (aerobic or facultative anaerobic) and *Clostridium* (strictly anaerobic). These bacteria form endospores, thick-walled structures that are released from vegetative cells upon cell lysis. Endospores contain less moisture than the corresponding vegetative cells and are significantly more resistant to disinfectants, heat or other environmental stresses compared to their vegetative counterparts because of their relatively dehydrated state and distinct structural and physiological properties (Leggett et al. 2012).

Spore-forming bacteria are ubiquitous in the environment (soil, animal feces, silage, and bedding materials) and as such, contamination of raw foods, including raw milk is not an infrequent event (Postollec et al. 2012). The aerobic spore-forming bacteria belonging to *Bacillus* spp. and related genera have been studied extensively by dairy microbiologists as they are of great importance both in terms of quality (spoilage) and safety (De Jonghe et al. 2010). Members of the *B. cereus* group (Bartoszewicz et al. 2008) as well as those of the *B. subtilis* group and *B. licheniformis* are commonly present in raw milk. Other aerobic sporeformers isolated from raw milk belong to the genera *Paenibacillus*, *Oceanobacillus*, *Brevibacillus*, *Lysinibacillus*, *Ureibacillus*, *Ornithinibacillus* and *Sporosarcina* (De Jonghe et al. 2008, Coorevits et al. 2008). Spore counts in raw milk are generally low (usually up to *ca.* $10^3$/mL) (Lukasova et al. 2001, te Giffel et al. 2002, Coorevits et al. 2008) and can be influenced by environmental, housing and feeding factors within dairy farms. The type of bedding material, the degree of contamination of the teats with soil and the milking equipment (Christiansson et al. 1999, Magnusson et al. 2007) or persistence strategies of spore-forming bacteria, such as the ability to adhere to stainless steel and form biofilms (Shaheen et al. 2010) can influence aerobic spore counts in milk.

Most of the endospores of spore-forming bacteria can survive milk pasteurization, and, upon certain conditions, they can germinate and grow during product fermentation or storage. In addition, some species such as *B. sporothermodurans* produce highly heat-resistant endospores, which can survive even UHT processing of milk (Pettersson et al. 1996, Klijn et al. 1997, Scheldeman et al. 2006). Heat-resistant aerobic spore-forming bacteria have been isolated from various sites within dairy farms (animal feed and milking equipment) including raw milk (Scheldeman et al. 2005).

Some aerobic spore-formers are known to cause food spoilage via the production of extracellular enzymes (proteases, lipases) and/or food poisoning though the production of toxins (food intoxications). *B. cereus* is probably the best known example, a frequent spoilage bacterium in (dairy) foods and the causative agent of two food poisoning syndromes (one of the emetic and one of the diarrheal type). *Bacillus* spp. present in raw milk other than *B. cereus* (Coorevits et al. 2008) have shown to possess spoilage potential or being capable of toxin production (McKillip 2000, De Jonghe et al. 2010).

*Paenibacillus* is a genus of facultative anaerobic, endospore-forming bacteria, originally included in the genus *Bacillus* and then reclassified as a separate genus (Ash et al. 1993). *Paenibacillus* spp. are now recognized as significant psychrotolerant spoilage organisms of pasteurized milk (Huck et al. 2007). In an extended pasteurized milk survey in the US, many genera of Gram-negative and Gram-positive bacteria (both *Bacillus* spp. and *Paenibacillus* spp.) were isolated from milk from all geographical regions sampled. Regarding the Gram-positive spore-formers, however, the authors recorded a shift in the predominant population of endospore-forming spoilage bacteria from *Bacillus* spp. to *Paenibacillus* spp. over the products' shelf-life. Hence *Paenibacillus* spp. were more frequently isolated towards the end of the shelf-life of high-temperature, short-time (HTST) pasteurized milk (i.e., beyond 14 d of storage at 6°C). The authors suggested that the low rate of isolation of *Paenibacillus* spp. during the early post-pasteurization days indicate that these bacteria may be present in low numbers in raw milk; they are nonetheless capable of growing at low temperatures to levels that limit the shelf-life of HTST pasteurized milk (Ranieri and Boor 2009). A great diversity of *Bacillus* spp. and *Paenibacillus* spp. exist in the dairy farm environment and can therefore contaminate raw milk. Aerobic endospore-forming bacteria are also isolated throughout the fluid milk processing line, i.e., from raw milk through packaged products (Scheldeman et al. 2004, Huck et al. 2007, 2008, Ivy et al. 2012).

The presence of thermophilic bacilli in dairy products undergoing intense heat treatment (such as milk powder) is usually the result of selection by the conditions (high temperatures) employed during manufacture. These bacteria can grow in sections of dairy plants where temperatures reach 40–65°C and readily form biofilms. Obligate thermophiles grow only at elevated temperatures (*ca.* 40–68°C) and therefore have limited potential to lead to spoilage at the usual (room or refrigeration) storage temperature of dairy products. *Anoxybacillus flavithermus* and *Geobacillus* spp. are examples of obligate thermophilic bacilli (Burgess et al. 2010). The facultative thermophilic bacilli, on the other hand, are members of the genus *Bacillus* (e.g., *B. licheniformis*, *B. coagulans*) and, depending on the strain, are capable of growing at both mesophilic and thermophilic temperatures. These facultative thermophilic bacilli can cause spoilage of pasteurized milk, cream or UHT milk. A comprehensive review on thermophilic bacilli and their importance in dairy processing, including their food spoilage potential, enumeration and identification methods has been published by Burgess et al. (2010).

*Clostridium* spp. are anaerobic spore-formers and, similar to aerobic spore-forming bacteria, they are widespread in the (dairy) environment. Clostridia have been isolated from several sources in dairy farms such as soil, silage, forage, hay and raw milk, with silage identified as a major source of raw-milk contamination. Frequent raw-milk clostridial isolates include *C. disporicum*, *C. tyrobutyricum* and *C. sporogenes* (Julien et al. 2008, Cremonesi et al. 2012). Garde et al. (2011) detected lactate-fermenting clostridial spores in 97% of the raw ovine milk samples examined. The Most Probable Number (MPN) counts ranged from 0.36 (detection limit of the MPN method) to 240 spores/mL with most of the milk samples having a spore count between 1 and 10 spores/mL. Spores of clostridia that germinate and grow fermentatively (converting lactic acid to butyric acid, carbon dioxide and hydrogen) towards the later stages of cheese maturation are the causative agents of the "late blowing" defect in cheeses (butyric acid fermentation) with long ripening times (Cocolin et al. 2004). *C. tyrobutyricum* is considered the primary cause of "late blowing" although additional species of clostridia such as *C. sporogenes* may play secondary roles in cheese defects (Klijn et al. 1995, Le Bourhis et al. 2007, Garde et al. 2011).

*Clostridium* spp. are also important in dairy microbiology as they are involved in food poisoning episodes. *C. perfringens* food poisoning is usually linked with spore germination and proliferation of vegetative cells after temperature abuse of heat-treated foods. Other pathogenic clostridia include *C. botulinum*, the causative agent of food-borne botulism, *C. baratii* and *C. butyricum*. *C. butyricum* has been implicated in outbreaks of food-borne, type-E botulism (Meng et al. 1997, Peck 2009). It has been reported that clostridial proliferation in freshly produced raw milk is improbable due to its highly positive Eh value (Goudkov and Sharpe 1965). Nonetheless, the botulinum neurotoxin is the most toxic substance known to date and the (raw) milk supply has been considered as a likely target for terrorist attacks in terms of deliberate contamination with botulinum toxin. *C. botulinum* neurotoxins type A and B and their corresponding complexes are inactivated by at least 99.99% and 95%, respectively during raw milk pasteurization at 72°C for 15 s (Weingart et al. 2010).

### *Psychrotrophic bacteria*

As mentioned previously, upon withdrawal from the mammary gland, raw milk contains microorganisms at low concentrations, even under ideal milking conditions. If raw milk is to be collected by the dairy industry for further processing, it needs to be cooled and maintained at low temperatures in order to prevent microbial growth. Therefore, refrigeration is applied during the time interval between milking at the farm and processing at the dairy plant as a means to keep microbial counts to low levels and prevent milk microbiological deterioration. Depending on the circumstances, this time period can vary from less than 24 hr to up to 2 or more days given that: a) raw-milk pick-up from the farms is not always done on a daily basis, particularly for small-sized cow farms or small-ruminant farms, and b) upon arrival to the dairy plant, raw milk is stored in silos for several hours (e.g., for quality control checks and in cases of late-evening deliveries) before processing.

The rates of enzymatic reactions are reduced at low temperatures and, in general, the growth of mesophilic microorganisms is significantly retarded or even inhibited at low temperatures (e.g., refrigeration). The principle physiological mechanisms responsible for microbial growth retardation/ inhibition at low temperatures involve cold-induced physicochemical changes in the cell membrane lipid bilayer, protein misfolding and ribosome instability. The refrigerated storage of raw milk, however, selects for microorganisms that are capable of acclimation and proliferation at low temperatures. Such microorganisms possess the genetic background and concomitant physiological mechanisms enabling adaptation and proliferation under chill stress (Jones and Inouye 1994, Graumann and Marahiel 1996, Berry and Foegeding 1997, Beales 2004). Mesophilic microorganisms that are also capable of proliferation at refrigeration temperatures (0–7°C) are called psychrotrophic. Psychrotrophic organisms are characterized by temperature growth optima and maxima in the mesophilic temperature range, yet they are able to withstand and proliferate, albeit slower, at refrigeration temperatures. Therefore psychrotrophic microorganisms possess a selective advantage over other raw-milk microorganisms which are unable to grow during cold storage of milk. Hence, upon prolonged storage of raw milk at refrigeration temperatures, psychrotrophic microorganisms gradually increase in numbers and, depending on the length and temperature of cold storage, they can eventually become the dominant microbial group of raw milk (Celestino et al. 1996).

Among other microbial groups of raw milk, most probably psychrotrophic microorganisms cause the greatest concern to the dairy industry nowadays, both in terms of quality and safety. Psychrotrops that contaminate raw milk during milking originate from the farm environment (e.g., soil or contaminated rinsing or cooling water), or from poorly cleaned and sanitized milking equipment (Thomas 1966, Christiansson et al. 1999). Psychrotrophic microorganisms can consist of molds, yeasts and bacteria (Cousin 1982). Psychrotrophic bacteria belonging to different genera have been isolated from raw or heat-treated milk. These include both Gram-negative (e.g., *Pseudomonas, Aeromonas, Serratia, Acinetobacter, Alcaligenes, Achromobacter, Enterobacter, Flavobacterium*) and Gram-positive (e.g., *Bacillus, Clostridium, Corynebacterium, Microbacterium, Micrococcus*) bacteria (Cousin 1982). *Pseudomonas* spp. are probably the best known and most studied among the Gram-negative psychrotrophic organisms. Growth of pseudomonads in raw milk occurs from the beginning of the dairy chain (farm tank), under both optimal and suboptimal storage temperature conditions (De Jonghe et al. 2011). Raw-milk isolates belonging to several additional bacterial genera (mainly Gram-negative), however, have been recently identified as psychrotrophic and some also possess lipolytic and/ or proteolytic traits, which may vary at the species or even strain level (Munsch-Alatossava and Alatossava 2006, Martins et al. 2006, Hantsis-Zacharov and Halpern 2007, Ercolini et al. 2009, Nörnberg et al. 2010).

The results of different studies focusing on psychrotrophs in raw milk are not directly comparable, as differences in the relative representation and populations of different psychrotrophs have been noted among studies, depending on factors such as milking practices, raw-milk storage temperatures and holding times and milk type (e.g., bovine vs. ovine) (Sanjuan et al. 2003). Different studies, however, help portray the often underestimated variety of raw-milk microorganisms capable of growing at refrigeration temperatures. In addition, the presence of unidentified psychrotrophic bacterial isolates has been recently reported in raw milk (Hantsis-Zacharov and Halpern 2007). It has been also demonstrated that several Gram-negative proteolytic psychrotrophic bacteria isolated from raw milk are capable of production of acylated homoserine lactones *in vitro*. It has been speculated, therefore, that quorum sensing mechanisms may be involved in the spoilage potential of such bacteria (Pinto et al. 2007) and a review of the literature suggests the involvement of quorum sensing mechanisms in spoilage of different food commodities (Ammor et al. 2008). In order to control the microbiological quality of raw milk, several treatments have been proposed and implemented, such as thermization (typically heating milk at 65°C for 10–20 s), addition of $CO_2$, or microfiltration (Roberts and Torrey 1988, Champagne et al. 1994, Singh et al. 2012).

The dynamics and shifts in the bacterial populations of raw milk during refrigerated storage were studied using molecular methods (Lafarge et al. 2004, Rasolofo et al. 2010, Raats et al. 2011). Using Temporal Temperature Gel Electrophoresis (TTGE) and Denaturing Gradient Gel Electrophoresis (DGGE) analyses, Lafarge et al. (2004) showed that a 24-h storage of raw cows' milk at 4°C results in considerable evolution of certain psychrotrophic bacterial populations, including pathogenic psychrotrophs. The authors also noted considerable variation in the bacterial dynamics between milk samples, indicating that the presence of different species and/or strains in raw milks prior to refrigeration may significantly affect their microbial balance during refrigerated storage. Rasolofo et al. (2010) studied the bacterial dynamics in raw milks using a combination of culture-dependent and molecular methods. Milk samples that had been treated by either addition of $CO_2$, thermization or microfiltration were monitored over 7 days of storage at 4 or 8°C. Dominant bacterial species in untreated, $CO_2$-treated and thermized milk samples at day 3 belonged to the genera *Staphylococcus, Streptococcus, Clostridium, Aerococcus, Facklamia, Corynebacterium, Acinetobacter* and *Trichococcus*. Dominant bacterial genera detected in micro-filtered milk were *Stenotrophomonas, Pseudomonas* and *Delftia*. *Pseudomonas* spp. dominated the bacterial population of untreated, $CO_2$-treated and micro-filtered milk samples at day 7. *Staphylococcus* spp. and *Delftia* spp. were the dominant bacterial genera in thermized milk. Raats et al. (2011) demonstrated that considerable evolution of bacterial communities occurs during cold storage of raw milk, but the taxonomic diversity decreased with storage time as a consequence of some microbial populations' dominance during refrigeration. In their study, raw milk samples were collected from bulk tanks of dairy farms and silo tanks of an associated industrial processing plant. According to their analyses Gram-positive bacteria (Bacilli, Clostridia, and Actinobacteria) prevailed in the milk from the farm tanks during cold incubation, whereas milk samples from the dairy plant were dominated by Gram-negative species belonging to *Gammaproteobacteria*, especially *Pseudomonadales*.

Reduced quality of raw and processed milks is frequently a consequence of the proliferation and metabolic activities of psychrotrophic bacteria. Although the Gram-negative psychrotrophs are effectively destroyed by pasteurization, some species of psychrotrophic bacteria can produce lipolytic and proteolytic enzymes during growth in raw milk. These enzymes are either not inactivated, or only marginally affected by the time/temperature schemes used in the dairy industry for the manufacture of processed dairy products (e.g., pasteurization, UHT processing). These lipases and proteases can lead to the development of off-odors in raw milk. Additionally, being heat-stable, they retain activity after the heat treatment of milk and therefore gradually degrade milk proteins and lipids, leading to noticeable off-odors and reducing dairy products' shelf-life (Griffiths et al. 1981, Sørhaug and Stepaniak 1997, Nörnberg et al. 2010). Therefore numerous published manuscripts have emphasized the importance of the initial microbiological quality of raw milk and the importance of the rapid and efficient cooling of milk upon milking in terms of milk and dairy products' quality (Banks et al. 1988, Griffiths et al. 1987, 1988, Zeng et al. 2007).

As stated above, some aerobic spore-forming bacteria belonging to the genus *Bacillus* (e.g., *B. cereus*) are psychrotrophic (Grosskopf and Harper 1974). Psychrotrophic strains of aerobic spore-formers are a major concern to the dairy industry because they are able to survive milk pasteurization and can subsequently proliferate in dairy products such as pasteurized milk stored at low temperatures, often leading to spoilage (Meer et al. 1991). For instance, in the absence of post-pasteurization re-contamination of milk by Gram-negative psychrotrophic bacteria (Eneroth et al. 2000), the shelf-life of pasteurized milk largely depends on the presence (levels) and spoilage potential of psychrotolerant spore-formers (Fromm and Boor 2004). Psychrotrophic aerobic spore-forming bacteria may be also significant in terms of food safety. Psychrotrophic strains of *B. cereus* producing enterotoxin have been identified (Van Netten et al. 1990). Also, the ability of certain *B. cereus* strains to produce toxins in milk stored at 8°C under aerated conditions has been demonstrated (Christiansson et al. 1989). Other well-known psychrotrophic pathogens (Schofield 1992) that have been isolated from raw milk are *Listeria monocytogenes*, *Yersinia enterocolitica* and *Aeromonas hydrophila*. *L. monocytogenes* and *Y. enterocolitica* have been identified as the causative agents in food-borne outbreaks associated with the consumption of contaminated raw milk, raw-milk cheeses or dairy products that had been contaminated post-pasteurization (Tacket et al. 1984, Lundén et al. 2004). Amongst these pathogens, undoubtedly, *L. monocytogenes* is of greatest concern for the dairy industry and its cold-adaptive mechanisms have been studied in detail (Angelidis et al. 2002, Tasara and Stephan 2006). The non-proteolytic group II *Clostridium botulinum* are psychrotrophic. The reported *C. botulinum* outbreaks associated with dairy products are rare, yet large, and have been associated with both commercial and home-prepared dairy products. However, the role of the psychrotrophic group II *C. botulinum* in dairy outbreaks is unclear due to the lack of epidemiological data (Lindström et al. 2010).

A great number of comprehensive review articles on psychrotrophic microorganisms of concern to raw milk and processed milk products can be found in the literature. These review articles cover aspects pertaining to the types (genera, species) of relevant organisms, their growth potential and growth kinetics at refrigeration temperatures, their spoilage potential through production of enzymes, public health considerations, their negative effects and methods of control in raw milk and dairy products. Whereas some of the information presented in older reviews may be outdated (e.g., due to the re-classification of certain microbial species to new or different genera), older review articles offer a nice historical overview of approaches practiced over the years to isolate, identify and characterize psychrotrophs and of approaches used or proposed in order to minimize their negative effects in raw milk and dairy products. Only a few of the many published review articles on psychrotrophs are cited here (Cousin 1982, Champagne et al. 1994, Shah 1994, Sørhaug and Stepaniak 1997).

### *Thermoduric bacteria*

The term "thermoduric" is used to denote bacteria (Gram-positive or Gram-negative) that are usually isolated from raw milk after pasteurization. Some authors use the term thermoduric to denote countable survivors following milder heat treatments such as milk thermization (e.g., 60–65°C for *ca.* 10 s). In official textbooks, thermoduric bacteria are defined as "microorganisms (vegetative cells or spores) that survive pasteurization treatment" (Frank and Yousef 2004). Thermoduric bacterial counts are estimated by heating milk at 62.8°C ± 0.5°C for 30 min to simulate batch pasteurization and survivors are counted using the SPC method. The presence of thermoduric bacteria in sufficiently high numbers in raw milk will result in small, though significant surviving populations after pasteurization. In the absence of post-treatment contamination, the microorganisms that survive pasteurization can affect the shelf-life of pasteurized milk. This is particularly the case for psychrotrophic thermoduric bacteria, such as those belonging to the genus *Bacillus* (Collins 1981).

Although some dairy microbiology textbooks include only a few, yet diverse bacterial genera under this category (*Bacillus* spp., *Clostridium* spp., *Corynebacterium* spp., *Kocuria* spp., *Lactobacillus* spp., *Micrococcus* spp., *Microbacterium* spp., *Rothia* spp.), it appears that members of additional bacterial genera could potentially meet the definition of "thermoduric". For instance, members of the genera *Staphylococcus* spp., *Pseudomonas* spp., and *Moraxella* spp., have been isolated from raw milk samples after heat treatment at 80°C for 10 min, i.e., a heat treatment more intense than laboratory pasteurization (Coorevits et al. 2008). Ranieri and Boor (2009) characterized the bacterial isolates obtained from 2% fat pasteurized milk samples processed at 18 fluid milk processing plants, representing five geographical regions across the US. Overall, 21 different bacterial genera were identified by 16S rDNA sequencing. The most frequently isolated Gram-positive genus was *Bacillus*; *Paenibacillus* was the second most frequently isolated Gram-positive genus. However, additional Gram-positive bacteria identified were *Staphylococcus*, *Leuconostoc*, *Enterococcus*, *Streptococcus*, *Brevibacillus*, *Corynebacterium*, *Lactococcus*, *Microbacterium*, *Micrococcus*, and *Oceanobacillus*. *Pseudomonas* was the most frequently isolated Gram-negative genus. Other Gram-negative bacteria identified belonged to the genera *Acinetobacter*, *Yersinia*, *Enterobacter*, *Shewanella*, *Aeromonas*, *Flavobacterium*, *Pantoea*, *Sphingobacterium* and some isolates belonged to the Enterobacteriaceae family (Ranieri and Boor 2009). It was not possible, however, to distinguish how many of these isolates were post-pasteurization contaminants.

As previously discussed, in the absence of post-pasteurization contamination, thermoduric spore-forming psychrotrophs (e.g., members of the genera *Bacillus* and *Paenibacillus*) often limit the shelf-life of refrigerated pasteurized milk.

### *Lactic acid bacteria*

Lactic acid bacteria (LAB) constitute one of the main groups of raw-milk microorganisms. The most common source of LAB in raw milk is the udder skin and teat canal, but LAB also colonize animal sites such as the vagina or the intestine. The term often used to denote LAB present in raw milk is autochthonous, or non-starter LAB, to distinguish them from the starter LAB, i.e., the known and characterized strains of LAB that are deliberately added to milk for the manufacture of fermented dairy products. Traditionally, the manufacture of fermented dairy products relied upon the presence and acidification activity of LAB present in raw milk. This practice is still used nowadays for the manufacture of some artisanal dairy products. The biopreservation effect of LAB relies on the fermentation of the major milk sugar (lactose) and the production of organic acids with a concomitant reduction of milk pH and therefore the creation of an acidic environment, unfavorable for the proliferation and/or survival of pathogenic microorganisms. The rate of acid production as well as the final pH is of vital importance for the safety of fermented dairy products. The production of several antimicrobial compounds such as bacteriocins, which are active against several food-borne pathogens, also contributes to the bio-preservative effects of LAB. Nowadays a vast majority of fermented dairy products is produced using well-defined (mixtures of) LAB strains.

Most LAB do not grow at refrigeration temperatures and souring of raw milk is highly unlikely under adequately refrigerated conditions. Elevated titratable acidity in milk nowadays is rare and usually an indication of raw-milk temperature abuse in the farm or during transport. In certain instances, the presence of adventitious LAB or the use of inappropriately selected (e.g., overactive) starter LAB, can lead to quality defects in fermented dairy products (e.g., bitterness, production of gas, or extreme acidity). On the other hand, underactive LAB, due to the presence of antibiotic residues or bacteriophages in raw milk, can delay or even halt the fermentation process.

LAB are usually classified based on: a) their optimum growth temperature into mesophilic (around 30°C) and thermophilic (around 43°C), and b) fermentation reactions/metabolic pathways into homofermentative (lactic acid is the primary and most abundant end-product of lactose fermentation) and heterofermentative (other compounds in addition to lactic acid are produced in significant amounts, such as acetic acid, or carbon dioxide). Raw-milk LAB belong to all major LAB genera (*Lactococcus, Lactobacillus, Streptococcus, Leuconostoc, Pediococcus, Enterococcus*). There are probably hundreds of published articles on the isolation, physiological and technological characterization of LAB isolated from raw bovine, caprine or ovine milk and raw-milk cheeses and one such study is cited here (Franciosi et al. 2009). Details on LAB used in the manufacture of dairy products are presented in Chapter 4 of this book.

### *Mastitis-causing organisms*

Mastitis (intra-mammary infection, IMI) is one of the costlier diseases of lactating animals. Mastitis alters milk composition, reduces milk secretion and has serious animal-welfare, economic and possibly public health implications. Depending on its clinical manifestation (clinical or subclinical), severity (e.g., acute, sub-acute, chronic), the epidemiology of the primary pathogen (contagious or environmental) and the specific etiology, mastitis results in variable degrees of animal suffering, losses in milk yield, increases of milk SCC and animal culling. Mastitis can be of infectious or traumatic etiology but the vast majority of cases are of infectious nature. Mastitis-causing pathogens include bacteria, fungi, viruses or algae. Bacterial pathogens (including mycoplasms) are responsible for the vast majority of infectious mastitis cases in lactating ruminants.

Contagious bovine mastitis is typically spread between cows during milking. The primary reservoir of contagious pathogens is the infected cows' udder quarter(s). Contagious mastitis is usually caused by *Staphylococcus aureus*, *Streptococcus agalactiae* and *Mycoplasma* spp. Environmental mastitis is usually the result of infection with coliform bacteria (*Escherichia coli*, *Klebsiella* spp., *Enterobacter* spp.), non-coliform Gram-negative bacteria such as *Serratia* spp., *Pseudomonas* spp., *Proteus* spp., Gram-positive bacteria such as *Enterococcus faecium* and *E. faecalis*, some *Bacillus* spp. and streptococci other than *Str. agalactiae* (i.e., "environmental streptococci" such as *Str. uberis*). *Prototheca* spp. are unicellular algae rarely reported to infect the mammary gland of dairy cattle (Marques et al. 2010). Environmental mastitis pathogens originate from the farm environment (e.g., bedding material). The etiology, control/prevention and treatment of the various types of bovine mastitis have been reviewed by several experts in the field (Watts 1988, Cullor 1993, Fox et al. 2005, Contreras and Rodríguez 2011, Fox 2012, Hogan et al. 2012, Schukken et al. 2012).

*Staphylococcus* spp. are the most prevalent pathogens responsible for IMIs in small ruminants. Other pathogens such as *Streptococcus* spp., members of the Enterobacteriaceae family, *Pseudomonas aeruginosa*, *Mannheimia haemolytica*, *Mycoplasma* spp., Corynebacteria and fungi can cause IMI in small ruminants, albeit at a lower incidence (Contreras et al. 2007). Coagulase-negative staphylococci (CNS) are the most prevalent pathogens causing subclinical mastitis in dairy ruminants. CNS can also produce persistent subclinical mastitis and significantly increase the SCC of milk. Contreras et al. (2007) have summarized the etiological, epidemiological and control aspects of mastitis in small ruminants.

The role of viruses and viral infections in bovine mastitis has been reviewed by Wellenberg et al. (2002). At least four bovine pathogenic viruses have been isolated from the milk of cows with clinical mastitis (bovine herpesvirus 1, bovine herpesvirus 4, foot-and-mouth disease virus and parainfluenza 3 virus). The authors concluded that viral infections can play a direct or more likely an indirect (e.g., via immune suppression, teat skin or mammary tissue lesions predisposing the animals to bacterial IMIs) role in the etiology of bovine mastitis and that the role of viruses warrants further investigation.

### *Human pathogenic bacteria*

The intestinal tract of lactating mammals is the source of several microbial species that can be pathogenic for humans. These pathogenic organisms are therefore frequently present in the environment of lactating animals as well as on the animal hide and udder skin and can contaminate raw milk during milking under non-hygienic conditions. Additionally, vegetative cells and spores of spore-forming bacterial pathogens and molds can be found in the environment and their transfer to raw milk can be airborne or via dirt. The lactating udder of animals with clinical or sub-clinical mastitis as well as biofilms formed at the inner surfaces of milking and storage equipment can also constitute sources of pathogenic microorganisms.

The consequences of human illnesses due to food-borne pathogens are very severe (Hoffmann et al. 2012). Judging from their involvement in the reported food-borne outbreaks, processed milk and dairy products appear to maintain a good safety record among other food categories (http://www.cdc.gov/foodsafety/outbreaks/multistate-outbreaks/outbreaks-list.html). The consumption of unpasteurized milk and dairy products made with raw milk, however, is not advisable (Leedom 2006, Cavirani 2008, LeJeune and Rajala-Schultz 2009, Oliver et al. 2009, Giacometti et al. 2012, Langer et al. 2012). In the US for instance, between 1973 and 1992, 46 raw-milk associated outbreaks were reported to the Centers for Disease Control and Prevention, with a median number of people becoming ill in these outbreaks equal to 19 (Headrick et al. 1998). Nonetheless, in some States as well as EU Member States the sale of raw milk for human consumption is allowed, e.g., in delicatessen stores or via self-service automatic vending machines. There are dozens if not hundreds of published articles dealing with the prevalence of specific pathogens or groups of pathogens in raw milk worldwide, and numerous pathogenic organisms have been found in or associated with raw milk over the years. The complete listing of these pathogens or a thorough description of their incidence in raw milk, epidemiological spread, symptoms and pathogenesis in humans, methods of intervention for prevention of milk contamination or pathogen inactivation, or listing of the relevant outbreaks (raw-milk outbreaks or outbreaks from the consumption of dairy products made with unpasteurized milk) would be unfeasible to cover in one chapter. Furthermore, due to differences in the sampling and analytical protocols used in the pathogen prevalence studies it is not possible to report "average" pathogen-specific prevalence estimates in raw milks. The reader is therefore more suited to consult recently published comprehensive review articles on these subjects. For instance, Oliver et al. (2005, 2009) reviewed the food-borne pathogens in milk and the dairy farm environment and have commented on their food safety and public health implications; the potential food safety hazards associated with the consumption of raw milk are addressed and data on the prevalence of food-borne pathogens in raw milk and raw milk-borne disease outbreaks in the US from 2000 until 2008 are summarized. In addition, under the section "Pathogens in milk", the second edition of the encyclopedia of dairy sciences (Fuquay et al. 2011) contains chapters dedicated to 14 pathogenic bacterial families, genera or species commonly associated with raw milk (*Bacillus cereus*, *Brucella* spp., *Campylobacter* spp., *Clostridium* spp., *Coxiella burnetii*, *Escherichia coli*, Enterobacteriaceae, *Enterobacter* spp., *Listeria monocytogenes*, *Mycobacterium* spp., *Salmonella* spp., *Shigella* spp., *Staphylococcus aureus* and *Yersinia enterocolitica*). The chapters have been written by highly-respected experts in the field and contain updated information on pathogens' characteristics, physiology, resulting milk-borne illness, toxins, outbreaks, incidence in dairy products including raw milk, potential sources and suggested control measures at the farm. Pathogens in dairy products are also covered in Chapter 3 in this book. The remaining of this section focuses on specific (emergent?) bacterial agents with known or suspected public health importance, recently shown to be occasionally present in raw milk.

*Mycobacterium avium* subsp. *paratuberculosis* (MAP) is the causative agent of paratuberculosis (Johne's disease), a chronic debilitating disease (infectious enteritis with cachexia) affecting ruminants and other animals worldwide. Johne's disease has been reported in many countries worldwide. Across the EU there are many published studies on the prevalence of MAP in cattle and a smaller number of studies on sheep and goats (Nielsen and Toft 2009). Infected animals are shedding the organism periodically in both feces and milk (Streeter et al. 1995). The presence of MAP in milk for human consumption poses concerns due to its possible association with Crohn's disease in humans (Grant 2005). The occurrence of MAP in bulk tank milk and individual milk samples at cattle dairy farms worldwide was reviewed by Okura et al. (2012). The authors reported a considerable variation of MAP in bulk tank milk and individual cows' milk. The overall MAP apparent prevalence and 95% CI based on PCR and culture were summarized to 0.1 (0.04–0.22) or 10% on a per cent basis in bulk tank milk samples, and 0.2 (0.12–0.32) or 20% on a per cent basis in individual milk samples.

Milk pasteurization is a heat treatment process for a given period of time that has been designed to inactivate the most heat-resistant, non-spore-forming pathogenic bacteria that may be present in raw milk, i.e., *Mycobacterium bovis* and *Coxiella burnetti*, the causative agents of tuberculosis and Q-fever in humans, respectively. The legal minimum time/temperature combination for pasteurized milk is the heating of milk at 72°C for 15 s (HTST) or 63°C for 30 min [low-temperature, long-time pasteurization (LTLT)] (Commission Regulation (EC) No. 853/2004). There are several published reports indicating survival of low numbers of MAP after pasteurization of either artificially contaminated or naturally contaminated milk. Therefore it has been suggested that, depending on the initial MAP load of raw milk, low numbers of MAP may occasionally survive HTST pasteurization (Grant 2006, Van Brandt et al. 2011b). Review articles have been published summarizing the available (and often contradictive) data regarding the prevalence of MAP in raw and pasteurized milks and other dairy products, as well as the effects of pasteurizing treatments on MAP survival in milk (Eltholth et al. 2009, Gill et al. 2011). Only a few studies have been published on the behavior of MAP during the production and storage of fermented milk products (cheddar cheese, yogurt, acidophilus milk and kefir). It appears that the type of fermented dairy product, the type of starter culture, the stage of MAP inoculation (pre- vs. post-fermentation inoculation), the MAP strain, and the pH of the finished product affect the (extent of) survival of MAP (Donaghy et al. 2004, Van Brandt et al. 2011a, Klanicova et al. 2012). More studies are warranted, however in this field.

*Helicobacter pylori* is the causative agent of gastric ulcers and other pathogenic conditions in humans. The presence of *H. pylori* in a small fraction of raw cows' milk samples examined was demonstrated via fluorescence *in situ* hybridization (Angelidis et al. 2011), although its presence had been proposed/suspected by the findings of earlier studies based on PCR (Fujimura et al. 2002, Quaglia et al. 2008). It has been hypothesized that sheep may constitute a reservoir of *H. pylori* (Dore et al. 2001). At present, the risk of human *H. pylori* infection via consumption of contaminated raw milk or dairy products manufactured with contaminated raw milk remains unknown.

*Arcobacter* spp. are Gram-negative, spiral-shaped bacteria belonging to the family Campylobacteraceae. *Arcobacter* spp. have been repeatedly isolated from feces of livestock animals including cows, whereas only a few investigations have reported their presence in feces of goats and sheep (Van Driessche et al. 2003, De Smet et al. 2011). *Arcobacter* spp. are considered emerging waterborne and food-borne human pathogens with infection symptoms similar to campylobacteriosis. Consumption and handling or raw or undercooked meats (mainly poultry) are potential sources of infections in humans (Lehner et al. 2005). The potential role of *Arcobacter* spp. in human disease, however, needs further evaluation. A few studies from different countries have reported the isolation of *Arcobacter* spp. at different frequencies (5.8–46%) from raw bovine milk (Scullion et al. 2006, Shah et al. 2012).

Generalizations in food microbiology/food safety should be avoided or at least made with caution. Nonetheless, from a practical point of view if raw milk is to be used for the manufacture of processed milks or other dairy products whose technology involves a heat treatment step equal to or more intense than that of pasteurization, the main food-safety concerns ought to be the potential presence of *S. aureus* enterotoxins, mycotoxins and possibly MAP.

### *Fungi (Yeasts and Molds)*

Yeasts and molds are eukaryotic microorganisms and their differentiation is often difficult due to their structural transitions during reproduction and/or environmental/growth conditions. Molds usually grow by forming elongated filaments (hyphae) resulting in the formation of mycelia. Molds are strictly aerobic organisms, whereas yeasts can grow both in the presence and absence of oxygen. Yeasts and molds are special in the sense that they can grow in environments characterized by high osmolality, increased acidity or low temperature. Fleet (1990) has summarized some of the properties of yeasts that are important for their growth and predominance in dairy products. These are the fermentation or assimilation of lactose, the production of extracellular proteolytic and lipolytic enzymes, the assimilation of lactic and citric acid, their potential for growth at low temperatures and their tolerance of elevated osmotic strength (high salt concentrations). Yeast populations in raw milk are typically less than $10^3$ cfu/mL and, unless bacterial growth is inhibited by the presence of antibiotic residues, they usually do not grow during refrigerated storage of milk, as they are overgrown by psychrotrophic bacteria (Cousin 1982). Yeasts may develop in spoiled raw milk as secondary flora, when bacterial growth has ceased and the pH of milk has dropped significantly.

Compared to the bacterial diversity, very few studies have looked into the fungal diversity of raw milks. The study of Vadillo Machota et al. (1987) is one of the few studies in the literature dedicated solely to the examination of yeasts and molds in raw milk. In this study, the examination of 103 tank milk samples in Spain revealed more than twenty different fungal genera, the most frequently isolated being *Geotrichum, Cladosporium, Penicillium, Aureobasidium* and *Aspergillus.* More recent studies have shown that common mold genera isolated from raw milk include *Penicillium* spp., *Aspergillus* spp., and *Eurotium* spp. (Vacheyrou et al. 2011), whereas common yeast genera/species include *Candida* spp., *Kluyveromyces* spp., as well as *Debaryomyces hansenii, Rhodotorula* spp. and *Cryptococcus* spp. (Callon et al. 2007, Mallet et al. 2012). The analysis of raw milk with molecular, DNA-based methods can help identify a wider variety of yeast species. For instance, Cocolin et al. (2002) studied the yeast biodiversity in raw cows' milk by using a) traditional culture methods, and b) PCR to amplify a region of the 26S rRNA gene from a DNA pool extracted directly from raw milk, followed by DGGE. The resulting bands were extracted and sequenced to identify yeast isolates. The combination of PCR-DGGE led to the identification of a number of additional yeast species, not isolated via the traditional cultural approach. Chen et al. (2009) reported on the biodiversity of yeasts in raw milk from dairies in China. The authors identified 11 different species of yeasts using an integrated approach including phenotypic and molecular (RAPD-PCR analysis and partial sequencing of the 26S rDNA) methods. A recently published article reported the results of a Canadian survey aiming on characterizing the fungal microflora of raw cows' milk (111 samples) and raw-milk cheeses collected from 19 dairy farms (Lavoie et al. 2012). Molecular identification analyses of the isolates using the ITS1-5.8S-ITS2 rDNA region showed that almost two-thirds of the fungal isolates were yeasts that could be assigned to 37 species/11 genera and the remaining isolates were molds that could be assigned to 33 species/25 genera. *Debaryomyces hansenii* was the most abundant among the yeast species (21% of the 339 yeast isolates) and was detected in the milk of 14 from the 19 farms. *Candida* was the most abundant (44% of the isolates) and most diverse (11 species) genus. Other frequently identified yeast genera were *Cryptococcus* spp. (10.9% of the yeast isolates) and *Pichia* spp. (8% of the yeast isolates). *Eurotium* was the most abundant mold genus (about one-quarter of the mold isolates) and was isolated from the milk of 13 from the 19 farms. *Lichtheimia* was the second most abundant mold genus (13.3% of the mold isolates) and was isolated from nine farms. The authors argued that only a fraction of the fungal species may have been actually identified. Nonetheless, the findings of their study indicate that the fungal profile of milk differs from farm to farm. In France, Delavenne et al. (2011) evaluated the fungal diversity in a smaller collection of raw milk samples (nine), which also included goat and ewe samples, in addition to cow milk samples. Following DNA extraction from milk samples, a semi-nested PCR method was used to amplify the ITS1 region of fungal DNA and PCR products were analyzed by ion-pair, reverse-phase, denaturing, high-performance liquid chromatography (D-HPLC). Fractions of each peak were retrieved and sequenced. The approach enabled the identification of 27 fungal species (18 yeast species belonging to 9 genera, *Candida*, *Cryptococcus*, *Debaryomyces*, *Geotrichum*, *Kluyveromyces*, *Malassezia*, *Pichia*, *Rhodotorula* and *Trichosporon* and 9 mold species belonging to 7 genera, *Aspergillus*, *Chrysosporium*, *Cladosporium*, *Engyodontium*, *Fusarium*, *Penicillium* and *Torrubiella*). The authors reported the highest fungal diversity in cow milk with a total recovery of 23 different species (five mold and 18 yeast), whereas only six (one mold and five yeast) and seven (one mold and six yeast) species were recovered in goat and ewe milk, respectively. The most common fungal species among the nine milk samples were *G. candidum*, *K. marxianus* and *Candida* spp. such as *C. parapsilosis* (Delavenne et al. 2011). In Italy, Corbo et al. (2001) analyzed 26 samples of raw cow, ewe, goat and water buffalo milk for the presence of yeasts using cultural and biochemical methods and reported the isolation of 36 yeast species.

The most frequently occurring yeasts belonged to the species *Trichosporon cutaneum* (15.2%), *Candida catenulata* (10.5%), *Yarrowia lipolytica* (8.6%), *C. zeylanoides* (4.8%) and *C. sake* (4.8%).

Yeasts and molds are of interest to dairy microbiologists for several reasons (Rohm et al. 1992, Jakobsen and Narvhus 1996, Leclercq-Perlat 2011). On the positive side, some species are used in dairy fermentations. An example is the use of yeasts residing in kefir grains for the production of Kefir, a fermented dairy product that undergoes a mixed lactic and alcoholic fermentation (Wang et al. 2012). Another example is the use of molds for the ripening of certain types of cheeses, e.g., the use of *Penicillium roquefortii* for the interior mold-ripened cheese Roquefort and *P. camemberti* for the surface mold-ripened cheese Camembert. On the other hand, being ubiquitous organisms (air, soil), yeasts and molds can easily contaminate raw milk. The contamination of dairy products with yeasts or molds is often the reason leading to quality defects and/or public health concerns. For instance, the accidental yeast contamination followed by their proliferation during the manufacture or ripening of dairy products leads to the generation of off-odors, as a result of extensive proteolysis of milk proteins and/or lipolysis of milk fats, respectively, or causes bloating of cheese containers due to the production of carbon dioxide. The surface contamination of yoghurt by yeasts or molds often leads to visible surface growth and product rejection. Yeast spoilage is a problem primarily in fermented milks and cheeses and the main defects caused by spoilage yeasts are the development of fruity or bitter flavors, discolorations and swelling of products or product containers due to gas production. Some of the yeast strains isolated from raw milks are able to grow at refrigeration temperatures and also able to produce proteinases and phospholipases when incubated in laboratory media under refrigeration temperatures (Melville et al. 2011).

Molds can act as spoilage agents mostly in cheeses, yogurt and sweetened condensed milk and spoilage is typically the result of airborne post-pasteurization contamination. In addition, some mold species are known to produce mycotoxins. Aflatoxins (AFs) are mycotoxins produced by some strains of the mold species *Aspergillus flavus, A. parasiticus* and *A. nominus*. From a public health perspective, AFs are probably the most important mycotoxins because of their highly toxic chronic or acute effects on human health; AFs are potent human carcinogens (Pitt 2000, Fujimoto 2011) and can be present in raw milk usually as a result of mold-growth and AF production in animal feedstuffs. There are four main AFs: $AFB_1$, $AFB_2$, $AFG_1$ and $AFG_2$. $AFB_1$ can be produced in animal feeds during production or storage. Upon ingestion by lactating animals the AFB toxins are converted to the AFM metabolites in the liver and *ca.* 0.9% of ingested $AFB_1$ is found in the milk as the hydroxylated metabolite $AFM_1$ (Tabata 2011). The occurrence of $AFM_1$ in milk, especially bovine milk, is of particular concern for public health because of the importance of cows' milk as a foodstuff for children and adults. In the EU, the maximum limit for $AFM_1$ in milk is set at 0.05 μg/Kg (Commission Regulation (EC) No 1881/2006). Chromatographic and ELISA methods are used for the determination of the AF content of raw milk (Tabata 2011). AFs are particularly heat tolerant. Although the results of studies on $AFM_1$ degradation during heat processing of milk and dairy products are not always consistent, most studies indicate that $AFM_1$ is not appreciably altered during the time/temperature combinations used in milk processing (pasteurization, sterilization). Hence, the presence of AFs in raw milk poses a severe risk to public health. Therefore raw milk containing AFs in levels exceeding the legal limits should never be used or further processed for animal or human consumption. Numerous surveys are published each year worldwide conveying the results of investigations regarding the presence and/or quantitative determination of AFs in raw milk or other dairy products (Galvano et al. 1996, Prandini et al. 2009). The construction and use of a vaccine for lactating dairy cows for the prevention of $AFB_1$ carry over in milk has been reported (Polonelli et al. 2011).

## *Viruses*

### *Human pathogenic viruses*

There are many pathogenic viruses for humans known to be transmitted via consumption of contaminated foods, and according to a literature review by Newell et al. (2010) food-borne viruses belong to at least 11 known viral families. Contamination of foods with most of these food-borne viruses often results from non-hygienic food-handling practices by human carriers. In addition, the actual involvement of foods in viral food-borne outbreaks is very difficult to approximate, because in many cases, following the initial food-borne incident ("seeding event"), viruses easily spread from one infected individual to another, without involvement of contaminated food sources. Among food-borne viruses, noroviruses and the hepatitis A virus have been the causative agents in well-documented causes of food-borne illness (Koopmans and Duizer 2004). Cliver (1997) stated that "all known food-borne viruses except the agent of tick-borne encephalitis are human specific and transmitted by a fecal-oral cycle".

There are some publications dated back in the seventies about the incidence/contamination of raw milk with animal viruses (zoonotic or not) and human viral outbreaks that have occurred as a consequence of consumption of contaminated milk or raw-milk cheeses (Kefford et al. 1979). It should be stressed, however, that in most cases the origin of viral contamination of raw milk in these instances was external, i.e., from environmental sources such as contaminated water or human carriers, and most of the implicated viruses (poliomyelitis virus or "infectious hepatitis", now called hepatitis A) were of human origin. More recently, Wellenberg et al. (2002) reviewed the role of viruses and viral infections in bovine mastitis. Animal viruses that have been isolated from raw bovine milk are the bovine herpesvirus 1, the bovine herpesvirus 4, the foot-and-mouth disease (FMD) virus, and the parainfluenza 3 virus. The authors concluded that viral infections can play a direct or indirect role in the etiology of bovine mastitis. It has been hypothesized that bovine herpesvirus 4 could represent a danger for human health. The virus has been shown to replicate in permissive human cells and protect non-permissive, persistently infected cells from apoptosis (Gillet et al. 2004). A subsequent study demonstrated that pasteurization was sufficient to completely inactivate the infectivity of $3.0 \times 10^6$ plaque-forming units of bovine herpesvirus 4 per mL of milk (Bona et al. 2005).

The literature pertaining to the incidence of zoonotic viruses in raw milk is scant. Some animal-origin viruses that are pathogenic to humans can be found in the raw milk of infected lactating mammals. Milk from rabid cows can contain the rabies virus (*Lyssavirus*), one of the members of the viral family Rhabdoviridae. In the US, two incidents of potential mass exposures to rabies through drinking unpasteurized milk have been reported (Centers for Disease Control and Prevention 1999). Tick-borne encephalitis is a zoonotic, potentially lethal neurological viral infection usually transmitted to humans by bites of ticks (*Ixodes ricinus* and *Ixodes persulcatus*) (Mansfield et al. 2009). The tick-borne encephalitis virus (TBEV) is a member of the genus *Flavivirus*. The TBEV can be found in the milk of cows and goats with encephalitis and consumption of unpasteurized milk from TBEV infected animals may constitute a secondary means of transmission to humans (Mansfield et al. 2009, Cisak et al. 2010, Caini et al. 2012). FMD affects all cloven-hoofed animals and is probably the most contagious among animal diseases. It is caused by an *Aphthovirus* of the family Picornaviridae. FMD can be rarely transmitted to humans in close contact with infected animals (Bauer 1997, Prempeh et al. 2001). The FMD virus is secreted into the milk of infected animals before the onset of clinical signs. Therefore the movement of contaminated milk during FMD outbreaks can contribute to the spread of the disease among susceptible animals (Donaldson 1997). Literature data on human infections by consumption of contaminated raw milk seem to be limited to a report (cited by Bauer 1997) describing the self-infection of veterinarians who deliberately drank raw milk from infected cows in the 19th century. Current minimum pasteurization standards of milk may not be adequate to completely eliminate FMD virus (Tomasula and Konstance 2004). Rift Valley Fever (RVF) is a viral zoonosis that affects sheep, goats, buffalos and cattle. RVF virus is a *Phlebovirus* of the family Bunyaviridae. Humans get infected by mosquito bites, especially during periods of intense epizootic activity, i.e., during ongoing epidemics among animals. It has been stated that "Drinking raw, unpasteurized milk from infected animals can also transmit RVF" (Balky and Memish 2003). However, to my knowledge there are no published reports documenting milk-borne RVF infections in humans. Foamy viruses are a subfamily of retroviruses. Cows infected with bovine foamy virus (BFV) shed BFV into the milk (Romen et al. 2007). Humans can be infected by simian foamy viruses. I am not aware of any published studies pertaining to the likelihood of BFV transmission to humans. Whether additional zoonotic animal viruses exist with the potential of milk-borne transmission to humans appears to be unsubstantiated at present.

#### *Lactic bacteriophage*

Strains of *Streptococcus thermophilus* and species of the genera *Lactococcus, Lactobacillus, Leuconostoc* and *Pediococcus* are widely used by the dairy industry as starter cultures for the manufacture of a variety of fermented dairy products. Bacteriophage are "bacteria-eating" viruses (viruses that infect bacteria) that require bacterial hosts to propagate. Bacteriophage are ubiquitous in nature and are found in all bacterial ecosystems. Bacteriophage specific for dairy starter cultures, i.e., phage attacking and inactivating LAB (lactic bacteriophage) have been recognized as a significant and often persistent problem for the dairy industry. Lactic bacteriophage are often present in raw milk. The presence of lactic bacteriophage in the milk used for milk fermentations (e.g., cheesemaking) leads to loss of fermentative capacity associated with starter culture lysis, which can significantly retard or even halt the fermentation process. Despite the development of a variety of counter-measures, such as the application of starter culture rotation schemes, improved sanitation strategies, and the use of bacteriophage-resistant starter strains, phage contamination during dairy products' manufacture continues to be the leading cause of failed or retarded fermentations. The dairy environment frequently serves as a phage reservoir, especially the incoming milk and lysogenic starter cultures. The technological importance of phage in the dairy industry has been reviewed by Lyne (2011). Sturino and Klaenhammer (2004) have reviewed the life cycles of bacteriophage as well as the defense strategies used by the dairy industry aiming at protecting starter cultures against phage-related problems.

## Indigenous (Natural) Antimicrobial Agents of Raw Milk

Raw milk contains several indigenous antimicrobial agents. Probably the best known antimicrobial agents are the immunoglobulins (i.e., pathogen-specific antibodies), the non specific defense agents lactoferrin and lysozyme and the enzyme lactoperoxidase (International Dairy Foundation 1991b, Korhonen et al. 2000). Several peptides and other organic compounds (e.g., free fatty acids) in milk have shown to possess antimicrobial properties. In addition, a wide variety of bacteriocins with variable antimicrobial spectra can be produced by LAB. In recent years, the presence of additional host-defense related minor proteins and peptides has been documented in cow's milk (Hettinga et al. 2011, Wheeler et al. 2012).

Bovine milk contains low levels of lysozyme (*ca.* 0.1 μg/mL) but its concentration increases during mastitis (1–2 μg/mL). The antibacterial role of lysozyme relies on the cleavage of the glycosidic bond between N-acetylmuramic acid and N-acetylglucosamine in bacterial cell-wall peptidoglycan (International Dairy Federation 1991b, Benkerroum 2008). Lactoferrin is a whey protein with iron chelating properties. Its principle antimicrobial action relies on depriving bacteria of iron, which is an essential element for bacterial growth. Its concentration in bovine milk (*ca.* 0.2 mg/mL) can increase several-fold during mastitis (International Dairy Federation 1991b). Recent studies have attributed additional beneficial properties to lactoferrin such as anti-cancer, immunomodulatory and antioxidant properties (García-Montoya et al. 2012). Lactoperoxidase (LP), i.e., the milk peroxidase, is one of the most abundant enzymes in bovine milk constituting *ca.* 0.5% of the whey proteins and one of the three components of the LP system of raw milk. LP is an oxidoreductase which catalyses the oxidation of thiocyanate ions (SCN-) that are present in milk as a result of the cows' diet into hypothiocyanate ions (SCNO-) at the expense of hydrogen peroxide ($H_2O_2$). Hypothiocyanate ions and other intermediates are the reactive products of the LP system of milk and are potent antimicrobials. The LP activity and thiocyanate content in raw milk can be affected by factors such as the individual animal, the lactating species, the animal feed, the time after milking and the stage of lactation (Althaus et al. 2001, Fonteh et al. 2002, Yaqub et al. 2012). The hydrogen peroxide needed to activate the LP system can be generated by the LAB of milk or by an indigenous enzymatic system (xanthine oxidase/hypoxanthine). Most typically, $H_2O_2$ is provided exogenously in order to activate the LP system and thus help preserve raw milk in situations where refrigerated storage of raw milk between milking and processing is not possible (Haddadin et al. 1996). The concentrations of the components of the LP system, the extent and range of its antimicrobial action and its applications in milk and dairy products have been presented in related review articles (Wolfson and Sumner 1993, Kussendrager and van Hooijdonk 2000, Seifu et al. 2005).

The raw-milk natural antimicrobial compounds exert a bacteriostatic effect for a time period following milking, the duration of which may depend on several factors including storage temperature. For instance, the duration of the antibacterial effect following activation of the LP system was found to be inversely related to the storage temperature of milk (International Dairy Federation 1988). The raw-milk natural antimicrobial compounds are inactivated at variable degrees upon milk pasteurization or more intense heat treatments. LP is relatively heat-stable and its inactivation has been reported to start at 70°C; its heat stability depends on pH, being less heat-stable under acidic conditions (Kussendrager and van Hooijdonk 2000). The inactivation of LP after heat treatment of milk at 85°C for 20 s is used as an index for the determination of milk that has undergone high-temperature processing (Extended Shelf-Life milk).

## Concluding Remarks

The determination of the microbiological flora of raw milks has been the subject of scientific research for more than one century. Depending on the nature of the organisms sought, traditionally milk samples (diluted or undiluted) are plated onto agar media (selective or nonselective) and are incubated (aerobically, under microaerophilic or even anaerobic conditions) at temperatures thought or known to be optimum for the growth of the target organism(s). Enrichment steps in semi-selective or selective broths are frequently used to enable preferential proliferation of the target organism(s) against the background microbial flora of raw milk. Other approaches and steps such as filtration and centrifugation have been used as a means of bacterial concentration, assisting subsequent detection steps. Following the isolation of any given microorganism, further identification strategies include the direct observation under the microscope (wet mounts and stained-preparations) in which characteristics such as the size, morphology and motility of the microorganism can be evaluated, as well as the presence of specific structures (e.g., capsules, flagella, spores). Additional laboratory tests are then conducted to narrow down the identity of unknown isolates, such as biochemical and serological tests. Biochemical assays most often target enzymatic activities that rely on the possession and expression of specific genes by the isolate. Biochemical tests include the utilization of sugars, the determination of proteolytic and lipolytic traits, types of energy metabolism, physiological growth boundaries (e.g., temperature, pH, osmolarity), tolerance to antimicrobial agents and other enzymatic activities.

The methods for bacterial identification and enumeration based on culture, microscopy and biochemistry have supported dairy microbiology for decades and have provided invaluable service to the scientific community and public health agencies. However, the phenotypic culture-based identification approaches are often hampered by difficulties. The morphology and/or motility of certain bacterial species under the microscope may vary depending on temperature and/or other culture conditions (e.g., osmolarity). The number of biochemical tests necessary for the identification of microbial isolates to the species level, can in some instances be very high, with a concomitant requirement in growth media, reagents, incubation times and labor. In addition, culture and phenotypic (biochemical) identification methods often fail to make reliable distinctions between isolates belonging to the same species because of variable expression of phenotypic characteristics or because of ambiguous or "intermediate" biochemical results. Routine bacteriological culture fails to identify viable, yet unculturable organisms or fastidious microorganisms requiring specific growth factors.

During the last decades the application of PCR revolutionized many aspects in microbiology by permitting rapid and selective cloning of specific target DNA (or RNA) regions among a heterogeneous collection of DNA (or RNA) sequences. The ability to amplify specific genetic fragments was successfully paired with other advances in nucleic acid research such as the ability for detailed sequencing, specific labeling, specific cutting (restriction) and analysis via electrophoretic separation. These and other principles have been used as the basis for the development of specialized genomic-based methods for microbial identification. Hence, in recent years researchers have progressively started to rely on molecular identification methods. These methods are generally faster, more sensitive and more robust. The application of molecular approaches has helped identify microbial species of technological, spoilage, food-safety, or clinical importance that were either present in very low numbers, uncultivable, or previously not associated with raw milk. Most often bacterial identification methods rely on unraveling the sequence of the gene encoding the 16S ribosomal subunit (16S rRNA), whereas the 26S rRNA is the most common target for eukaryotic cell identification. The 16S rRNA gene possesses conserved regions found in all prokaryotes as well as regions whose sequence is hyper-variable and in most cases specific enough to make identifications to the species or even the sub-species level. The sequencing of the gene encoding the RNA-polymerase beta-subunit is another alternative target.

Numerous molecular approaches (both culture-dependent and culture independent) have been developed for the analysis of the microbial composition of raw milk and raw-milk cheeses and only a few recent examples are cited here (Callon et al. 2007, Giannino et al. 2009, Ajitkumar et al. 2012, Bhatt et al. 2012, Deperrois-Lafarge and Meheut 2012). The choice of a specific technique and approach depends upon the aim of the investigation and in particular on whether the research aims on elucidating the general microbial diversity of an ecosystem, identifying specific microorganisms, or both. In addition, molecular techniques can provide semi-quantitative or quantitative output. Quigley et al. (2011) have presented an extensive review of DNA-based technologies and molecular approaches that are available for the analysis of the microbial composition of raw milk and raw milk cheeses, covering technical aspects, advantages and disadvantages. Most experts nowadays seem to agree that it is probably most beneficial to apply polyphasic approaches, i.e., both culture-dependent and culture independent methods when assessing the diversity of the raw milk microbiota.

## References Cited


Ajitkumar, P., H.W. Barkema and J. De Buck. 2012. Rapid identification of bovine mastitis pathogens by high-resolution melt analysis of 16S rDNA sequences. Vet. Microbiol. 155: 332–340.

Althaus, R.L., M.P. Molina, M. Rodríguez and N. Fernández. 2001. Analysis time and lactation stage influence on lactoperoxidase system components in dairy ewe milk. J. Dairy Sci. 84: 1829–1835.

Amagliani, G., A. Petruzzelli, E. Omiccioli, F. Tonucci, M. Magnani and G. Brandi. 2012. Microbiological surveillance of a bovine raw milk farm through multiplex real-time PCR. Foodborne Path. Dis. 9: 1–6.

Ammor, M.S., C. Michaelidis and G.-J.E. Nychas. 2008. Insights into the role of quorum sensing in food spoilage. J. Food Prot. 71: 1510–1525.

Angelidis, A.S., L.T. Smith and G.M. Smith. 2002. Elevated carnitine accumulation by *Listeria monocytogenes* impaired in glycine betaine transport is insufficient to confer wild-type cryotolerance in milk whey. Int. J. Food Microbiol. 5: 1–9.

Angelidis, A.S., I. Tirodimos, M. Bobos, M.S. Kalamaki, D.K. Papageorgiou and M. Arvanitidou. 2011. Detection of *Helicobacter pylori* in raw bovine milk by fluorescence *in situ* hybridization. Int. J. Food Microbiol. 151: 252–256.

Anonymous. Pasteurized Milk Ordinance. 2011. U.S. Department of Health and Human Services. http://www.fda.gov/downloads/Food/Food Safety/Product-Specific Information/ Milk Safety/NationalConferenceonInterstateMilkShipmentsNCIMSModelDocuments/ UCM291757.pdf. Accessed Oct. 29 2012.

Ash, C., F.G. Priest and M.D. Collins. 1993. Molecular identification of rRNA group 3 bacilli (Ash, Farrow, Wallbanks and Collins) using a PCR probe test. Antonie van Leeuwenhoek 64: 253–260.

Balkhy, H.H. and Z.A. Memish. 2003. Rift Valley Fever: an uninvited zoonoses in the Arabian peninsula. Int. J. Antimicrob. Agents 21: 153–157.

Banks, J.M., M.W. Griffiths, J.D. Phillips and D.D. Muir. 1988. A comparison of the effects of storage of raw milk at 2°C and 6°C on the yield and quality of Cheddar cheese. Food Microbiol. 5: 9–16.

Bartoszewicz, M., B.M. Hansen and I. Swiecicka. 2008. The members of the *Bacillus cereus* group are commonly present contaminants of fresh and heat-treated milk. Food Microbiol. 25: 588–596.

Bauer, K. 1997. Foot-and-mouth disease as zoonoses. Arch. Virol. 13(suppl.): 95–97.

Beales, N. 2004. Adaptation of microorganisms to cold temperatures, weak acid preservatives, low pH, and osmotic stress: A review. Comp. Rev. Food Sci. Food Saf. 3: 1–20.

Benkerroum, N. 2008.Antimicrobial activity of lysozyme with special relevance to milk. Afr. J. Biotechnol. 7: 4856–4867.

Berry, E.D. and P.M. Foegeding. 1997. Cold adaptation and growth of microorganisms. J. Food Prot. 12: 1583–1594.

Bhatt, V.D., V.B. Ahir, P.G. Koringa, S.J. Jakhesara, D.N. Rank, D.S. Nauriyal, A.P. Kunjadia and C.G. Joshi. 2012. Milk microbiome signatures of subclinical mastitis-affected cattle analysed by shotgun sequencing. J. Appl. Microbiol. 112: 639–650.

Bolzoni, G., A. Marcolini and G. Varisco. 2000. Evaluation of the Bactoscan FC. 1 Accuracy, comparison with the Bactoscan 8000 and somatic cell effect. Milchwissenschaft 55: 67–70.

Bolzoni, G., A. Marcolini and G. Varisco. 2001. Evaluation of the Bacoscan FC. 2. Stability, repeatability, carry-over and linearity. Milchwissenschaft 56: 318–321.

Bona, C., B. Dewals, L. Wiggers, K. Coudijzer, A. Vanderplasschen and L. Gillet. 2005. Pasteurization of milk abolishes bovine herpesvirus 4 infectivity. J. Dairy Sci. 88: 3079–3083.

Boor, K.J., D.P. Brown, S.C. Murphy, S.M. Kozlowski and D.K. Bandler. 1998. Microbiological and chemical quality of raw milk in New York state. J. Dairy Sci. 81: 1743–1748.

Burgess, S.A., D. Lindsay and S.H. Flint. 2010. Thermophilic bacilli and their importance in dairy processing. Int. J. Food Microbiol. 144: 215–225.

Caini, S., K. Szomor, E. Ferenczi, Á.S. Gáspár, Á. Csohán, K. Krisztalovics, Z. Molnár and J.K. Horváth. 2012. Tick-borne encephalitis transmitted by unpasteurized cow milk in western Hungary, September to October 2011. Euro. Surveil. 17: pii=20128.

Callon, C., F. Duthoit, C. Deblés, M. Ferrand, Y. Le Frileux, R. De Crémoux and M.-C. Montel. 2007. Stability of microbial communities in goat milk during a lactation year: Molecular approaches. Syst. Appl. Microbiol. 30: 547–560.

Cassoli, L.D., P.F. Machado, A.C.D.O. Rodrigues and A. Coldebella. 2007. Correlation study between standard plate count and flow cytometry for determination of raw milk total bacterial count. Int. J. Dairy Tech. 60: 44–48.

Cavirani, S. 2008. Cattle industry and zoonotic risk. Vet. Res. Commun. 32(suppl. 1): S19–S24.

CDC Centers for Disease Control and Prevention. 1999. Mass treatment of humans who drank unpasteurized milk from rabid cows-Massachusetts, 1996–1998. Morb. Mortal. Wkly Rep. 48: 228–229.

CDC Centers for Disease Control and Prevention. 2014. http://www.cdc.gov/foodsafety/outbreaks/multistate-outbreaks/outbreaks-list.html.

Celestino, E.L., M. Iyer and H. Roginski. 1996. The effects of refrigerated storage on the quality of raw milk. Aust. J. Dairy Tech. 51: 59–63.

Champagne, C.P., R.R. Laing, D. Roy and A.A. Mafu. 1994. Psychrotrophs in dairy products: Their effects and their control. Crit. Rev. Food Sci. Nutr. 34: 1–30.

Chen, L.-S., Y. Ma, J.-L. Maubois, L.-J. Chen, Q.-H. Liu and J.-P. Guo. 2009. Identification of yeasts from raw milk and selection for some specific antioxidant properties. Int. J. Dairy Tech. 63: 47–54.

Christiansson, A., A.S. Naidu, I. Nilsson, T. Wadsrtöm and H.-E. Pettersson. 1989. Toxin production by *Bacillus cereus* dairy isolates in milk at low temperatures. Appl. Environ. Microbiol. 55: 2595–2600.

Christiansson, A., J. Bertilsson and B. Svensson. 1999. *Bacillus cereus* spores in raw milk: Factors affecting the contamination of milk during the grazing period. J. Dairy Sci. 82: 305–314.

Cisak, E., A. Wójcik-Fatla, V. Zając, J. Sroka, A. Buczek and J. Dutkiewicz. 2010. Prevalence of tick-borne encephalitis virus (TBEV) in samples of raw milk taken randomly from cows, goats and sheep in eastern Poland. Ann. Agric. Environ. Med. 17: 283–286.

Cliver, D.O. 1997. Virus transmission via food. Food Technol. 51: 71–78.

Cocolin, L., D. Aggio, M. Manzano, C. Cantoni and G. Comi. 2002. An application of PCR-DGGE analysis to profile the yeast populations in raw milk. Int. Dairy J. 12: 407–411.

Cocolin, L., N. Innocente, M. Biasutti and G. Comi. 2004. The late blowing in cheese: a new molecular approach based on PCR and DGGE to study the microbia1 ecology of the alteration process. Int. J. Food Microbiol. 90: 83–91.

Collins, E.B. 1981. Heat resistant psychrotrophic microorganisms. J. Dairy Sci. 64: 157–160.

Commission Regulation No 853/2004 of the European Parliament and of the Council of 29 November 2004 laying down specific hygiene rules for food of animal origin. Official J. Eur. Union L 139.

Commission Regulation No 1881/2006 of 19 December 2006 setting maximum levels for certain contaminants in foodstuffs. Official J. Eur. Union L 364.

Contreras, A., D. Sierra, A. Sánchez, J.C. Corrales, J.C. Marco, M.J. Paape and C. Gonzalo. 2007. Mastitis in small ruminants. Small Rumin. Res. 68: 145–153.

Contreras, A.G. and J.M. Rodríguez. 2011. Mastitis: Comparative etiology and epidemiology. J. Mammary Gland Biol. Neoplasia 16: 339–356.

Coorevits, A., V. De Jonghe, J. Vandroemme, R. Reekmans, J. Heyrman, W. Messens, P. De Vos and M. Heyndrickx. 2008. Comparative analysis of the diversity of aerobic spore-forming bacteria in raw milk from organic and conventional dairy farms. Syst. Appl. Microbiol. 31: 126–140.

Corbo, M.R., R. Lancioti, M. Albenzio and M. Sinigaglia. 2001. Occurrence and characterization of yeasts isolated from milks and dairy products of Apulia region. Int. J. Food Microbiol. 69: 147–152.

Cousin, M.A. 1982. Presence and activity of psychrotrophic microorganisms in milk and dairy products: a review. J. Food Prot. 45: 172–207.

Council Directive 96/23/EC on measures to monitor certain substances and residues thereof in live animals and animal products and repealing Directives 85/358/EEC and 86/469/EEC and Decisions 89/187/EEC and 91/664/EEC. Official J. Eur. Communities L 125/10.

Council Regulation (EEC) No. 2377/90 of 26 June 1990 laying down a Community procedure for the establishment of maximum residue limits of veterinary medicinal products in foodstuffs of animal origin. Official J. Eur. Commun. L224/1.

Cremonesi, P., L. Vanoni, T. Silvetti, S. Morandi and M. Brasca. 2012. Identification of *Clostridium beijerinckii, Cl. butyricum, Cl. sporogenes, Cl. tyrobutyricum* isolated from silage, raw milk and hard cheese by a multiplex PCR assay. J. Dairy Res. 79: 318–323.

Cullor, J.S. 1993. The control, treatment and prevention of the various types of bovine mastitis. Vet. Med. Food Anim. Pract. 88: 571–579.

Cullor, J.S. 1997. HACCP (Hazard Analysis Critical Control Points): is it coming to the dairy? J. Dairy Sci. 80: 3349–3352.

Davidson, P.M., L.A. Roth and S.A. Gambrel-Lenarz. 2004. Coliform and other indicator bacteria. pp. 187–226. *In*: H.M. Wehr and J.F. Frank [eds.]. Standard Methods for the Examination of Dairy Products. 17th edn. American Public Health Association, Washington, DC, USA.

De Jonghe, V., A. Coorevits, J. Vandroemme, J. Heyrman, L. Herman, P. De Vos and M. Heyndrickx. 2008. Intraspecific genotypic diversity of *Bacillus* species from raw milk. Int. Dairy J. 18: 496–505.

De Jonghe, V., A. Coorevits, J. De Block, E. Van Coillie, K. Grijspeerdt, L. Herman, P. De Vos and M. Heyndrickx. 2010. Toxinogenic and spoilage potential of aerobic spore-formers isolated from raw milk. Int. J. Food Microbiol. 136: 318–325.

De Jonghe, V., A. Coorevits, K. Van Hoorde, W. Messens, A. Van Landschoot, P. De Vos and M. Heyndrickx. 2011. Influence of storage conditions on the growth of Pseudomonas species in refrigerated raw milk. Appl. Environ. Microbiol. 77: 460–470.

Delavenne, E., J. Mounier, K. Asmani, J.-L. Jany, G. Barbier and G. Le Blay. 2011. Fungal diversity in cow, goat and ewe milk. Int. J. Food Microbiol. 151: 247–251.

Delavenne, E., J. Mounier, F. Déniel, G. Barbier and G. Le Blay. 2012. Biodiversity of antifungal lactic acid bacteria isolated from raw milk samples from cow, ewe and goat over one-year period. Int. J. Food Microbiol. 155: 185–190.

Deperrois-Lafarge, V. and T. Meheut. 2012. Use of the *rpoB* gene as an alternative to the V3 gene for the identification of spoilage and pathogenic bacteria species in milk and milk products. Lett. Appl. Microbiol. 55: 99–108.

Desmasures, N., F. Bazin and M. Guéguen. 1997a. Microbiological composition of raw milk from selected farms in the Camembert region of Normandy. J. Appl. Microbiol. 83: 53–58.

Desmasures, N., W. Opportune and M. Guéguen. 1997b. *Lactococcus* spp., yeasts and *Pseudomonas* spp. on teats and udders of milking cows as potential sources of milk contamination. Int. Dairy J. 7: 643–646.

De Smet, S., L. De Zutter and K. Houf. 2011. Small ruminants as carriers of the emerging foodborne pathogen *Arcobacter* on small and medium farms. Small Rumin. Res. 97: 124–129.

Donaghy, J.A., N.L. Totton and M.T. Rowe. 2004. Persistence of *Mycobacterium paratuberculosis* during manufacture and ripening of cheddar cheese. Appl. Environ. Microbiol. 70: 4899–4905.

Donaldson, A.I. 1997. Risks of spreading foot and mouth disease through milk and dairy products. Rev. Sci. Tech. Off. Int. Epiz. 16: 117–124.

Dore, M.P., A.R. Sepulveda, H. El-Zimaity, Y. Yamaoka, M.S. Osato, K. Mototsugu, A.M. Nieddu, G. Realdi and D.Y. Graham. 2001. Isolation of *Helicobacter pylori* from sheep—Implications for transmission to humans. Am. J. Gastroenterol. 96: 1396–1401.

Elmoslemany, A.M., G.P. Keefe, I.R. Dohoo and B.M. Jayarao. 2009. Risk factors for bacteriological quality of bulk tank milk in Prince Edward Island dairy herds. Part 2: Bacterial count-specific risk factors. J. Dairy Sci. 92: 2644–2652.

Elmoslemany, A.M., G.P. Keefe, I.R. Dohoo, J.J. Wichtel, H. Stryhn and R.T. Dingwell. 2010. The association between bulk tank milk analysis for raw milk quality and on-farm management practices. Prev. Vet. Med. 95: 32–40.

Eltholth, M.M., V.R. Marsh, S. Van Winden and F.J. Guitian. 2009. Contamination of food products with *Mycobacterium avium paratuberculosis*: a systematic review. J. Appl. Microbiol. 107: 1061–1071.

Eneroth, A., S. Ahrné and G. Molin. 2000. Contamination of milk with Gram-negative spoilage bacteria during filling of retail containers. Int. J. Food Microbiol. 57: 99–106.

Ercolini, D., F. Russo, I. Ferrocino and F. Villani. 2009. Molecular identification of mesophilic and psychrotrophic bacteria from raw cow's milk. Food Microbiol. 26: 228–231.

Eur-Lex website, http://eur-lex.europa.eu/homepage.html.

Fleet, G.H. 1990. Yeasts in dairy products. J. Appl. Bacteriol. 68: 199–211.

Fonteh, F.A., A.S. Grandison and M.J. Lewis. 2002. Variations of lactoperoxidase activity and thiocyanate content in cows' and goats' milk throughout lactation. J. Dairy Res. 69: 401–409.

Fox, L.K., J.H. Kirk and A. Britten. 2005. Mycoplasma mastitis: A review of transmission and control. J. Vet. Med. B 52: 153–160.

Fox, L.K. 2012. Mycoplasma mastitis causes, transmission and control. Vet. Clin. Food Anim. 28: 225–237.

Franciosi, E., L. Settanni, A. Cavazza and E. Poznanski. 2009. Biodiversity and technological potential of wild lactic acid bacteria from raw cows' milk. Int. Dairy J. 19: 3–11.

Frank, J.F. and A.E. Yousef. 2004. Tests for groups of microorganisms. pp. 227–247. *In*: H.M. Wehr and J.F. Frank [eds.]. Standard Methods for the Examination of Dairy Products. 17th edn. American Public Health Association, Washington, DC, USA.

Fricker, M., B. Skånseng, K. Rudi, B. Stessl and M. Ehling-Schulz. 2011. Shift from farm to dairy tank milk microbiota revealed by a polyphasic approach is independent from geographical origin. Int. J. Food Microbiol. 145: S24–S30.

Fromm, H.I. and K.J. Boor. 2004. Characterization of pasteurized fluid milk shelf-life attributes. J. Food Sci. 69: 207–214.

Fujimoto, H. 2011. Mycotoxins: Classification, occurrence and determination. *In*: J.W. Fuquay, P.F. Fox and P.H.L. McSweeney [eds.]. Encyclopedia of Dairy Sciences, 2nd edition. Academic Press, New York.

Fujimura, S., T. Kawamura, S. Kato, H. Tateno and A. Watanabe. 2002. Detection of *Helicobacter pylori* in cow's milk. Lett. Appl. Microbiol. 35: 504–507.

Fuquay, J.W., P.F. Fox and P.L.H. McSweeney. 2011. Encyclopedia of dairy sciences, second edition. Elsevier, Academic Press, New York.

Galvano, F., V. Galofaro and G. Galvano. 1996. Occurrence and stability of aflatoxin $M_1$ in milk and milk products: A worldwide review. J. Food Prot. 59: 1079–1090.

García-Montoya, I.A., T.S. Cendón, S. Arévalo-Gallegos and Q. Rascón-Cruz. 2012. Lactoferrin a multiple bioactive protein: An overview. Biochim. Biophys. Acta 1820: 226–236.

Garde, S., R. Arias, P. Gaya and M. Nuñez. 2011. Occurrence of *Clostridium* spp. in ovine milk and Manchego cheese with late blowing defect: Identification and characterization of isolates. Int. Dairy J. 21: 272–278.

Giacometti, F., A, Serraino, G. Finazzi, P. Daminelli, M.N. Losio, P. Bonilauri, N. Arrigoni, A. Garigliani, R. Mattioli, S. Alonso, S. Piva, D. Florio, R. Riu and R.G. Zanoni. 2012. Foodborne pathogens in in-line milk filters and associated on-farm risk factors in dairy farms authorized to produce and sell raw milk in northern Italy. J. Food Prot. 75: 1263–1269.

Giannino, M.L., M. Aliprandi, M. Feligini, L. Vanoni, M. Brasca and F. Fracchetti. 2009. A DNA array based assay for the characterization of microbial community in raw milk. J. Microbiol. Meth. 78: 181–188.

Gilchrist, J.E., J.E. Campbell, C.B. Donnelly, J.T. Peeler and J.M. Delaney. 1973. Spiral plate method for bacterial determination. Appl. Microbiol. 25: 244–252.

Gill, J.J., P.M. Sabour, J. Gong, H. Yu, K.E. Leslie and M.W. Griffiths. 2006. Characterization of bacterial populations recovered from the teat canals of lactating dairy and beef cattle by 16S rRNA gene sequence analysis. FEMS Microbiol. Ecol. 56: 471–481.

Gill, C.O., L. Saucier and W.J. Meadus. 2011. *Mycobacterium avium* subsp. *Paratuberculosis* in dairy products, meat, and drinking water. J. Food Prot. 74: 480–499.

Gillet, L., F. Minner, B. Detry, F. Farnir, L. Willems, M. Lambot, E. Thiry, P.-P. Pastoret, F. Schynts and A. Vanderplasschen. 2004. Investigation of the susceptibility of human cell lines to bovine herpesvirus 4 infection: Demonstration that human cells can support a nonpermissive persistent infection which protects them against tumor necrosis factor alpha-induced apoptosis. J. Virol. 78: 2336–2347.

Goudkov, A.V. and M.E. Sharpe. 1965. Clostridia in dairying. J. Appl. Bacteriol. 28: 63–73.

Grant, I.R. 2005. Zoonotic potential of *Mycobacterium avium* subsp. *paratuberculosis*: the current position. J. Appl. Microbiol. 98: 1282–1293.

Grant, I.R. 2006. *Mycobacterium avium* subsp. *paratuberculosis* in foods: current evidence and potential consequences. Int. J. Dairy Tech. 59: 112–117.

Graumann, P. and M.A. Marahiel. 1996. Some like it cold: response of microorganisms to cold shock. Arch. Microbiol. 166: 293–300.

Griffiths, M.W., J.D. Phillips and D.D. Muir. 1981. Thermostability of proteases and lipases from a number of species of psychrotrophic bacteria of dairy origin. J. Appl. Bacteriol. 50: 289–303.

Griffiths, M.W., J.D. Phillips and D.D. Muir. 1987. Effect of low-temperature storage on the bacteriological quality of raw milk. Food Microbiol. 4: 285–291.

Griffiths, M.W., J.D. Phillips, I.G. West and D.D. Muir. 1988. The effect of extended low-temperature storage of raw milk on the quality of pasteurized and UHT milk. Food Microbiol. 5: 75–87.

Grosskopf, J.C. and W.J. Harper. 1974. Isolation and identification of psychrotrophic sporeformers in milk. Milchwissenschaft 29: 467–470.

Haddadin, M.S., S.A. Ibrahim and R.K. Robinson. 1996. Preservation of raw milk by activation of the natural lactoperoxidase systems. Food Cont. 7: 149–152.

Hantsis-Zacharov, E. and M. Halpern. 2007. Culturable psychrotrophic bacterial communities in raw milk and their proteolytic and lipolytic traits. Appl. Environ. Microbiol. 73: 7162–7168.

Hassan, A.N. and J.F. Frank. 2011. Microorganisms associated with milk. pp. 447–457. *In*: J.W. Fuquay, P.F. Fox and P.H.L. McSweeney [eds.]. Encyclopedia of Dairy Sciences, 2nd edition. Academic Press, New York.

Headrick, M.L., S. Korangy, N.H. Bean, F.J. Angulo, S.F. Altekruse, M.E. Potter and K.C. Klontz. 1998. The epidemiology of raw-milk-associated foodborne disease outbreaks in the United States, 1973 through 1992. Am. J. Pub. Health. 88: 1219–1221.

Hettinga, K., H. van Valenberg, S. de Vries, S. Boeren, T. van Hooijdonk, J. van Arendonk and J. Vervoort. 2011. The host defense proteome of human and bovine milk. PLoS One 6: e19433: 1–8.

Hill, B.M. 1991. Direct microscopic count method. IDF Bullet. 256: 17–20.

Hill, B., B. Smythe, D. Lindsay and J. Zealand. 2012. Microbiology of raw milk in New Zealand. Int. J. Food Microbiol. 157: 305–308.

Hoffmann, S., M.B. Batz and J.G. Morris, Jr. 2012. Annual cost of illness and quality-adjusted life year losses in the United States due to 14 foodborne pathogens. J. Food Prot. 75: 1292–1302.

Hogan, J.S., K.L. Smith, K.H. Hoblet, D.A. Todhunter, P.S. Schoenberger, W.D. Hueston, D.E. Pritchard, G.L. Bowman, L.E. Heider, B.L. Brockett and H.R. Conrad. 1989. Bacterial counts in bedding materials used on nine commercial dairies. J. Dairy Sci. 72: 250–258.

Hogan, J. and K.L. Smith. 2012. Managing environmental mastitis. Vet. Clin. Food Anim. 28: 217–224.

Huck, J.R., N.H. Woodcock, R.D. Ralyea and K.J. Boor. 2007. Molecular subtyping and characterization of psychrotolerant endospore-forming bacteria in two New York state fluid milk processing systems. J. Food Prot. 70: 2354–2364.

Huck, J.R., M. Sonnen and K.J. Boor. 2008. Tracking heat-resistant, cold-thriving fluid milk spoilage bacteria from farm to packaged product. J. Dairy Sci. 91: 1217–1228.

Hutchison, M.L., D.J.I. Thomas, A. Moore, D.R. Jackson and I. Ohnstad. 2005. An evaluation of raw milk microorganisms as markers of on-farm hygiene practices related to milking. J. Food Prot. 68: 764–772.

IDF International Dairy Federation. 1988. Code of practice for preservation of raw milk by lactoperoxidase system. IDF Bulletin No. 234.

IDF International Dairy Federation. 1991a. Methods for assessing the bacteriological quality of raw milk from the farm. IDF Bulletin No. 256.

IDF International Dairy Federation. 1991b. Significance of the indigenous antimicrobial agents of milk to the dairy industry. IDF Bulletin No. 256: 2–19.

International Organization for Standardization. 4833: 2003. Microbiology of food and animal feeding stuffs—Horizontal method for the enumeration of microorganisms—Colony-count technique at 30 degrees C. International Organization for Standardization. Geneva, Switzerland.

International Organization for Standardization 707 | IDF 050 2008. Milk and milk products —Guidance on sampling. International Organization for Standardization. Geneva, Switzerland.

International Organization for Standardization 6887-5 2010. Microbiology of food and animal feeding stuffs—Preparation of test samples, initial suspension and decimal dilutions for microbiological examination—Part 5: Specific rules for the preparation of milk and milk products. International Organization for Standardization. Geneva, Switzerland.

Ivy, R.A., M.L. Ranieri, N.H. Martin, H.C. den Bakker, B.M. Xavier, M. Wiedmann and K.J. Boor. 2012. Identification and characterization of psychrotolerant sporeformers associated with fluid milk production and processing. Appl. Environ. Microbiol. 78: 1853–1864.

Jakobsen, M. and J. Narvhus. 1996. Yeasts and their possible beneficial and negative effects on the quality of dairy products. Int. Dairy J. 6: 755–768.

Jackson, E.E., E.S. Erten, N. Maddi, T.E. Graham, J.W. Larkin, R.J. Blodgett, J.E. Schlesser and R.M. Reddy. 2012. Detection and enumeration of four foodborne pathogens in raw commingled silo milk in the United States. J. Food Prot. 75: 1382–1393.

Jayarao, B.M. and L. Wang. 1999. A Study on the prevalence of Gram-negative bacteria in bulk tank milk. J. Dairy Sci. 82: 2620–2624.

Jones, P.G. and M. Inouye. 1994. The cold-shock response—a hot topic. Mol. Microbiol. 11: 811–818.

Julien, M.-C., P. Dion, C. Lafrenière, H. Antoun and P. Drouin. 2008. Sources of clostridia in raw milk on farms. Appl. Environ. Microbiol. 74: 6348–6357.

Kefford, B., R. Borland and A.J. Sinclair. 1979. Viral and coxiella contamination of milk. Austr. J. Dairy Tech. 34: 102–105.

Klanicova, B., I. Slana, P. Roubal, I. Pavlik and P. Kralik. 2012. *Mycobacterium avium* subsp. *paratuberculosis* survival during fermentation of soured milk products detected by culture and quantitative real time PCR methods. Int. J. Food Microbiol. 157: 150–155.

Kliem, K.E. and D.I. Givens. 2012. Dairy products in the food chain: their impact on health. Ann. Rev. Food Sci. Tech. 2: 21–36.

Klijn, N., F.F. Nieuwenhof, J.D. Hoolwerf, C.B. van der Waals and A.H. Weerkamp. 1995. Identification of *Clostridium tyrobutyricum* as the causative agent of late blowing in cheese by species-specific PCR amplification. Appl. Environ. Microbiol. 61: 2919–2924.

Klijn, N., L. Herman, L. Langeveld, M. Vaerewijck, A.A. Wagendorp, I. Huemer and A.H. Weerkamp. 1997. Genotypical and phenotypical characterization of *Bacillus sporothermodurans* strains, surviving UHT sterilization. Int. Dairy J. 7: 421–428.

Koopmans, M. and E. Duizer. 2004. Foodborne viruses: an emerging problem. Int. J. Food Microbiol. 90: 23–41.

Korhonen, H., P. Marnila and H.S. Gill. 2000. Milk immunoglobulins and complement factors. Brit. J. Nutr. 84 Suppl. 1: S75–S80.

Kussendrager, K.D. and A.C.M. van Hooijdonk. 2000. Lactoperoxidase: physic-chemical properties, occurrence, mechanism of action and applications. Br. J. Nutr. 84: S19–S25.

Lachowsky, W.M., W.B. McNab, M. Griffiths and J. Odumeru. 1997. A comparison of the Bactoscan 8000S to three cultural methods for enumeration of bacteria in raw milk. Food Res. Int. 30: 273–280.

Lafarge, V., J.-C. Ogier, V. Girard, V. Maladen, J.-Y. Leveau, A. Gruss and A. Delacroix-Buchet. 2004. Raw cow milk bacterial population shifts attributable to refrigeration. Appl. Environ. Microbiol. 70: 5644–5650.

Langer, A.J., T. Ayers, J. Grass, M. Lynch, F.J. Angulo and B.E. Mahon. 2012. Nonpasteurized dairy products, disease outbreaks, and State laws-United States, 1993–2006. Emerg. Inf. Dis. 18: 385–391.

Lavoie, K., M. Touchette, D. St-Gelais and S. Labrie. 2012. Characterization of the fungal microflora in raw milk and specialty cheeses of the province of Quebec. Dairy Sci. Technol. DOI 10.1007/s13594–011-0051-4.

Le Bourhis, A.G., J. Doré, J.-P. Carlier, J.-F. Chamba, M.-R. Popoff and J.-L. Tholozan. 2007. Contribution of *C. beijerinckii* and *C. sporogenes* in association with *C. tyrobutyricum* to the butyric fermentation in Emmental type cheese. Int. J. Food Microbiol. 113: 154–163.

Leedom, J.M. 2006. Milk of nonhuman origin and infectious diseases in humans. Clin. Inf. Dis. 43: 610–615.

Leggett, M.J., G. McDonnell, S.P. Denyer, P. Setlow and J.-Y. Maillard. 2012. Bacterial spore structures and their protective role in biocide resistance. J. Appl. Microbiol. 113: 485–498.

Lehner, A., T. Tasara and R. Stephan. 2005. Relevant aspects of *Arcobacter* spp. as potential foodborne pathogen. Int. J. Food Microbiol. 102: 127–135.

LeJeune, J.T. and P.J. Rajala-Schultz. 2009. Unpasteurized milk: A continued public health threat. Clin. Inf. Dis. 49: 93–100.

Leclercq-Perlat, M.-N. 2011. Camembert, Brie, and Related Varieties. pp. 773–782. *In*: J.W. Fuquay, P.F. Fox and P.H.L. McSweeney [eds.]. Encyclopedia of Dairy Sciences, 2nd edition. Academic Press, New York.

Lievaart, J.J., J.P.T.M. Noordhuizen, E. van Beek, C. van der Beek, A. van Risp, J. Schenkel and J. van Veersen. 2005. The Hazard Analysis Critical Control Point's (HACCP) concept as applied to some chemical, physical and microbiological contaminants of milk on dairy farms. A prototype. Vet. Quarterly 27: 21–29.

Lindström, M., J. Myllykoski, S. Sivelä and H. Korkeala. 2010. *Clostridium botulinum* in cattle and dairy products. Crit. Rev. Food Sci. Nutr. 50: 281–304.

Lukasova, J., J. Vyhnalkova and Z. Pacova. 2001. *Bacillus* species in raw milk and in the farm environment. Milchwissenschaft 56: 609–611.

Lundén, J., R. Tolvanen and H. Korkeala. 2004. Human listeriosis outbreaks linked to dairy products in Europe. J. Dairy Sci. (E. supp.): E6–E11.

Lyne, J. 2011. Technological importance in the dairy industry. *In*: J.W. Fuquay, P.F. Fox and P.H.L. McSweeney [eds.]. Encyclopedia of Dairy Sciences, 2nd edition. Academic Press, New York.

Magnusson, M., A. Christiansson and B. Svensson. 2007. *Bacillus cereus* spores during housing of dairy cows: Factors affecting contamination of raw milk. J. Dairy Sci. 90: 2745–2754.

Mallet, A., M. Guéguen, F. Kauffmann, C. Chesneau, A. Sesboué and N. Desmasures. 2012. Quantitative and qualitative microbial analysis of raw milk reveals substantial diversity influenced by herd management practices. Int. Dairy J. 27: 13–21.

Mansfield, K.L., N. Johnson, L.P. Phipps, J.R. Stephenson, A.R. Fooks and T. Solomon. 2009. Tick-borne encephalitis virus—a review of an emerging zoonoses. J. Gen. Virol. 90: 1781–1794.

Marchand, S., J. De Block, V. De Jonghe, A. Coorevits, M. Heyndrickx and L. Herman. 2012. Biofilm formation in milk production and processing environments; influence on milk quality and safety. Comp. Rev. Food Sci. Food Safety 11: 133–147.

Martin, N.H., M.L. Ranieri, S.C. Murphy, R.D. Ralyea, M. Wiedmann and K.J. Boor. 2011. Results from raw milk microbiological tests do not predict the shelf-life performance of commercially pasteurized fluid milk. J. Dairy Sci. 94: 1211–1222.

Marques, S., E. Silva, J. Carvalheira and G. Thompson. 2010. Phenotypic characterization of Mastitic *Prototheca* spp. isolates. Res. Vet. Sci. 89: 5–9.

Martins, M.L., C.L.O. Pinto, R.B. Rocha, E.F. de Araújo and M.C.D. Vanetti. 2006. Genetic diversity of Gram-negative, proteolytic, psychrotrophic bacteria isolated from refrigerated raw milk. Int. J. Food Microbiol. 111: 144–148.

McKillip, J.L. 2000. Prevalence and expression of enterotoxins in *Bacillus cereus* and other *Bacillus* spp., a literature review. Antonie van Leeuwenhoek 77: 393–399.

Meer, R.R., J. Baker, F.W. Bodyfelt and M.W. Griffiths. 1991. Psychrotrophic *Bacillus* spp. in fluid milk products: A review. J. Food Prot. 54: 969–979.

Melville, P.A., N.R. Benites, M. Ruz-Peres and E. Yokoya. 2011. Proteinase and phospholipase activities and development at different temperatures of yeasts isolated from bovine milk. J. Dairy Res. 78: 385–390.

Meng, X., T. Karasawa, K. Zou, X. Kuang, X. Wang, C. Lu, C. Wang, K. Yamakawa and S. Nakamura. 1997. Characterization of a neurotoxigenic *Clostridium butyricum* strain isolated from the food implicated in an outbreak of food-borne type E botulism. J. Clin. Microbiol. 35: 2160–2162.

Morse, P.M., H. Jackson, C.H. McNaughton, A.G. Leggatt, G.B. Landerkin and C.K. Johns. 1968. Investigation of factors contributing to the bacterial count of bulk tank milk. II. Bacteria in milk from individual cows. J. Dairy Sci. 51: 1188–1191.

Murphy, S.C. and K.J. Boor. 2000. Trouble-shooting sources and causes of high bacteria counts in raw milk. Dairy Food Environ. Sanit. 20: 606–611.

Munsch-Alatossava, O. and T. Alatossava. 2006. Phenotypic characterization of raw milk-associated psychrotrophic bacteria. Microbiol. Res. 161: 334–346.

Nada, S., D. Ilija, T. Igor, M. Jelena and G. Ruzica. 2012. Implication of food safety measures on microbiological quality of raw and pasteurized milk. Food Control 25: 728–731.

Newell, D.G., M. Koopmans, L. Verhoef, E. Duizer, A. Aidara-Kane, H. Sprong, M. Opsteegh, M. Langelaar, J. Threfall, F. Scheutz, J. van der Giessen and H. Kruse. 2010. Food-borne diseases—The challenges of 20 years ago still persist while new ones continue to emerge. Int. J. Food Microbiol. 139: S3–S15.

Nielsen, S.S. and N. Toft. 2009. A review of prevalences of paratuberculosis in farmed animals in Europe. Prev. Vet. Med. 88: 1–14.

Nickerson, S.C. and R.M. Akers. 2001. Anatomy. pp. 1680–1689. *In*: J.W. Fuquay, P.F. Fox and H. Roginski [eds.]. Encyclopedia of Dairy Sciences. Academic Press, New York.

Nörnberg, M.F.B.L., R.S.C. Friedrich, R.D.N. Weiss, E.C. Tondo and A. Brandelli. 2010. Proteolytic activity among psychrotrophic bacteria isolated from refrigerated raw milk. Int. J. Dairy Tech. 63: 41–46.

Okura, H., N. Toft and S.S. Nielsen. 2012. Occurrence of *Mycobacterium avium* subsp. *paratuberculosis* in milk at dairy cattle farms: A systematic review and meta-analysis. Vet. Microbiol. 157: 253–263.

Oliver, S.P., B.M. Jayarao and R.A. Almeida. 2005. Foodborne pathogens in milk and the dairy farm environment: Food safety and public health implications. Foodborne Path. Dis. 2: 115–129.

Oliver, S.P., K.J. Boor, S.C. Murphy and S.E. Murinda. 2009. Food safety hazards associated with consumption of raw milk. Foodborne Path. Dis. 6: 793–806.

Oliver, S.P., D.A. Patel, T.R. Callaway and M.E. Torrence. 2009b. Developments and future outlook for preharvest food safety. J. Anim. Sci. 87: 419–437.

Omiccioli, E., G. Amagliani, G. Brandi and M. Magnani. 2009. A new platform for real-time PCR detection of *Salmonella* spp., *Listeria monocytogenes*, and *Escherichia coli* O157 in milk. Food Microbiol. 26: 615–622.

O'Toole, D.K. 1983. Methods for the direct and indirect assessment of the bacterial content of milk. J. Appl. Bacteriol. 55: 187–201.

Paape, M.J., B. Poutrel, A. Contreras, J.C. Marco and A.V. Capuco. 2001. Milk somatic cells and lactation in small ruminants. J. Dairy Sci. 84(E. Suppl.): E237–E244.

Paduch, J.-H., E. Mohr and V. Krömker. 2012. The association between teat end hyperkeratosis and teat canal microbial load in lactating dairy cattle. Vet. Microbiol. 158: 353–359.

Pantoja, J.C.F., D.J. Reinemann and P.L. Ruegg. 2009. Associations among milk quality indicators in raw bulk milk. J. Dairy Sci. 92: 4978–4987.

Peck, M.W. 2009. Biology and genomic analysis of *Clostridium botulinum*. Adv. Microb. Phys. 55: 183–265.

Pettersson, B., F. Lembke, P. Hammer, E. Stackebrandt and F.G. Priest. 1996. *Bacillus sporothermodurans*, a new species producing highly heat-resistant endospores. Int. J. Syst. Bacteriol. 46: 759–764.

Pinto, U.M., E. de S. Viana, M.L. Martins and M.C.D. Vanetti. 2007. Detection of acylated homoserine lactones in gram-negative proteolytic psychrotrophic bacteria isolated from cooled raw milk. Food Control 18: 1322–1327.

Pitt, J.I. 2000. Toxigenic fungi and mycotoxins. Brit. Med. Bull. 56: 184–192.

Polonelli, L., L. Giovati, W. Magliani, S. Conti, S. Sforza, A. Calabretta, C. Casoli, P. Ronzi, E. Grilli, A. Gallo, F. Masoero and G. Piva. 2011. Vaccination of lactating dairy cows for the prevention of aflatoxin $B_1$ carry over in milk. PLoS One 6: e26777. doi:10.1371/journal.pone.0026777.

Postollec, F., A.-G. Mathot, M. Bernard, M.-L. Divanac'h, S. Pavan and D. Sohier. 2012. Tracking spore-forming bacteria in food: From natural biodiversity to selection by processes. Int. J. Food Microbiol. 158: 1–8.

Prandini, A., G. Tansini, S. Sigolo, L. Filippi, M. Laporta and G. Piva. 2009. On the occurrence of aflatoxin M1 in milk and dairy products. Food Chem. Toxicol. 47: 984–991.

Prempeh, H., R. Smith and B. Müller. 2001. Foot and mouth disease: the human consequences. Brit. Med. J. 322: 565–566.

Quaglia, N.C., A. Dambrosio, G. Normanno, A. Parisi, R. Patrono, G. Ranieri, A. Rella and G.V. Celano. 2008. High occurrence of *Helicobacter pylori* in raw goat, sheep and cow milk inferred by *glmM* gene: A risk of food-borne infection? Int. J. Food Microbiol. 124: 43–47.

Quigley, L., O. O'Sullivan, T.P. Beresford, R.P. Ross, G.F. Fitzgerald and P.D. Cotter. 2011. Molecular approaches to analyzing the microbial composition of raw milk and raw milk cheese. Int. J. Food Microbiol. 150: 81–94.

Raats, D., M. Offek, D. Minz and M. Halpern. 2011. Molecular analysis of bacterial communities in raw cow milk and the impact of refrigeration on its structure and dynamics. Food Microbiol. 28: 465–471.

Ramsahoi, L., A. Gao, M. Fabri and J.A. Odumeru. 2011. Assessment of the application of an automated electronic milk analyzer for the enumeration of total bacteria in raw goat milk. J. Dairy Sci. 94: 3279–3287.

Ranieri, M.L. and K. Boor. 2009. Bacterial ecology of high-temperature, short-time pasteurized milk processed in the United States. J. Dairy Sci. 92: 4833–4840.

Rasolofo, E.A., D. St-Gelais, G. LaPointe and D. Roy. 2010. Molecular analysis of bacterial population structure and dynamics during cold storage of untreated and treated milk. Int. J. Food Microbiol. 138: 108–118.

Raynal-Ljutovac, K., A. Pirisi, R. de Crémoux and C. Gonzalo. 2007. Somatic cells of goat and sheep milk: Analytical, sanitary, productive and technological aspects. Small Rumin. Res. 68: 126–144.

Rendos, J.J., R.J. Eberhart and E.M. Kesler. 1975. Microbial populations of teat ends of dairy cows, and bedding materials. J. Dairy Sci. 58: 1492–1500.

Roberts, R.F. and G.S. Torrey. 1988. Inhibition of psychrotrophic bacterial growth in refrigerated milk by addition of carbon dioxide. J. Dairy Sci. 71: 52–60.

Rohm, H., F. Eliskases-Lechner and M. Bräuer. 1992. Diversity of yeasts in selected dairy products. J. Appl. Bacteriol. 72: 370–376.

Romen, F., P. Backes, M. Materniak, R. Sting, T.W. Vahlenkamp, R. Riebe, M. Pawlita, J. Kuzmak and M. Löchelt. 2007. Serological detection systems for identification of cows shedding bovine foamy virus via milk. Virol. 364: 123–131.

Ruegg, P.L. 2003. Practical food safety interventions for dairy production. J. Dairy Sci. 86(E. suppl.): E1–E9.

Sanjuan, S., J. Rúa and M. Garcia-Armesto. 2003. Microbial flora of technological interest in raw ovine milk during 6°C storage. Int. J. Dairy Tech. 56: 143–148.

Scheldeman, P., K. Goossens, M. Rodriguez-Diaz, A. Pil, J. Goris, L. Herman, P. De Vos, N.A. Logan and M. Heyndrickx. 2004. *Paenibacillus lactis* sp. nov. isolated from raw and heat-treated milk. Int. J. Syst. Evol. Microbiol. 54: 885–891.

Scheldeman, P., A. Pil, L. Herman, P. De Vos and M. Heyndrickx. 2005. Incidence and diversity of potentially highly heat-resistant spores isolated at dairy farms. Appl. Environ. Microbiol. 71: 1480–1494.

Scheldeman, P., L. Herman, S. Foster and M. Heyndrickx. 2006. *Bacillus sporothermodurans* and other highly heat-resistant spore formers in milk. J. Appl. Microbiol. 101: 542–555.

Schofield, G.M. 1992. Emerging food-borne pathogens and their significance in chilled foods. J. Appl. Bacteriol. 72: 267–273.

Sculion, R., C.S. Harrington and R.H. Madden. 2006. Prevalence of *Arcobacter* spp. in raw milk and retail raw meats in northern Ireland. J. Food Prot. 69: 1986–1990.

Schukken, Y., M. Chuff, P. Moroni, A. Gurjar, C. Santisteban, F. Welcome and R. Zadoks. 2012. The "other" gram-negative bacteria in mastitis. Vet. Clin. Food Anim. 28: 239–256.

Seifu, E., E.M. Buys and E.F. Donkin. 2005. Significance of the lactoperoxidase system in the dairy industry and its potential applications: a review. Trends Food Sci. Tech. 16: 137–154.

Shah, A.H., A.A. Saleha, M. Murugaiyah, Z. Zunita and A.A. Memon. 2012. Prevalence and distribution of *Arcobacter* spp. in raw milk and retail raw beef. J. Food Prot. 75: 1474–1478.

Shah, N.P. 1994. Psychrotrophs in milk: a review. Milchwissenschaft 49: 432–437.

Shaheen, R., B. Svensson, M.A. Andersson, A. Christiansson and M. Salkinoja-Salonen. 2010. Persistence strategies of *Bacillus cereus* spores isolated from dairy silo tanks. Food Microbiol. 27: 347–355.

Sierra, D., A. Sánchez, A. Contreras, C. Luengo, J.C. Corrales, C. de la Fe, I. Guirao, C.T. Morales and C. Gonzalo. 2009. Effect of storage and preservation on total bacterial counts determined by automated flow cytometry in bulk tank goat milk. J. Dairy Sci. 92: 4841–4845.

Singh, P., A.A. Wani, A.A. Karim and H.-C. Langowski. 2012. The use of carbon dioxide in the processing and packaging of milk and dairy products: A review. Int. J. Dairy Tech. 65: 161–177.

Sørhaug, T. and L. Stephaniak. 1997. Psychrotrophs and their enzymes in milk and dairy products: Quality aspects. Trends Food Sci. Tech. 8: 35–41.

Souza, F.N., M.G. Blagitz, C.F.A.M. Penna, A.M.M.P. Della Libera, M.B. Heinemann and M.M.O.P. Cerqueira. 2012. Somatic cell count in small ruminants: Friend or foe? Small Rumin. Res. 107: 65–75.

Sperber, W.H. 2005. HACCP does not work from farm to table. Food Control 16: 511–514.

Streeter, R.N., G.F. Hoffsis, S. Bech-Nielsen, W.P. Shulaw and D.M. Rings. 1995. Isolation of *Mycobacterium paratuberculosis* from colostrum and milk of subclinically infected cows. Am. J. Vet. Res. 56: 1322–1324.

Sturino, J.M. and T.R. Klaenhammer. 2004. Bacteriophage defense systems and strategies for lactic acid bacteria. Adv. Appl. Microbiol. 56: 331–378.

Suhren, G. and J. Reichmuth. 2000. Interpretation of quantitative microbiological results. Milchwissenschaft 55: 18–22.

Tabata, S. 2011. Mycotoxins: Aflatoxins and related compounds. *In*: J.W. Fuquay, P.F. Fox and P.H.L. McSweeney [eds.]. Encyclopedia of Dairy Sciences, 2nd edition. Academic Press, New York.

Tacket, C.O., J.P. Narain, R. Sattin, J.P. Lofgren, C. Konigsberg Jr., R.C. Rendtorff, A. Rausa, B.R. Davis and M.L. Cohen. 1984. A multistate outbreak of infections caused by *Yersinia enterocolitica* transmitted by pasteurized milk. J. Am. Med. Assoc. 251: 483–486.

Tasara, T. and R. Stephan. 2006. Cold stress tolerance of *Listeria monocytogenes*: A review of molecular adaptive mechanisms and food safety implications. J. Food Prot. 69: 1473–1484.

te Giffel, M.C., A. Wagendorp, A. Herrewegh and F. Driehuis. 2002. Bacterial spores in silage and raw milk. Antonie van Leeuwenhoek 81: 625–630.

Thomas, S.B. 1966. Sources, incidence and significance of psychrotrophic bacteria in milk. Milchwissenschaft 21: 270–275.

Thompson, D.I. 1960. A plate loop method for determining viable counts of raw milk. J. Milk Food Tech. 23: 167–171.

Tomáška, M., G. Suhren, O. Hanuš, H.-G. Walte, A. Slottová and M. Hofericová. 2006. The application of flow cytometry in determining the bacteriological quality of raw sheep's milk in Slovakia. Lait. 86: 127–140.

Tomasula, P.M. and R.P. Konstance. 2004. The survival of foot-and-mouth disease virus in raw and pasteurized milk and milk products. J. Dairy Sci. 87: 1115–1121.

Tormo, H., C. Agabriel, C. Lopex, D. Ali Haimoud Lekhal and C. Roques. 2011. Relationship between the production conditions of goat's milk and the microbial profiles of milk. Int. J. Dairy Sci. 6: 13–28.

Uraz, G. and S. Çitak. 1998. The isolation of *Pseudomonas* and other Gram(–) psychrotrophic bacteria in raw milks. J. Basic Microbiol. 2: 129–134.

Vacheyrou, M., A.-C. Normabd, P. Guyot, C. Cassagne, R. Piarroux and Y. Bouton. 2011. Cultivable microbial communities in raw cow milk and potential transfers from stables of sixteen French farms. Int. J. Food Microbiol. 146: 253–262.

Vadillo Machota, S., M.J. Paya Vicens, M.T. Cutuli de Simon and S. Suarez Fernandez. 1987. Raw milk mycoflora. Milchwissenschaft 42: 20–22.

Van Brandt, L., K. Coudijzer, L. Herman, C. Michiels, M. Hendrickx and G. Vlaemynck. 2011a. Survival of *Mycobacterium avium* ssp. *paratuberculosis* in yoghurt and in commercial fermented milk products containing probiotic cultures. J. Appl. Microbiol. 110: 1252–1261.

Van Brandt, L., I. Van der Plancken, J. De Block, G. Vlaemynck, E. Van Coillie, L. Herman and M. Hendrickx. 2011b. Adequacy of current pasteurization standards to inactivate *Mycobacterium paratuberculosis* in milk and phosphate buffer. Int. Dairy J. 21: 295–304.

Van Driessche, E., K. Houf, J. Van Hoof, L. De Zutter and P. Vandamme. 2003. Isolation of *Arcobacter* species from animal feces. FEMS Microbiol. Lett. 229: 243–248.

Van Netten, P., A. Van de Moosdijk, P. Van Hoensel, D.A.A. Mossel and I. Perales. 1990. Psychrotrophic strains of *Bacillus cereus* producing enterotoxin. J. Appl. Bacteriol. 69: 73–79.

Van Schothorst, M. and T. Kleiss. 1994. HACCP in the dairy industry. Food Control 5: 162–166.

Vasavada, P.C. 1993. Rapid methods and automation in dairy microbiology. J. Dairy Sci. 76: 3101–3113.

Verdier-Metz, I., V. Michel, C. Deblès and M.-C. Montel. 2009. Do milking practices influence the bacterial diversity of raw milk? Food Microbiol. 26: 305–310.

Verdier-Metz, I., G. Gagne, S. Bornes, F. Monsallier, P. Veisseire, C. Deblès-Paus and M.-C. Montel. 2012. Cow teat skin, a potential source of diverse microbial populations for cheese production. Appl. Environ. Microbiol. 78: 326–333.

Wang, S.-Y., K.-N. Chen, Y.-M. Lo, M.-L. Chiang, H.-C. Chen, J.-R. Liu and M.-J. Chen. 2012. Investigation of microorganisms involved in biosynthesis of the kefir grain. Food Microbiol. 32: 274–285.

Watts, J.L. 1988. Etiological agents of bovine mastitis. Vet. Microbiol. 16: 41–66.

Weingart, O.G., T. Schreiber, C. Mascher, D. Pauly, M.B. Dorner, T.F.H. Berger, C. Egger, F. Gessler, M.J. Loessner, M.-A. Avondet and B.G. Dorner. 2010. The case of botulinum toxin in milk: Experimental data. Appl. Environ. Microbiol. 76: 3293–3300.

Wellenberg, G.J., W.H.M. van der Poel and J.T. Van Oirschot. 2002. Viral infections and bovine mastitis: a review. Vet. Microbiol. 88: 27–45.

Wheeler, T.T., G.A. Smolenski, D.P. Harris, S.K. Gupta, B.J. Haigh, M.K. Broadhurst, A.J. Molenaar and K. Stelwagen. 2012. Host-defense-related proteins in cows' milk. Animal 6: 415–422.

White, D.G., R.J. Harmon, J.E.S. Matos and B.E. Langlois. 1989. Isolation and identification of coagulase-negative *Staphylococcus* species from bovine body sites and streak canals of nulliparous heifers. J. Dairy Sci. 72: 1886–1892.

Wolfson, L.M. and S.S. Sumner. 1993. Antibacterial activity of the lactoperoxidase system: A review. J. Food Prot. 56: 887–892.

Yaqub, T., S. Sadaf, N. Mukhtar, H. Zaneb, M.Z. Shabbir, A. Ahmad, A. Aslam, K. Ashraf, N. Ahmad and H.U. Rehman. 2012. The influence of time elapsed after milking and lactation stage on the lactoperoxidase system of milk in Sahiwal cattle and Nili-Ravi buffaloes. Int. J. Dairy Tech. 65: 360–364.

Zdanowicz, M., J.A. Shelford, C.B. Tucker, D.M. Weary and M.A.G. von Keyserlingk. 2004. Bacterial populations on teat ends of dairy cows housed in free stalls and bedded with either sand or sawdust. J. Dairy Sci. 87: 1694–1701.

Zeng, S.S., S.S. Chen, B. Bah and K. Tesfai. 2007. Effect of extended storage on microbiological quality, somatic cell count, and composition of raw goat milk on a farm. J. Food Prot. 70: 1281–1285.

CHAPTER 3

# Dairy Pathogens: Characteristics and Impact

*Photis Papademas** and *Maria Aspri*

## INTRODUCTION

Milk and dairy products can be an ideal medium for the growth of a variety of microorganisms, both pathogenic and spoilage. Even though milk and milk products are amongst the safest food worldwide and account only a small percentage of all food-borne diseases, they have an inherent potential for causing illness as they are potentially the source of a very broad range of microbial, chemical, and physical hazards. The presence of food-borne pathogens in milk and milk products is due to direct contact with contaminated sources in the dairy farm environment, to excretion from the udder of an infected animal and during processing and handling.

Some important pathogens for milk and milk products include *Salmonella* spp., *Listeria monocytogenes, Staphylococcus aureus, Campylobacter* spp., pathogenic *Escherichia coli, Cronobacter sakazakii* and will be discussed in detail in this chapter.

Department of Agricultural Sciences, Biotechnology and Food Science, Cyprus University of Technology, P.O. Box 50329, 3603, Limassol, Cyprus.
Email: mg.aspri@edu.cut.ac.cy
* Corresponding author: photis.papademas@cut.ac.cy

**Figure 1.** Sources of Contamination of Milk and Dairy Products (After: Oliver et al. 2005)

## *Listeria monocytogenes*

### *Introduction*

*Listeria monocytogenes*, has been recognized as a significant food-borne pathogen, and has become one of the most studied microorganisms in the last decades (Ryser and Marth 2004). The organism was first identified as a causative agent for animal illness in 1926, when Murray, Webb, and Swann isolated *L. monocytogenes* as the cause of a septicemic disease affecting rabbits and guinea pigs in their laboratory at Cambridge in England. This strain was named Bacterium monocytogenes, as it was observed to infect monocytes of the blood.

### *Characteristics*

According to *Bergey's Manual of Systematic Bacteriology (1994) Listeria* genus includes: *L. monocytogenes, L. ivanovii, L. innocua, L. seeligeri, L. welshimeri* and *L. gray. L. monocytogenes* is pathogenic for humans and animals, and *L. ivanovii* is mainly pathogenic for animals, primarily sheep. Other species are considered to be non-pathogenic. It has thirteen serotypes, but the most common causes of disease are 1/2a, 1/2b, and 4b.

*L. monocytogenes is* a gram-positive, micro-aerophilic, non–spore-forming cocoid to rod shape, measuring 0.4–0.5 µm in diameter and 0.5–2 µm in length. The tumbling motility by means of peritrichous flagella is characteristic of *Listeria* and often used as a conventional marker for *Listeria* identification. The degree of motility is temperature dependent, for example

**Table 1.** Pathogens associated with milk and dairy products.

| **Milkborne Pathogens** | | |
|---|---|---|
| **Organism** | **Disease Characteristics** | **Associated Foods** |
| *Salmonella* spp. | Typhoid like fever, headache, abdominal pain, diarrhea, nausea, vomiting, constipation, malaise, chronic symptoms (arthritis) | Raw milk, dairy products Others: Eggs, meat fish, poultry, shrimp, salad dressings (especially home-made that contain raw eggs) |
| *Listeria monocytogenes* | Nausea, vomiting, diarrhea, fever, influenza like symptoms, meningitis, encephalitis. In pregnant women can lead to spotaneous abortion, premature delivery, or still birth | Fresh soft cheeses, raw milk, inadequately pasteurised milk, Others: Ready to eat deli-meals, hot dogs, raw and smoked fish |
| *Staphylococcus aureus* | Nause, vomiting, abdominal cramps, retching, prostration | Milk and dairy products Others: Meat and meat products, salads (chicken, potato, egg and makaroni), cream pastries |
| *Campylobacter jejuni* | Diarrhea, abdominal pain and cramps, fever, vomiting, headache | Raw milk Others: Raw and undercooked poultry, contaminated water, shellfish |
| *Bacillus cereus* | Diarrheal toxin: Watery diarrhea, abdominal cramps, nausea Emetic toxing: nausea and vomiting, diarrhea may be present | Milk, dairy products Others: Meats, vegetables, rice, pasta, soups, fish |
| *Escherichia coli* | Watery diarrhea, abdominal cramps, low-grade fever, nausea, malaise | Raw milk Others: Contaminated water, undercooked ground beef, raw apple juice and cider, alfalfa sprouts, cut melons |
| *Cronobacter sakazakii* | Meningitis, Enteritis | Powdered Infant formula |
| *Mycobacterium avium* ssp. *paratuberculosis (MAP)* | paratuberculosis, fever, chills, weight loss | Raw milk, pasteurised milk (few cells might survive) |
| *Shigella* spp. | Abdominal pains and cramps, diarrhea, fever, vomiting, stools may contain blood, pus or mucus | Milk and dairy products Others: bakery products (cream-filled pastries), sandwitches, raw vegetables and salads (potato, tuna, chicken, macaroni) |
| *Yersinia enterocolytica* and *Y. pseudoturbeculosis* | Fever, abdominal pain, diarrhea and vomiting | Raw milk, Others: oysters and fish, Meat |

it shows mobility when growth temperature is between 20 and 25°C but reduces on 37°C. *L. monocytogenes* is not a fastidious organism and grows well in most common nutrient media, including brain heart infusion broth

(BHI), trypticase-soy broth with 0.6% yeast extract (TSBYE), Luria broth (LB), etc. *L. monocytogenes* is catalase positive, oxidase negative and ferments rhamnose, dextrose, esculin, and maltose, but does not ferment xylose and manitol. These biochemical properties are used for the differentiation of *L. monocytogenes* from other members of the genus *Listeria.*

The organism's ability to grow and reproduce in harsh conditions makes it a food-borne pathogen of great concern. Optimum growth temperature is 30–37°C, although *L. monocytogenes* grows reasonably well at temperatures over a wide range of temperature from 0–42°C. The growth rate reduces as the growth temperature decreases. The pH range for growth is 5–9 with an optimum pH of 6–8. However, these values depend on the acidulant, strain, and temperature. *L. monocytogenes* is also quite salt tolerant being able to grow in 10% sodium chloride. The organism is ubiquitous in the environment. It has been isolated from fresh and salt water, soil, sewage sludge, decaying vegetation and silage.

### *Food-borne Illness*

*Listeria monocytogenes* is the causative agent for the disease Listeriosis, one of the most important foods borne illness of humans. Listeriosis is a zoonotic illness that affects both animal and humans and is transmitted via three main routes: direct contact with animals, cross-infection of new-born babies in hospital and food-borne infection (Ryser and Marth 2004). The latter two sources result in the majority of listeriosis cases in humans. As previously noted, *L. monocytogenes* is widely distributed in the environment. Thus humans can come into contact with this pathogen through a variety of sources including animals, meats, milk, dairy products, seafood, and plants as well as insects, air, dust, dirt, feces and other humans. Also, it can be transmitted by cross-infection of newborn babies in hospitals and via food vehicles.

Pregnant women, newborns, elderly and immunocompromised individuals are more susceptible to the disease and experience a more severe illness. Listeriosis is a very serious and often fatal infection with a fatality rate that often reaches as high as 30–40%, even though the morbidity of listeriosis is relatively low (Liu 2006).

Symptoms of listeriosis range from flu-like vomiting and diarrhea to septicemia, meningitis and meningoencephalitis. Listeriosis refers to the more serious life-threatening illness while gastroenteritis is the mild illness experienced by healthy adults.

In pregnant women, it most commonly features as an influenza-like illness with fever, headache and occasional gastrointestinal symptoms, but there may be associated transplacental foetal infection which can result in abortion, still birth, premature labour or birth of a severely ill infant.

*L. monocytogenes* is one of the few bacteria capable of crossing the placenta and gain direct access to the fetus. Newborn babies may also acquire infection after birth from the mother or from other infected infants. Listeriosis in the newborn can be an early-onset syndrome, which occurs at birth or shortly afterwards, or a late-onset disease appearing several days to weeks after birth. Early-onset illness results from utero infection, possibly through the aspiration of infected amniotic fluid, and is characterised by pneumonia, septicaemia and widely disseminated granulomas. In immunocompromised and elderly adults, the illness typically involves infection of the tissues surrounding the brain (meningitis) and infection of the bloodstream (septicemia). Healthy adults are thought rarely to suffer from listeriosis. However, high levels of this organism can cause symptoms in healthy individuals that are similar to flu-like symptoms of vomiting, nausea and diarrhea.

The infective dose (ID) of *L. monocytogenes* is very unclear. The minimum dose of bacteria required to cause clinical infection depends on several factors that include the virulence mechanism of the microorganism, immune status of the host, the concentration of the pathogen in the contaminated food, as well as the food and the amount of food consumed. Based on documented outbreaks of listeriosis, foods that contained around $10^2$–$10^4$ cfu/g have been responsible in various outbreaks (Ooi and Lorber 2005).

### Detecting L. monocytogenes

There is a variety of conventional and rapid methods currently available for the detection and identification of *L. monocytogenes* in food samples and specimens from animal listeriosis. The most commonly used culture reference methods for the detection of *Listeria* in foods are the ISO 11290 standards (International Organization for Standardization 1996). In the United States of America (USA) two main standards are used as reference methods to isolate *L. monocytogenes* from foods. One of the protocols was developed by the US Food and Drug Administration (FDA) to isolate *Listeria* spp. from dairy products, seafood, and vegetables (Hitchins 2003). The US Department of Agriculture (USDA) developed another method to isolate the organism from meat and poultry products as well as from environmental samples (USDA 2013). All of the methods require an enrichment step to increase the viable cell count to detectable numbers. Enrichment steps are followed by plating onto agars containing selective/differential agents (Jeyaletchumi et al. 2010). Oxford agar was the medium of choice of the International Organization for Standardization (ISO) until 1998, when it was replaced by PALCAM agar. In 2004, it was replaced by a chromogenic agar, ALOA (Agar Listeria Ottaviani Agosti), which allows the differentiation of *L. monocytogenes* and *L. innocua*. The detection in the first two plating media is based on the hydrolysis of aesculin which does not differentiate between species. In ALOA, the detection is based on the phosphatidylinositol phospholipase C enzyme (PI-PLC) activity present in *L. monocytogenes* and in *L. ivanovii*, and b-glucosidase activity detected by a chromogenic substrate (Gasanov et al. 2005).

The confirmation of suspected colonies can be performed by testing a limited number of biochemical markers such as hemolytic activity (CAMP test) and sugar fermentation patterns. Figure 2 summarises the ISO 11290 method for the detection of *L. monocytogenes*. Rapid methods for the identification of *L. monocytogenes* include enzyme-linked immunosorbent assay (ELISA) kits, nucleic acid assay kits, and Polymerase chain reaction (PCR) (Jemmi and Stephan 2006). For a detailed account of the molecular methods employed for the identification of bacteria in dairy products see Chapter 6.

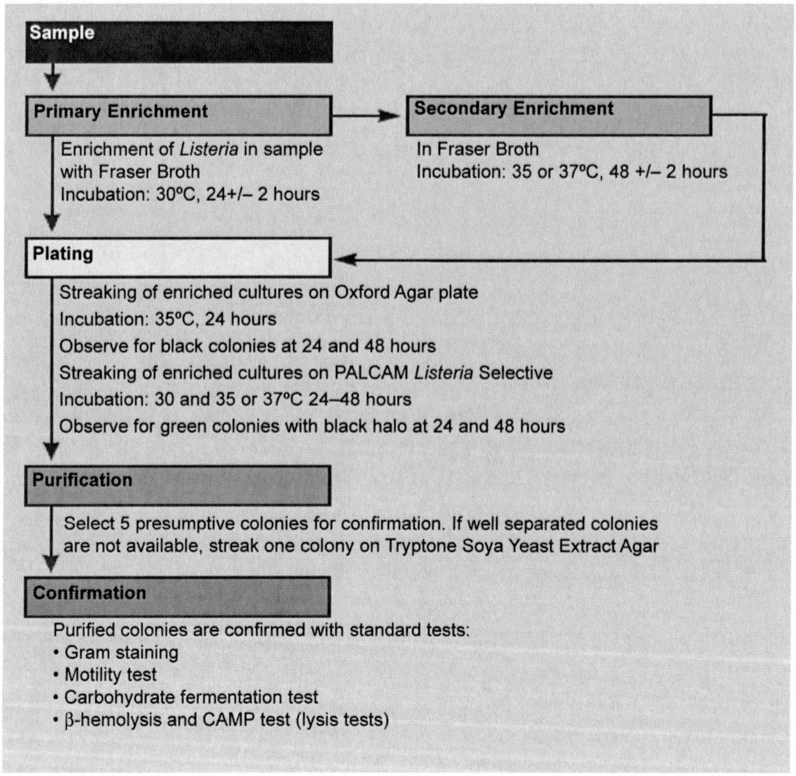


**Figure 2.** ISO 11290 for detection of *L. monocytogenes* in food samples (After: Scharlau 2007).

*Color image of this figure appears in the color plate section at the end of the book.*

### *Association with Milk and Milk Products*

Dairy products such as raw and pasteurized milk and soft cheeses have been associated with a number of major outbreaks of listeriosis (Kasalica 2011), since Listeria population may be extremely acid and salt resistant.

*L. monocytogenes'* association with soft cheeses is due to the cheese ripening process. It survives poorly in unriped soft cheeses such as cottage cheese but well in products such as Camembert and Brie (especially if made from raw milk). During this ripening process microbial utilization of lactate and release of amines increase the surface pH allowing *Listeria* spp. to multiply. Hard cheese such as Parmesan (Parmigiano Reggiano) does not favour growth (Yousef and Marth 1990) while in white-brined cheeses, i.e., Feta, the acidic environment of the cheese (pH 4.6) will inhibit the growth of *Listeria* spp. but the pathogen will survive even for 90 days (Papageorgiou and Marth 1989). It was also reported by Belessi et al. (2008) that the growth of *L. innocua* in Feta cheese was assisted by the presence of fungi. The findings indicated that the growth of fungi on the surface of Feta cheese and yogurt may compromise the safety of these products by enhancing survival of the bacterium. Particularly, when fungi increase the pH of Feta cheese, *L. innocua* demonstrates better survival and prolonged storage may raise concerns for the development of acid-resistant Listeria populations.

In high-scalding cheeses, i.e., Mozzarella, Halloumi the bacteria will not survive the stretching/cooking temperatures. For fresh Halloumi cheese (pH 6.2) post-scalding contamination may result in a favorable environment for *Listeria* spp. growth therefore, careful handling of the cheese is crucial.

Gougouli et al. (2008) showed that in ice cream the freezing conditions did not cause a severe stress in *L. monocytogenes* cells capable of leading to a significant "additional" lag phase during the subsequent growth of the pathogen at chilling conditions. Additionally they observed that under freezing conditions, no significant changes in the population of the pathogen were observed throughout a 90-d storage period for either of the inoculum levels tested. The above statements highlight the importance of post-pasteurisation hygiene conditions to be employed in all dairy factories.

## *Salmonella* spp.

### *Introduction*

*Salmonella* are one of the most important food-borne pathogens and have been recognized for over 100 years as the cause of illnesses ranging from mild to severe food poisoning (gastroenteritis), and even more severe typhoid (enteric fever), paratyphoid, bacteraemia, septicaemia and a variety of associated longer-term conditions (sequelae).

### *Characteristics*

*Salmonella* is facultative anaerobe, gram negative flagellated rod-shaped bacterium belonging to the family of Enterobacteriacae. The size of the rods is about 2–3 x 0.4–0.6 µm in size. They are non-spore forming, oxidase negative and catalase positive and most members of the genus are motile by peritrichous flagella. They are generally: able to reduce nitrate to nitrite, able to grow on citrate as sole carbon source, capable of producing acid and gas from glucose, able to produce hydrogen sulfide on triple sugar iron and decarboxylate lysine and ornithine, and able to hydrolyze indole and urea.

Strains of *Salmonella* are antigenically distinguishable by agglutination (formation of aggregates/clumping) reactions with homologous antisera and the combination of antigens possessed by each strain, referred to as the antigenic formula, is unique to each *Salmonella* serotype. These are classified in two species, *S. enterica* and *S. bongori*. *S. enterica* is further divided into six subspecies, *Salmonella enterica* subsp. *arizonae*, *Salmonella enterica* subsp. *diarizonae*, *Salmonella enterica* subsp. *enterica*, *Salmonella enterica* subsp. *houtenae*, *Salmonella enterica* subsp. *indica*, and *Salmonella enterica* subsp. *salamae* (Tindall et al. 2005).

Optimum growth for salmonellae occurs at 35–37°C (mesophilic), but they can grow at temperature from 5°C to 46°C, depending on serotype (El-Gazzar and Marth 1992). Although optimal growth pH is between 6.5 and 7.5, salmonellae can grow in more acidic environments (pH 4). They do not require sodium chloride for growth, but can grow in the presence of 0.4 to 4%. They are sensitive to heat and often killed at temperature of 70°C or above (pasteurization temperature). They require high water activity ($a_w$) between 0.99 and 0.94 but can survive very well in dry foods ($a_w < 0.2$) (Pui et al. 2011).

### *Food-borne Illness*

Illness caused by Salmonella serotypes known as salmonellosis. Infection is initiated by consumption of raw or undercooked contaminated animal food (common source of infection for humans) or water containing fecal material. In addition to transmission by food, *Salmonella* can also be spread via the environment in a number of ways, indirectly infecting humans and other animals, as it can survive for long periods of time under both wet and dry conditions. Transmission can also occur through direct contact with ill or in apparently infected farm animals or ill pets such as cats and dogs. The outcome of this infection largely depends on the serotype and type of host. Serotypes Typhimurium and Enteritidis can cause disease in humans, cattle, poultry, sheep, pigs, horse and wild rodents.

There are two clinical types of human salmonellosis: enteric fever (a severe, life-threatening illness) following infection with *S. typhi* or *paratyphi* and the more common food-borne illness syndrome caused by nontyphoid Salmonella species. In both cases, the responsible microorganisms enter the body via the oral route.

Salmonellosis generally is a self-limiting acute gastroenteritis similar to intestinal influenza, and is believed to be grossly underreported. Onset of non-typhoidal salmonellosis typically occurs 12 to 36 hours after ingestion of the contaminated food characterized by nausea and vomiting; symptoms which tend to subside within a few hours. Symptoms include nausea, vomiting, chills, fever, abdominal pain or cramps, and headache, is typically followed by diarrhea (El-Gazzar and Marth 1992). The illness usually lasts 4 to 7 days and most persons recover without antibiotic treatment. The elderly, infants and those with underlying chronic illness or immuno-compromised individuals are more likely to have a severe illness.

In some cases septicemia can occur as a complication of gastroenteritis which can be fatal in immunocompromised hosts. Prolonged septicemic infections can result in localized tissue and organ infections, especially in those previously damaged or diseased. The severity and duration of symptoms depends upon the concentration of the pathogens in the food, the physiological composition and the type of the food consumed the susceptibility of the host, and the virulence of the pathogen. As few as 15 cells can cause illness.

*S. typhi* is responsible for bacteremia-related enteric fever referred to as typhoid fever. Onset typically occurs within 8 to 15 days, and sometimes as long as 30 to 35 days. Symptoms include fever, headache, malaise, anorexia, and congestion of the mucous membranes, especially of the upper respiratory tract. The high mortality rate of *S. typhi* (10%) compared to other *Salmonella* spp. can be reduced with the prompt administration of antibiotics.

### Detecting *Salmonella* spp.

Culture and colony counting methods, Polymerase Chain Reaction (PCR) as well as immunology-based methods are the most common tools used for pathogens detection including *Salmonella* detection. They involve counting of bacteria, DNA analysis and antigen-antibody interactions, respectively. These methods are often combined together to yield more robust results.

Detection of *Salmonella* in foods by conventional culture methods consist of four steps 1) non selective pre-enrichment, 2) selective enrichment in different media, 3) plating on selective and indicative media, and 4) confirmation through biochemical and serological tests (ISO 6579: 2002). The culture method is time-consuming and labor intensive, requiring a minimum of 4–6 days.

Pre-enrichment in a non selective medium needed to increase the recovery of sub lethally damaged cells of salmonella. Buffered peptone water and lactose broth are two of the most widely used media for pre-enrichment.

The second step is enrichment in selective media. This increases the number of salmonella to a level where detection on selective agar plates is possible, while at the same time restricted the growth of other microorganisms present by selective agents in the media. The most commonly used media for selective enrichment are Rappaport-Vassiliades soy broth, selenite cysteine broth, and tetrathionate broth. Modified semisolid Rappaport-Vassiliades is another selective media particularly useful for detecting salmonella in feces and environmental samples.

From the selective enrichment broths, cultures are streaked on selective solid media in order to obtain isolates colonies. Most standard methods recommend employing two media in parallel. Commonly employed media include brilliant green, xylose lysine Tergitol-4, bismuth sulfite, Hektoen enteric, and xylose lysine deoxycholate agars. Recently, chromogenic agars have also been used.

Finally, presumptive Salmonella colonies from selective media are subcultured on nonselective plates in order to get well-isolated colonies that can be used for confirmation by biochemical and serological analysis. Most common biochemical reactions for salmonella confirmation include triple sugar iron, mannitol, urea, ornithin decarboxylase, and lysine decarboxylase. A serological verification by determining the antigenic composition is performed. The O and H antigens are determined by agglutination testing using polyvalent antisera. Miniaturized tests such as API 20 E (bioMerieux) and BBLTM EnterotubeTM II (BD Diagnostics) have been developed for confirmation and they present an efficient and labour-saving alternative. Figure 3 summarises the International Organization for Standardization method for the detection of Salmonella.

Following confirmation of the identity of *Salmonella*, it is important for food surveillance purposes and the investigation of outbreaks, to subtype or 'fingerprint' *Salmonella* serotypes. There are a variety of techniques available now that can allow confident traceability of strains in factory environments. These include biotyping, serotyping (including variation in H antigens), phage typingantibiotic resistance patterns (resistotyping), various molecular typing methods including pulsed-field gel electrophoresis (PFGE) and ribotyping.

### Association with Milk and Milk Products

Outbreaks of human salmonellosis have been linked to the consumption of unpasteurized milk or milk products. In addition, inadequate pasteurizarion

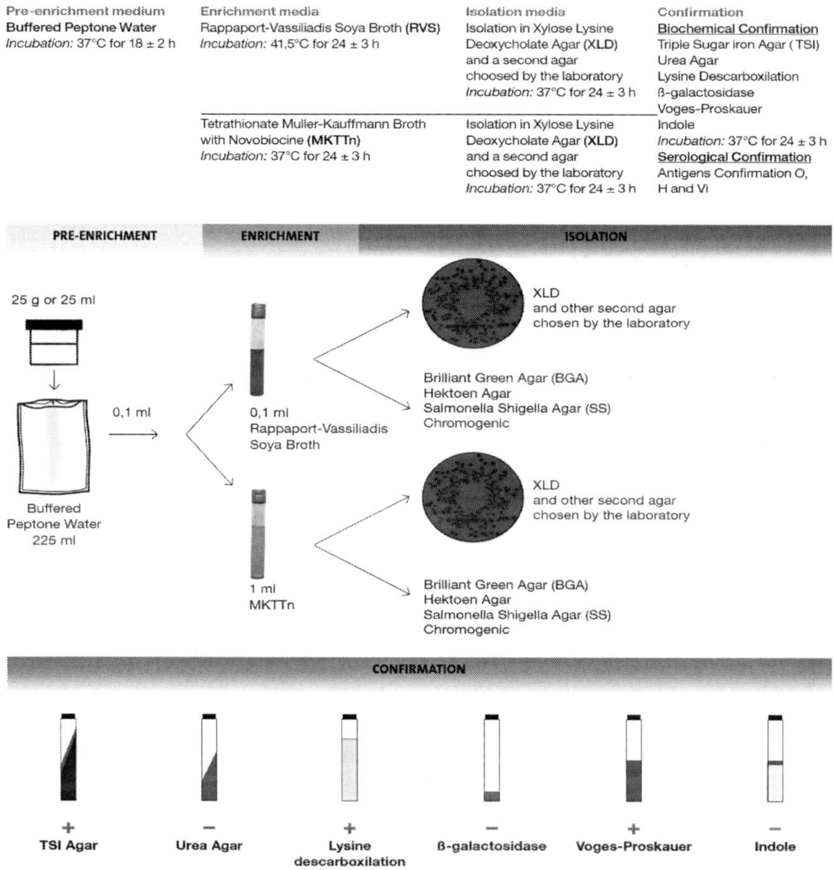


**Figure 3.** ISO 6579 Detection of Salmonella in Food (After: Scharlau 2007).

*Color image of this figure appears in the color plate section at the end of the book.*

or post process contamination have occasionally resulted in milk and milk products that tested positive for *Salmonella* spp. Entry of *Salmonella* spp. via contaminated raw milk into dairy food processing plants can lead to persistence of this pathogen in biofilms, and subsequent contamination of processed milk products and exposure of consumers to the pathogen.

Presumably, the largest outbreak of Salmonellosis ever reported in the USA occurred in Illinois (Ryan et al. 1987), where an antimicrobial-resistant strain of *Salmonella typhimurium* yielded 16,000 culture-confirmed cases traced to two brands of pasteurized 2% milk produced by a single dairy plant. The authors reported that the study of stored isolates showed it had caused clusters of salmonellosis during the previous ten months that may have been related to the same plant, suggesting that the strain had persisted in the plant and repeatedly contaminated milk after pasteurization.

## *Staphylococcus aureus*

### *Introduction*

*Staphylococcus aureus* was first described in 1880 by the Scottish surgeon, Sir Alexander Ogston from pus in a knee joint abscess. The first description of food-borne illness dates back 1884, when a spherical organism in cheese caused a large food-poisoning outbreak in Michigan in the United States. Since then we know that S. aureus is a very frequent microorganism in humans and animals with approximately 20–50% of persons being long-term carriers, mainly as a part of the normal skin flora and in anterior nares of the nasopharynx, but also the throat and hair.

### *Characteristics*

*Staphylococcus aureus* subsp. *aureus* (*S. aureus*) belongs to the genus *Staphylococcus* and to the family *Staphylococcaceae,* which includes the lesser known genera of *Gemella, Jeotgalicoccus, Macrococcus,* and *Salinicoccus.* The name of the organism is derived from Greek words *staphyle* (a bunch of grapes) and *coccus* (grain or berry).

*S. aureus* are spherical, coccus shaped Gram positive bacteria of 0.5–1.5 µm in diameter that appear in pairs, short chains, or bunched, grape-like clusters. They are non motile, non-spore forming facultative anaerobes which ferment most of the sugars except raffinose and salicin producing lactic acid during fermentation. They are catalase and coagulase positive and oxidase negative.

In general, *S. aureus* is mesophile with growth temperature range from 7 to 48°C, with an optimum temperature for growth of 35°C. The pH range for growth is between 4.5 and 9.3, with the optimum between pH 7.0 and 7.5 (Singh and Prakash 2010).

With regard to water activity ($a_w$), the staphylococci are unique in being able to grow at lower levels than other non-halophilic bacteria. Growth has been demonstrated at as low as 0.8 Aw under ideal conditions. These low Aw conditions are too low for the growth of many competing organisms. Thus, it is highly tolerant to salts and sugars and has historically been responsible for food-borne disease outbreaks illnesses from contaminated hams (salt) and pies (sugar).

### *Food-borne Illness*

Staphylococcal food poisoning, also known as staphyloenterotoxicosis or staphyloenterotoxemia, is the name of the condition caused by the ingestion of SEs (Le Loir et al. 2003). Staphylococcal food poisoning occurs not as the result of the ingestion of the organism itself, but through ingestion of one or more of the 14 staphylococcal enterotoxins (SE A-N) produced by some strains of *S. aureus*. With 30–50% of the population carrying *S. aureus* in their nostrils and on skin and hair, foods can become contaminated quite readily prior to or following heat treatment during processing and handling.

Staphylococcal enterotoxins are heat-stable exoproteins consisting from 236 to 296 aminoacids with a molecular mass of 25–35 kDa. Upon hydrolysis, 18 amino acids are present, mostly aspartic acid, glutamic acid, lysine and tyrosine.

There are five different types of classical enterotoxins (SEA-SEE) which are distinct in antigen reaction. Recently, new types of enterotoxins and enterotoxin-like types (SEG-SEV) have been described in *S. aureus*. There are several enterotoxins but only, SE A, B, C1, C2, C3, D, E, G, H, I, and J have been identified (Argudin et al. 2010). The most common toxin and the one that usually involved in staphylococcal outbreaks is Enterotoxin A and data from staphylococcal outbreaks indicate that less than 1 µg of toxin can result in illness.

The onset of symptoms including nausea, vomiting, retching, diarrhea, and abdominal cramps, normally develops within 1 to 6 hours following ingestion of the contaminated food. A toxin dose of less than 1.0 µg in contaminated food will produce symptoms of staphylococcal intoxication. This toxin level is reached when *S. aureus* populations exceed $10^6$ cfu. However, in highly sensitive people a dose of 100–200 ng is sufficient to cause illness. Depending on individual susceptibility to the toxin, the amount of toxin ingested and the general health of the affected individual, symptoms may also include headaches, cold sweats, and rapid pulse, transient changes in blood pressure, prostration and dehydration. Recovery generally takes one to two days rarely resulting in complications or hospitalization.

### *Detecting S. aureus*

The most successful and widely used medium for *S. aureus* is called Baird Parker. This contains tellurite, glycine and lithium chloride as selective agents, pyruvate and egg yolk to assist the recovery of damaged cells. Reduction of the tellurite by *S. aureus* gives characteristic shiny, jet-black colonies which are surrounded by a zone of clearing resulting from hydrolysis of the egg yolk lipovitellenin and often with an inner cloudy zone due to the precipitation of fatty acid salts. The appearance of the colonies on Baird-Parker agar gives a presumptive identification of *S. aureus* which is confirmed through the production of coagulase (Asperger and Zangerl 2003).

Molecular techniques based on the polymerase chain reaction (PCR) have been developed using specific primers directed against a variety of sequences in the *S. aureus* genome including genes for thermostable nuclease and enterotoxins. There are also kits available for detection based on nucleic acid hybridization. Enterotoxins can be detected based on immunological methods following extraction from the food matrix, and there are numerous kits available using maenzyme linked immunoassay or latex agglutination.

### *Association with Milk and Milk Products*

Outbreaks of staphylococcal poisoning have been linked to milk and milk products for over 100 years with *S. aureus* emerging as a major milk-borne pathogen. Milk and dairy products constitute 1–9% (mean 4.8%) of all *S. aureus* outbreaks in Europe. Most outbreaks linked to the use of raw milk due to mastitic dairy cows. Although *S. aureus* inactivated by standard heat treatments, food-borne outbreaks related to pasteurized products occur as the result of staphylococcal enterotoxin production prior to heat treatment or as a result of post-pasteurization contamination. An excellent, comprehensive report published by the European Food Safety Authority (EFSA 2003) describes the Staphylococcal enterotoxins (SE) in dairy products. The report states that the conditions favouring SE production in milk and dairy products are (a) Liquid milk is an excellent medium for the growth and SE production (favourable pH, water activity and nutrients) and at temperatures 7–43°C, numbers of enterotoxinogenic *S. aureus* may increase rapidly even when the initial levels are low, (b) in cheese manufacturing enterotoxinogenic *S. aureus* can multiply and produce enterotoxin during the first stages of production when the pH of the curd is higher than 5.0 and the competing LAB bacteria have not reach high number. This favorable for multiplication of *S. aureus* period extends, in the different types of cheeses from several (5–10) to 48 hours at the maximum (c) experiments for enterotoxins production in pasta filata cheeses, internal mould ripened cheeses and processed cheeses failed to prove the presence of enterotoxin whatever the size of the inoculum of enterotoxinogenic *S. aureus* was used, (d) on the contrary, whey cheeses and imitation cheeses appears to be a favourable environment for growth of *S. aureus* and even small inoculum ($10^2$–$10^3$ cfu/g) can result in enterotoxin production, (e) In cream production critical stage is considered the period of cream ripening if the process is conducted in favorable for *S. aureus* growth, and (f) in milk powder and ice cream production critical stages are the pre-drying and pre-freezing periods in which *S. aureus* can multiply if temperature is favorable.

## *Cronobacter sakazakii*

### *Introduction*

*Cronobacter* spp. (*Enterobacter sakazakii*) is an opportunistic food-borne pathogen that has been linked with serious infections in infants through dried infant milk formula, which causes bacteraemia and meningitis and is associated with necrotizing enterocolitis (NEC). Most cases of *Cronobacter* infections have been detected among newborn and very young infants. The first reported cases attributed to this organism occurred in 1958 in England and resulted in the death of two infants (Gurtler et al. 2005).

### *Characteristics*

It is a member of the family Enterobacteriaceae, and belongs to the genus Cronobacter. This organism was known as yellow pigmented Enterobacter cloacae until 1980 after which it was renamed as *E. sakazakii* (Farmer et al. 1980). The new name *E. sakazakii* was proposed based on differences between *E. sakazakii* and *E. cloacae* in DNA-DNA hybridization, some biochemical traits, production of yellow-pigmented colonies and antibiotic susceptibility. Recently, *E. sakazakii* was reclassified as a genus *Cronobacter* compromising five spp.: *C. sakazakii, C. malonaticus, C. tuicensis, C. muytjensii* and *C. dublinensis* (Iversen et al. 2008).

*C. sakazakii* is a Gram-negative rod bacterium of 0.3 to 10 μm long by 1.0 to 6.0 μm wide that are motile by peritrichou flagellated. It is facultative anaerobic, oxidase negative, nonspore forming, nonacid-fast bacterium.

The temperature growth range of *Cronobacter* is 6–47°C with an optimal range of 37–43°C (Iversen and Forsythe 2003). However, some strains are inhibited at temperatures above 44°C and some strains have the ability to grow at 5°C. In the Enterobacteriaceae family, *Cronobacter* spp. is amongst the most thermo-tolerant. *Cronobacter* spp. have the ability to survive in acidic environments with pH levels as low as 3. Of the members of the Enterobacteriaceae family, *Cronobacter* spp. also appears to be well adapted to dry stress.

### *Food-borne Illness*

*C. sakazakii* is an emerging human pathogen with mortality ranges from 40 to 80% among infected infants, and those who survive the infection usually develop irreversible neurological sequelae. *C. sakazakii* has been implicated in severe forms of neonatal infections, such as meningitis, bacteraemia, sepsis and necrotizing enterocolitis. *C. sakazakii* can cause also systemic respiration, cardiovascular and neurologic symptoms such as destruction of the frontal lobes of the brain, seizures, spastic quadriplegia, hypothermia, fever, Cheyne-Stokes respirations, bradycardia, poor feeding, irritability, jaundice, grunting respirations, instability of body temperature, hemorrhagic cerebral necrosis, meningo encephalitis, necrotic softened brain, cyst formation, liquefaction of cerebral white matter and severe neurologic complications (Arsalam et al. 2013). Low birth-weight neonates 45 (< 2.5 Kg) and infants < 28 days of age are at higher risk compared to more mature infants (Iversen and Forsythe 2003).

Sources of contamination include floor drains, air, vacuum, canister, broom bristles, room heater, electrical control box, transition socks and condensate in a dry product processing plant (Shaker et al. 2007). Cronobacter sakazakii infections are treated with antibiotics such as carbapenems or antipseudomonal penicillins (i.e., mezlocillin, piperacillin, piperacillin/tazobactam, ticarcillin, and ticarcillin/clavulanate) (Dumen 2010).

### *Detecting C. sakazakii*

The isolation of *C. sakazakii,* from dried foods requires a series of steps to resuscitate stressed cells that would otherwise not be cultured. Methods for isolation and enumeration of *C. sakazakii* from powdered formula have improved in the last years. These methods include the ISO/TS 22964:2006, the US-FDA(2002), and the revised method of the US-FDA which combines both a PCR assay and two newly developed chromogenic agars for detection (Yan et al. 2012).

Table 2 compares the available methods for the detection of *C. sakazakii.* According to US Food and Drug Administartion (FDA) protocol enrichment of the PIF samples in water overnight is required. This is followed by a second enrichment step in Enterobacteriaceae enrichment (EE) broth for up to 24 h. Then samples are plated on violet red bile glucose agar overnight and presumptive colonies purified onto TSA. The resulting yellow-pigmented colonies are selected for biochemical tests using analytical profile index (API) 20E test strips. On the other hand, the International Organization for Standardization method consists of pre-enriching the PIF samples in buffered peptone water at 37°C overnight. Then the samples are enriched in modified lauryl sulfate (addition of vancomycin) at 44°C overnight. Following the second enrichment, the samples are plated on to chromogenic agar and incubate at 25°C for a period of 48–72 h.

**Table 2.** Protocols for detection of *C. sakazakii*.

| Procedure | FDA (Original) | ISO 22964 | FDA (revised) |
|---|---|---|---|
| **Pre-enrichment** | Make 1:10 (w/v) of sample in distilled water, incubate overnight at 36°C | Make 1:10 (w/v) of sample in BPW, incubate at 37°C for 18 ± 2 h | Make 1:10 (w/v) of sample in BPW, incubate at 36°C for 6 h |
| **Selective Enrichment** | Transfer 10 ml pre-enrichment to 90ml EE broth, incubate overnight at 36°C | Transfer 100 µl pre-enrichment to 10 ml mLST/Vancomycin medium incubate at 44°C for 24 ± h | |
| **Selection/Isolation** | Make an isolation streak and spread plate from each EE broth onto VRBG agar, incubated overnight at 36°C | Streak from the cultures mLST/ Vancomycin medium one loopfull on the chromogenic agar and incubate at 44°C for 24 ± 2 h | Centrifuge 40 ml samples, 3000 g x 10 min and resuspend pellet in 200 µl PBS; Spread 100 µl onto chromogenic media and incubate overnight at 36°C |
| **Confirmation** | Pick 5 presumptive positive colonies and streak onto TSA and incubate overnight at 25°C | Select five typical colonies and streak on TSA agar, incubate at 25°C for 48 ± 4 h | Pick two typical colonies from each chromogenic media and confrmed with real-time PCR, API20E , Rapid ID 32E |
| **Identification** | Yellow colonies are confirmed with the API20E test kit | Select one colony from each TSA plate for biochemical characterization | |
| **Detection time (days)** | 5 | 6 | 3 |

Currently molecular detection methods such as polymerase chain reaction (PCR), real-time PCR, immunoassays and DNA microarray-based assays are used for its identification.

### *Association with Milk and Milk Products*

*C. sakazakii* is considered a ubiquitous microorganism it can be found in a wide variety of foods and in waters and several areas including hospitals and houses. While this organism has been isolated from a variety of foods such cheese, fermented bread, tofu, sour tea, cured meats, minced beef and sausage meat, food factories and environments, most infections are associated with rehydrated infant milk formula (RIMF). Powered infant milk formula (PIMF) is basically a non-sterile product which can be, once rehydrated, a good medium for microorganisms. Therefore the presence of *C. sakazakii* in PIMF is mainly due to post-processing, environmental contamination, the addition of contaminated ingredients during powder production or is due to colonization by *C. sakazakii* of utensils such as bottles, brushes and spoons used in PIMF preparation (Shaker et al. 2007).

## *Bacillus cereus*

### *Introduction*

There are several Bacillus species that have been responsible for food-borne illness, although the only one that is frequently involved is *Bacillus cereus.* Hauge (1955) was the first to establish *B. cereus* as a food poisoning organism.

### *Characteristics*

The *Bacillus cereus* group is a very homogeneous cluster within the genus Bacillus and consists of six species: *B. cereus, B. thuringiensis, N. anthracis, B. mycoides, N. pseudomycoides* and *B. weihenstephanensis* (Ehling-Schulz et al. 2004). *B. cereus* is a Gram-positive, catalase positive motile, facultative, aerobic sporeformer. Its cells are rod-shaped, hence the name "Bacillus" ("rod"). The name "cereus" ("wax") was given because its colonies have a waxy appearance on agar plates. Dimensions of vegetative cells are typically 1.0–1.2 µm in diameter by 3.0–5.0 µm in length, with rounded or square ends and are often arranged in pairs or chains (Arnesen et al. 2008). The temperature range for the growth of *B. cereus* has been reported to be between 8–55°C with the optimum being in the range of 28–35°C. The range of pH for growth of *B. cereus* has been reported to be 4.9–9.3 with the optimum being 6.0–7.0, and a water activity ($a_w$) greater than 0.94.

### Food-borne Illness

*Bacillus cereus* is an opportunistic pathogen commonly isolated from soil and causes two types of food poisoning syndromes, emesis and diarrhea (Granum and Lund 1997).

The diarrheal type is caused by heat-labile enterotoxins produced by *B. cereus* in the intestine after ingesting large numbers of cells and is mostly associated with proteinaceous foods such as meat. Three different enetrotoxins have been characterised, namely haemolysin BL (Hbl), non-haemolytic enetrotoxin (Nhe) and cytotoxin K (CytK) (Lund et al. 2000). It is characterised by an incubation period of 8–16 h before the onset of watery diarrhea, abdominal pain and occasionally nausea and vomiting. Generally, the symptoms resolve within 12–24 hours, however in some rare cases symptoms perpetuate and can eventually lead to death.

The emetic syndrome on the other hand is caused by consumption of pre-fromed toxin mostly in farinaceous food items, particularly fried or cooked rice and pasta (Ehling-Schulz et al. 2004). In most outbreaks, these foods are stored after preparation, in conditions (room temperature) that allow rapid growth and toxin production. The emetic syndrome is caused by cereulide, a small cyclic non-ribosomally synthesized heat-stable dodecadepsipeptide. The syndrome is characterised by an incubation period of 0.5–5 h and is accompanied by symptoms such as vomiting, nausea, malaise and occasional diarrhea. Generally, the symptoms are relatively mild and disappear within 24 hours.

Usually it takes $10^5$–$10^7$ cells in total for the diarrheal syndrome and $10^5$–$10^8$ cells per gram of food to cause the emetic syndrome (Ceuppens et al. 2013). The wide range of infective dose is partly due to the consumption of spores which can survive the acidic environment in the stomach and partly due to the ability of different strains to produce different amounts of enterotoxins. Therefore, any food containing more than $10^3$ *B. cereus* per gram cannot be considered completely safe for consumption.

### Detecting B. cereus

Selective media is primarly used to isolate *B. cereus* by direct plating. MYP (Mannitol-yolk-polymyxin) and PEMBA (Polymyxin-pyruvate-egg yolk mannitol-bromothymol blue agar) are the most widely used selective media for isolating *B. cereus*. Both media are based on the diagnostic features of *B. cereus* of lecitithin hydrolysis and inability to ferment mannitol. *B. cereus* forms blue and pink colonies on PEMBA and MYP respectively, surrounded by a halo of lecithin hydrolysis. Polymyxin acts as the selective agent by inhibiting the growth of competitive organisms (Bottone 2010). In addition a chromogenic agar BCM developed which relies on the phosphotidylinositol phospholipase C hydrolyase enzyme can be used for the detection of *B. cereus*.

Presumptive colonies are confirmed by biochemical tests; include anaerobic utilization of glucose, Voges-Proskauer, L-tyrosine decomposition, nitrate reduction and growth in 0.0001% lysozyme.

Commercial kits (immunoassays) are available for the detection of the diarrheal enterotoxin (Andersson et al. 2004). The Oxoid Bacillus cereus RPLA enterotoxin detection kit (Oxoid Ltd., Basingstoke, UK) detects the HblC (L2) component, whereas the Tecra Bacillus diarrheal enterotoxin visual immunoassay kit (Bioenterprises pty, Australia) mainly detects the NheA protein. Neither assay is quantitative.

### *Association with Milk and Milk Products*

In the dairy industry, *B. cereus* is an important cause of microbe-related problems. It is practically impossible to completely avoid the presence of *B. cereus* in raw milk samples, because *B. cereus* spores are ubiquitously present in the farm environment: in soil, on cattle-feed and in cattle-faeces. From these sources, the spores easily contaminate the udders and the raw milk. Furthermore, *B. cereus* spores are very hydrophobic and attach to the surfaces of processing equipment of the dairy industry. Spores attached to processing equipment may germinate, multiply and re-sporulate, resulting in a continuous source of contamination. In addition, its spore-forming and psychrotrophic properties and the insufficiency of pasteurization to kill the spores enable *B. cereus* to grow and produce toxins in pasteurized milk at refrigeration. Besides causing food-borne illness, *B. cereus* is also responsible for the spoilage of pasteurized milk and its products resulting in off-flavors, sweet curdling and bitter cream.

## *Campylobacter* spp.

### *Introduction*

*Campylobacter* spp. has been known as a veterinary problem from the early twentieth century. *Campylobacter jejuni* subsp. *jejuni* (referred to as *C. jejuni* throughout this chapter) is a major cause of bacterial human gastroenteritis in both developed and developing nations. It is one of the most important milk-borne pathogens that has come to rival or even surpass Salmonella as an etiological agent of human gastroenteritis worldwide.

### Characteristics

The name Campylobacter is derived from the Greek word "Kampylos", which means curved. There are 11 species in the genus *Campylobacter* including *C. fetus, C. hyointestinalis, C. sputorum, C. jejuni, C. coli, C. lari, C. upsaliensis, C. mucosalis, C. concisus, C. curvus,* and *C. rectus* and belongs to the delta-epsilon group of proteobacteria, which also comprises Helicobacter, Arcobacter, *Sulfurospirillum* , the species *Bacteroides ureolyticus* and Wolinella.

*C. jejuni* cells are pleomorphic and can be curved, spiral or occasionally straight rods (splender vibiod cells) that are 0.2 to 0.8 µm wide and 0.5 to 5 µm long, which form an "S" or a "V" shape when two or more bacterial cells are grouped together. They are gram negative and non sporeforming. The cells are highly motile and have a corkscrew-like motion by means of an unsheathed polar flagellum at one or both ends of the cell. *C. jejuni* are microaerophilic, requiring an oxygen concentration of 3–15% and capnophilic with a carbon dioxide concentration of 3–5%. They are catalase and oxidase positive and urease negative. Temperature range for growth is 37–42°C, with an optimum temperature of 41.5°C, but they cannot survive cooking or pasteurization temperature (Silva et al. 2011). The organism is sensitive to heat, drying and acidic conditions (pH of less than 4.9 with optimal growth at pH 6.5 to 7.5) and salinity.

### Food-borne Illness

*C. jejuni* is estimated to be responsible for 90% of campylobacteriosis cases in humans, and *C. coli* for the remaining 10% (Janseen et al. 2008). *C. jejuni* infection is generally sporadic and in contrast to infections caused by other food-borne pathogens such as *Salmonella* and *E. coli* O157, outbreaks are rarely reported. The reported outbreaks are usually associated with drinking contaminated/ untreated water, consumption of raw milk and contaminated food and through contact with domestic animals.

In addition, *C. jejuni* infection is linked to traveller's diarrhea. In temperate regions there is a peak in the number of cases in summer and to a lesser extent autumn and spring with the most at risk groups in both developed and developing countries being children, the elderly and the immunocompromised.

The most common symptoms are acute gastroenteritis, cramping abdominal pain, fever, and more rarely, vomiting and headaches (Hariharan et al. 2004). Diarrhea often develops shortly after onset of abdominal pain and varies from mild, non-inflammatory, watery symptoms to severe and bloody. In otherwise healthy individuals, infection last for approximately 4 days with an incubation period of 1–10 days. The infection dose for humans may be as few as 500–10000 cells depending on vehicle of ingested materials, virulence of the strain and the susceptibility of the individual. Normally the *C. jejuni* infection is a self-limiting disease, but in severe cases, macrolide antibiotic (erythromycin) or fluoroquinolones (ciprofloxacin) is the choice of treatment.

### *Detecting Campylobacter* spp.

*Campylobacter* spp. is typically fragile bacteria that are difficult to culture and maintain in laboratory due to the special requirements for growth temperature and gaseous environment.

For food and other samples where a low number/injured *Campylobacter* spp. cells are suspected to be present in a highly mixed background flora, an enrichment step is necessary prior to isolation on selective agar plates. There are many different enrichment broths, some of the most frequently employed enrichment broths are Bolton, Preston, Park-Sanders and Exeter. Since *Campylobacter* is sensitive towards peroxides, radical scavengers like horse/sheep blood and charcoal are often included in these enrichment broths, as well as growth promoting reagents like ferrous sulphate, sodium metabisulphite and sodium pyruvate (FBP) (Silva et al. 2011). Media commonly used with antibiotics include cefoperazone, amphotericin B, polymixin B, cycloheximide, rifampicin, trimethoprim lactate and vancomycin. Furthermore, culturing is performed at approx. 42°C in a microaerobic atmosphere (5% $O_2$, 10% $CO_2$, and 85% $N_2$).

Following enrichment, or directly from samples with presumed high numbers of *Campylobacter*, the samples are spread on selective agar plates. Some of the most common ones are: modified charcoal cefoperazone deoxycholate (mCCDA), Skirrow, Karmali, Preston, Abeyta-Hunt-Bark (AHB), Campy-cefex and Butzler. According to ISO 10272, two selective agars with different selective principles must be used in parallel in order to increase the yield. Agar plates are incubated at 42°C for 24–48 h in a microaerophilic atmosphere and confirmatory tests on characteristic colonies. Typical colonies on selective agar media are smooth, convex and glistening with a distinct edge or flat, translucent, shiny, and spreading with an irregular edge. They are colorless to light cream or grayish with diameter range from pinpoint to 5 mm.

Identification of *Campylobacter* spp. presumptive colonies is performed by sub-culturing five colonies from selective media onto non-selective media. These are examined microscopically regarding their morphology and motility. Furthermore, a number of tests can be performed to confirm the identification and determine the species; growth at 25, 37 and 42°C, catalase, oxidase, glucose utilisation, and hippurate hydrolysis (Levin 2007).

Regarding milk samples The United States Food and Drug Administration (US-FDA) recommends a method for isolation of *Campylobacter* spp. (Fig. 4). Many rapid methods have been developed for isolation and detection of *Campylobacter* in recent years, such as latex agglutination test, polymerase chain reaction (PCR) technique, enzyme-linked immunosorbent assay (ELISA) and immunomagnetic separation (IMS) method.

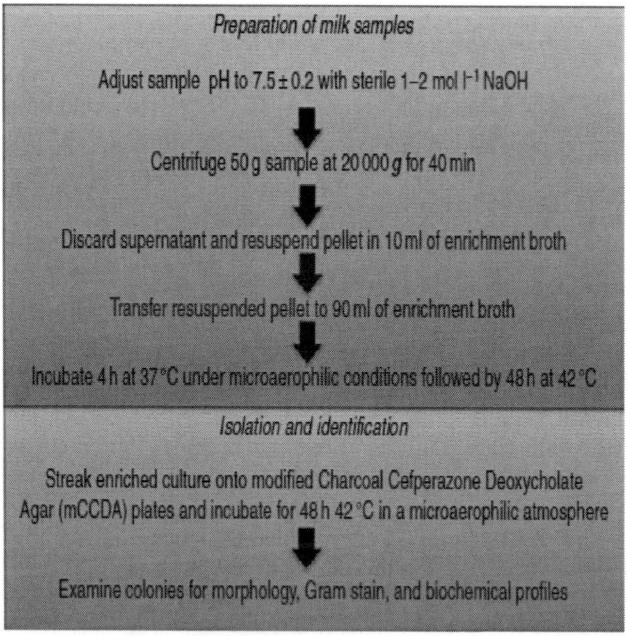


**Figure 4.** Procedure for Isolation and Identification of *Campylobacter* spp. from Milk (After: United States Food and Drug Administration).

*Color image of this figure appears in the color plate section at the end of the book.*

### Association with Milk and Milk Products

Reported cases of campylobacteriosis (Heuvelink et al. 2009, Longenberger et al. 2013) are linked with the consumption of raw milk rather than processed dairy products. Both studies stress the point that *Campylobacter jejuni* identified as the causal microorganism is isolated from raw milk therefore, dairy establishments that sell raw milk directly to the consumer should be aware of this pathogen.

On the other hand, El Sharoud (2009) reported low prevalence of Campylobacter in Egyptian dairy products (raw milk and fresh Domiati cheese) but a higher than expected survival rate. This suggests that this food-borne strain of *C. jejuni* may develop adaptive strategies that aid survival under food preservation conditions, which contradicts with what is known about this pathogen as a stress-sensitive organism.

## *Yersinia* spp.

### *Introduction*

The genus Yersinia is named after Alexander Yersin, a Swiss bacteriologist who first isolated the plaque bacillus in Hong Kong in 1894, while investigating a catastrophic epidemic of bubonic plaque that killed an estimated 60000 people.

### *Characteristics*

*Y. enterocolitica* are included in the genus *Yersinia*, which are classified into the family *Enterobacteriaceae*, a group of gram-negative, oxidase-negative, catalase positively, non spore-forming and facultatively anaerobic bacteria. *Y. enterocolitica* are rods or coccobacilli of 0.5–0.8 x 1–3 μm in size. Of the 11 species, only 3 species of Yersinia are recognised to be pathogenic to humans, the plague bacillus, Yersinia pestis, and two that cause gastroenteritis; *Y. enterocolitica* and *Y. pseudotuberculosis* (Bottone 1997). Strains belonging to *Y. enterocolitica* are urease positive and can be differentiated from *Y. pseudotuberculosis* with a positive result for fermentation of sucrose, and negative reactions for rhamnose and melibiose fermentation. Most of the strains are motile at 25°C but non-motile at 37°C. It can be isolated from many sources, such as food, drinking water, sewage, environmental water and human clinical samples with a world-wide distribution.

*Y. enterocolitica* can grow over a pH range of 4–10, generally with an optimum pH of 7–8 depending on the acidulant used, environmental temperature, composition of the medium, and growth phase of the bacteria. It is a psychotropic bacterium therefore it has the ability to grow at temperatures below 4°C, but the optimum growth temperature is 28–30°C.

### *Food-borne Illness*

Human clinical infections with this species ensue after ingestion of the microorganisms in contaminated food or water or by direct inoculation through blood transfusion (Sabina et al. 2011). Infection with *Y. enterocolitica* can cause a variety of symptoms depending on the age of the person infected. Infection with *Y. enterocolitica* occurs most commonly in young children, under seven years old and the infection is more common in the winter (Bottone 1997).

It is a self limiting enterocolitis with an incubation period of 1–11 days. Common symptoms include diarrhea and/or vomiting, fever and acute abdominal pain caused by mesenteric lymphadenitis, and it is often clinically indistinguishable from acute appendicitis. Sometimes post-infections, more specifically extra-intestinal sequelae, such as reactive arthritis, erythema nodosum, erythema multiforme, scarlatiniform exanthemata and septicemic types deserve particular clinical attention.

### Detection

Methods to isolate *Yersinia* from food are problematic due to competition and over-growth by food microflora. Enrichment procedures usually exploit the psychrotrophic character of Y. *enterocolitica* by incubating at 4°C for 7–21 days (Fredriksson-Ahomaa and Korkeala 2003). The most commonly used enrichment media are phosphate buffered saline (PBS) or phosphate-buffered saline with sorbitol and bile salts (PSB) or tryptone soya broth (TSB). The disadvantage of cold enrichment for prolonged periods can increase the isolation rate other psychotrophics organism present in the sample.

*Y. enterocolitica* grows in most routine media including blood, chocolate, MacConkey (MAC), in which they produce colorless colonies as it can't ferment lactose, and *Salmonella-Shigella* (SS) agar, but these media are low selective, as *Y. enterocolitica* strains grow slowly and of overgrowth by other enteric bacteria. However, the best results for the selective isolation of *Y. enterocolitica* from food samples give the Cefsulodin-Irgasan-Novobiocin (CIN) agar. In addition to the antibiotics, the medium contains deoxycholate and crystal violet as selective agents and mannitol as a fermentable carbon source. After incubation at 28°C for 24 hours *Y. enterocolitica* forms colonies with a deep red centre (bull's eye) with a sharp border surrounded by a translucent zone.

Confirmation of identity of *Y. enterocolitica* may be made using commercial identification kits. Rapid methods for detection of *Y. enterocolitica* have also been developed such as polymerase chain reaction (PCR) methods.

### Association with Milk and Milk Products

*Y. entercolitica* has been isolated from raw cow's milk in Mexico City (Bernardino-Varo et al. 2013) while other authors (Ackers et al. 2000) have reported that the same pathogen was associated with pasteurized bottled milk sold from a local dairy. This incident was connected to postpasteurization contamination, indicating the need for controls and training/awareness of dairy farmers. Tacket et al. (1984) stated that outbreaks of enteric disease caused by pasteurized milk are rare; although the ability of *Y. enterocolitica* to grow in milk at refrigeration temperatures makes pasteurized milk a possible vehicle for virulent *Y. enterocolitica*.

## *Shigella* spp.

### Introduction

The genus Shigella was first reported by the Japanese microbiologist Kiyoshi in 1898. It was classified into four species based on the O antigen; *Shigella dysenteriae, Shigella flexneri, Shigella boydii* and *Shigella sonnei* which were also known as *Shigella* subgroups A, B, C and D respectively.

### Characteristics

Shigella is a genus of gammaproteobacteria and belongs to the family of Enterobacteriaceae. They share common characteristcis and genetic relatedness with members of the genus Escherichia and in particular with enteroinvasive *E. coli*.

It is a small, non-spore forming, gram negative rod with a diameter of 0.3 to 1 µm and a length of 1 to 6 µm. It is catalase positive, oxidase negative, facultative anaerobe and is non-motile because of the absence of H antigens. It is non-capsulated and possesses the K and O antigens. O antigen (somatic antigen) is useful in serological identification to classify the four species. K antigen is the capsule antigen which occasionally interferes with O antigen determination. The Shiga toxin, also called as verotoxin, is produced by *Shigella dysenteriae* type 1. The toxin has a molecular weight of 68 kDa. It is a multisubunit protein made up of an A subunit (32 kDa), responsible for toxic action of the protein and five molecules of the B subunit (7.7 kDa), responsible for binding to a specific cell receptor. *Shigella* is a typical mesophile and is able to grow at temperatures ranging from 12°C to 48°C (optimum 37°C), at a pH range of 5.0 to 7.3 (Zaika and Philis 2005).

### Food-borne Illness

*Shigella* spp. are intracellular bacterial pathogens that inhibit the gastrointestinal tract of humans and are the causative agent of shigellosis (Niyogi 2005). Shigellosis is a global human health problem that is estimated to affect 80–165 million individuals annually. It is the most important cause of bloody diarrhea worldwide, especially in developing countries with poor hygienic conditions (Jennison and Verma 2004). Shigellosis is exclusively a human disease and is usually acquired through contaminated food and water sources. Shigella infects via the oral-faecal route and infection is transmissible with as few as a 100 microorganisms, partly due to the ability of the bacterium to survive the highly acidic environment of the stomach such as gastric secretions. Shigellosis is also frequently acquired from consumption of raw vegetables harvested in fields where sewage was used as fertilizer, as well consumption of oysters that have been contaminated with sewage. Transmission also occurs by accidentally drinking contaminated water in the swimming pools.

The most common symptoms of shigellosis range from water diarrhea to severe dysentery. Severe dysentery is characterized by fever, abdominal pains, nausea, vomiting and acute permanent bloody and mucoid diarrhea (Phalipon and Sansonetti 2007). The condition may be asymptomatic in some cases or severe, especially in children. Symptoms can last from 3 to 14 days.

In the absence of effective treatments, patients with shigellosis can develop secondary complications such as septicemia and pneumonia.

The effectiveness of the transmission of *Shigella* is due to the fact that *Shigella* is highly infectious, the 10 to 200 organisms that are sufficient to cause infection is significantly lower than that reported for most other enteric pathogens such as for *Vibrio* spp. and *Salmonella* spp. which require at least 104 to 105 organisms to cause infection.

### ***Detecting Shigella* spp.**

Due to the lack of interest in Shigella as food-borne pathogens, laboratory protocols for the isolation and identification are undeveloped. A pre-enrichment step in Selenite-F(SF), or Tetrathionate (TT) broth or gram-negative broth is required for the isolation of Shigella.

The common selective/differential agar media used for the recovery of *Shigella* are MacConkey (MAC), Xylose Lysine Deoxycholate (XLD), Hektoen (HEK) and *Salmonella-Shigella* (SS) and Deoxycholate Citrate Agar (DCA). It has typical nonlactose fermenting characteristic colonies in lactose enriched media such as on MAC, DCA and SS agar. *Shigella* is resistant to bile salts and this characteristic is usually useful in the selective media. Colonies on the MacConkey and DCA agar appear to be large, 2 to 3 mm in diameter, translucent and colourless (non-lactose fermenting). Whereas, on the XLD agar, colonies appear to be much smaller (1 to 2 mm diameter) and red in colour as lysine is decarboxylated producing alkaline end products which raises the pH and cause the agar to turn into deep red colour.

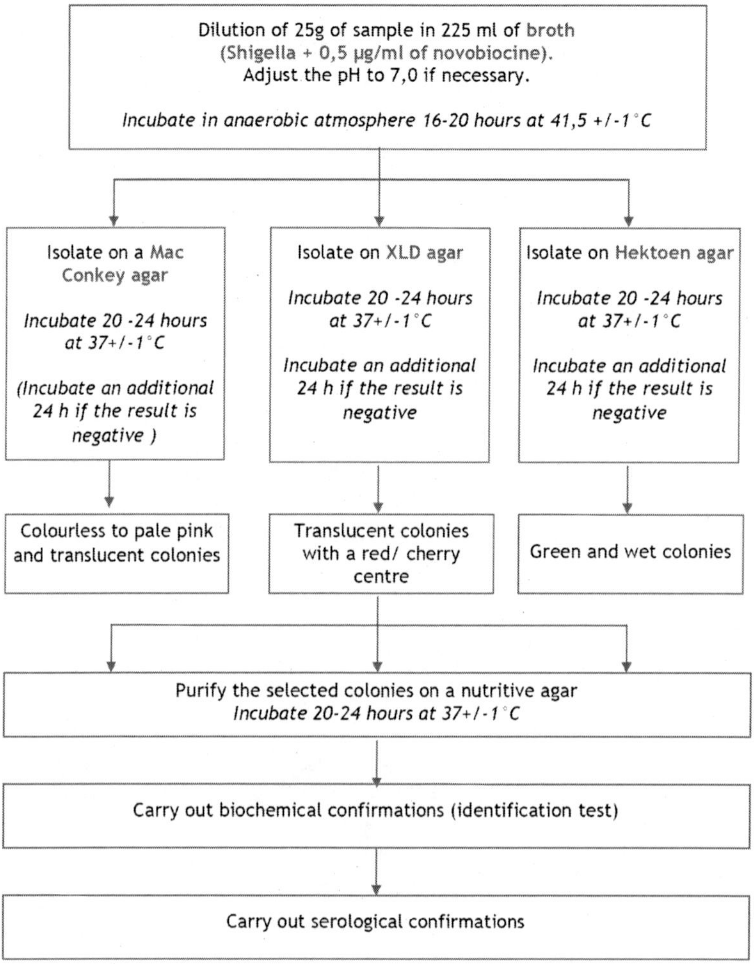


**Figure 5.** Detection of *Shigella* spp. in Food Samples (After: AES Chemunex 2008).

*Color image of this figure appears in the color plate section at the end of the book.*

## *Association with Milk and Milk Products*

*Shigella* spp. is better related to raw milk but in developing countries where hygiene and sanitation facilities are often poor the pathogen may contaminate food products too. The recent findings from a survey carried out in Egypt (Ahmed and Shimamoto 2014) from 1600 food samples (50:50 dairy and meat) showed that *Shigella* spp., along with *E. coli 0157:H7* and *S. enterica* serovar *typhimurium*, were isolated at a rate of 1.7%, 3.4% and

4.3%, respectively. The results may show a low prevalence but since the bacterium may cause dysentery it should not be underestimated.

## *Escherichia coli*

### *Introduction*

*Escherichia coli*, originally called "*Bacterium coli* commune", was first isolated from children feces in 1885 by the German bacteriologist Theodor Escherich. Nowadays it is the one of the most understood bacteria and is a common inhabitant of the gastrointestinal tract of humans and animals.

### *Characteristics*

*Escherichia coli* are Gram-negative, non-sporeforming bacilli and belong to the family of Entrobacteriacae. They are approximately 1.1–1.5 µm in wide and 2.0–6.0 µm in length and occur as single straight rods. *E. coli* is a facultative anaerobe, catalase positive, oxidase negative and is capable of reducing nitrates to nitrites. When growing fermentatively on glucose or other carbohydrates, it produces acid and gas (mainly $H_2$ and $CO_2$). The classic differential test in order to separate *E. coli* from *Shigella* and *Salmonella* is the ability of *E. coli* to ferment lactose, which the latter two genera fail to do.

Most *E. coli* strains are typical mesophiles, capable of growing over a wide range in temperature from 7–50°C with an optimum temperature of 37°C. *E. coli* can grow within a pH range of 5.5–8.0 with best growth occurring at neutrality.

### *Food-borne Illness*

Even though *E. coli* is a group of harmless bacteria that are most often used as indicator organisms for fecal contamination and breaches in hygiene, there are some strains that have acquired virulence factors that has allowed them to cause serious disease especially diarrhea. Pathogenic Escherichia coli is categorized into the following six categories according to their virulence properties, clinical manifestations, pathogenic mechanisms, clinical syndromes and O:H serogroups: enteropathogenic *E. coli* (EPEC), enterotoxigenic *E. coli* (ETEC), enterohaemorrhagic *E. coli* (EHEC), enteroinvasive *E. coli* (EIEC), diffusely adherent *E. coli* (DAEC), and enteroaggregative *E. coli* (EAEC) (Jafari et al. 2012).

### *Enteropathogenic E. coli (EPEC)*

EPEC were the first group of *E. coli* recognised as a causative agent of diarrheal illness in humans. EPEC are a major cause of acute or chronic enteritis and infant diarrhea in developing countries where sanitation and water quality may be poor Other symptoms include vomiting, fever, malaise and dehydration and appear usually 12–36 h after ingestion of the organism.

### *Enterotoxigenic E. coli (ETEC)*

ETEC were first recognised as a causative agent of diarrhea in the 1960s and 1970s. ETEC strains produce toxins which are heat-labile (LT) and/or heat-stable (ST) and they are responsible for secretory diarrhea in both humans and animals. Illness caused by ETEC usually occurs 12 and 36 h after ingestion of the organism and usually persists for 2–3 days and symptoms include acute watery diarrhea that may be mild and of short duration or more severe such as cholera characterized by watery stools accompanied by vomiting and severe stomach pains. The infective dose of ETEC for adults has been estimated to be at least 108 cells; but the young, the elderly and the infirm may be susceptible to lower levels.

### *Enterohaemorrhagic E. coli (EHEC)*

EHEC strains are implicated in food-borne diseases principally due to ingestion of uncooked minced meat and raw milk. They are more commonly referred to as verocytotoxigenic (VTEC) as they produce a toxin which has a cytotoxic effect on vero cell lines. These strains produce shiga like toxins. *E. coli O157:H7* is the prototype of this group. The symptoms of infection from this group of organisms includes watery diarrhea which may develop into bloody diarrhea (haemorrhagic colitis), severe abdominal pain and Haemolytic uraemic syndrome (HUS), a cause of acute renal failure, and thrombotic thrombocytopenic purpura (TTP) (Kaper 1998). The pathogen has an incubation period of 1–14 days with illness duration of 5–7 days. The infectious dose for O157:H7 is estimated to be 10–100 cells; but no information is available for other EHEC serotypes (Teunis et al. 2004).

### *Enteroinvasive E. coli (EIEC)*

EIEC (enteroinvasive *E. coli*) cause a broad spectrum of human's diseases. They are biochemically, genetically and pathogenetically closely related to *Shigella* spp. However the infective dose appears to be higher than that of Shigella and required at least $10^6$ cells to cause illness in health adults

EIEC cause invasive inflammatory colitis and dysentery with a clinical presentation (blood and mucous stools accompanied by fever and severe cramps). Other symptoms are headache, fever, and cramping. Onset of illness occurs about 24 h post ingestion.

### *Diffusely adherent E. coli (DAEC)*

DAEC are a major cause of urinary tract infections worldwide, but its role as a causative agent of diarrhea is not completely understood.

### *Enteroaggregative E. coli (EAEC)*

EAEC is the most recently identified and described diarrhoeagenic *E. coli* (Huang et al. 2006). This bacterium was described in 1987, and identified in a child from Chile with persistent diarrhea (Nataro 2005). EAEC cause watery diarrhea by adhering to the epithelium of terminal ileum and colon in a characteristic aggregative pattern followed by a damage/ secretion stage.

## *Detecting E. coli*

Pathogenic *E. coli* represents a phenotypically diverse group of pathogens and no single method or approach can be used to detect or isolate all of the pathotypes of concern. Therefore different methods have been developed for the detection and isolation of certain pathotypes.

Pathogenic *E. coli* strains that ferment lactose and are not adversely affected by elevated temperatures (e.g., 44°C) can be isolated using standard procedures for *E. coli*. Most pathogen detection in foods is performed on a 25 g sample which is enriched in 225 ml of a suitable enrichment broth. In the Food and Drug Administration Bacteriological Analytical Manual (FDA BAM) method the recommended procedure for pathogenic *E. coli* is to pre-enrich the sample in 225 ml brain heart infusion (BHI) broth at 35°C for 3 hours to facilitate resuscitation of sub-lethally injured cells. The entire pre-enrichment is transferred to 225 ml of tryptone phosphate (TP) broth and incubated at 44°C for 20 hours, after which time an aliquot of enriched broth is plated onto eosin-methylene blue (EMB) agar and MacConkey agar plates. These are incubated at 37°C for 24 hours. Further identification includes biochemical tests, serotyping and examination for key virulence associated genes.

Another biochemical feature of *E. coli* is β-glucuronidase activity, therefore a fluorogenic or chromogenic glucuronide is incorporated into a conventional media and enzyme activity is detected by the production of colour or flurorescence.

Regarding the isolation of *E. coli* 0157 the reference method ISO 16654:2001 is used. This method is based on an enrichment step in modified TSB with novobiocin (mTSBn) at 41.5°C for an initial period of 6 hours and then for a further period of 12 to 18 hours. This is followed by a separation and concentration step and then isolation on chromogenic media such as CT-SMAC (Farrokh et al. 2013).

### *Association with Milk and Milk Products*

Shiga toxin producing *E. coli* (STEC) are an important cause of human food-borne outbreaks. The consumption of raw milk dairy products may be an important route of STEC infection. There are 2 STEC strains of serotypes O157:H7 and O26:H11, and Miszczycha et al. 2014, reported that the survival of *E. coli O26:H11* was 13 times greater than that of *E. coli O157:H7* at the end of digestion in contaminated raw milk cheeses. *E. coli O157:H7* has also been isolated from dairy farms, even though a survey in Italy reported from Conedera et al. 2004 showed that none of the dairy products collected (2948 including mozzarella cheese, pasteurized and raw bovine and ovine milk) had tested positive for VTEC *E. coli 0157:H7*. A very comprehensive review on Shiga-toxin-producing Escherichia coli (STEC) and their significance in dairy production has been published by Farrokh et al. (2013), where the authors review the survival of STEC during cheese making and in particular the factors affecting the growth of STEC during cheesemaking (i.e., salt, pH, temperature, water activity), behaviour of STEC during cheesemaking.

As a conclusion, the International Dairy Federation in its IDF Factsheet (2012) states that in the processing industry, milk pasteurization is recommended, since pasteurization will eliminate pathogens including STEC. Good hygienic practices at the processing facility are also to be applied in order to prevent post-pasteurization contamination. At farm level, strict hygiene practices should be followed during the production and transport of raw milk. Additional hygiene practices are to be implemented in the production of raw milk cheese and other dairy products.

## *Mycobacterium* spp.

### *Introduction*

Mycobacterium species are a large, varied group of organisms, which include some significant pathogens (such as *M. tuberculosis, M. bovis, M. avium* subsp. *paratuberculosis,* and *M. leprae*) along with many free-living non-pathogens. *Mycobacterium tuberculosis,* the causative agent of tuberculosis (TB, also known as the "white plague"), was identified by Robert Koch in 1882. *Mycobacterium avium* subsp. *paratuberculosis* (MAP) causes paratuerculosis in animals (Johne's disease) and could have a role in human diseases like Crohn's (Marchetti et al. 2013). Though it requires a suitable cell medium for growth, MAP can survive in the environment well beyond its presence in an animal host, creating drastic problems for farmers trying to eradicate Johne's disease from their herds.

### Characteristics

The genus *Mycobacterium* is a member of the Actinobacteria. MAP, an obligate intracellular pathogen, is a small (0.5 x 1.5 μm), rod shaped bacteria which grow in circular colonies reaching about 1–2 mm in diameter. They are usually found to be off-white or yellow depending on the medium, and have the ability to adapt easily into many environments. A common feature of mycobacteria is the chemical composition of their cell walls which have a high lipid content, that avidly retains Carbol fuchsin dye even in the presence of acidic alcohol (acid fast staining) (Glickman and Jacobs 2001) .

### Food-borne Illness

MAP and its association (or not) to human Crohn's Disease is a debate which started in 1984, when a Mycobacterium similar to the agent of Johne's disease was isolated and cultivated from the intestinal tissue of three patients with Crohn's disease. Crohn's disease is a chronic inflammatory bowel disease which currently is of unknown cause. The most widely accepted hypothesis is that Crohn's disease occurs as a result of an abnormal immune response within the gastrointestinal mucosa. Grant (2006) reports that transmission to humans via consumption of animal-derived foods is a distinct possibility as milk, other dairy products, beef and water have been identified as possible food vehicles of transmission. To date, viable MAP has been cultured from raw cows', sheep and goats' milk, retail pasteurized cows' milk, and some retail cheeses in several countries during recent studies. MAP has not been isolated from retail beef to date, although limited testing has been carried out.

The public health consequences, if any, of low numbers of viable MAP being periodically consumed by susceptible individuals are uncertain. Moreover, an association between MAP and Crohn's disease is not proven, but neither can it be discounted on the basis of current evidence.

### *Detecting Mycobacterium aviums* subsp. *paratuberculosis*

It is generally accepted that molecular techniques (i.e., PCR basis) should be employed for the detection of MAP from a wide range of samples, while antibodies aginst MAP could be tested by commercially available ELISA tests for indirect confirmation test. Liapi et al. (2013) examined 192 MAP ELISA-positive sheep and goats from Cyprus using faecal culture and genotype MAP isolates using IS1311 PCR and restriction endonuclease analysis (IS1311 PCR-REA) with HinfI restriction enzyme. The same methods, i.e., ELISA and RT-PCR, were also used by Marchetti et al. (2013) in order to quantify MAP in bulk tanker milk (BTM). Botsaris et al. (2010) also reported rapid methods for the determination of MAP in milk and cheese.

### *Association with Milk and Milk Products*

Marchetti et al. (2013) reported that the transmission of MAP to humans most likely occurs via contaminated milk and milk products. MAP can survive low-temperature holding (63°C for 30 min) and high temperature-short time (HTST) (72°C for 15 sec) pasteurization, some surveys showed the presence of MAP in commercially pasteurized milk purchased at retail. The effect of homogenisation of heat inactivation of MAP in milk was studied by Hammer et al. (2014) showing that none of the 3 homogenization modes applied showed a statistically significant additional effect on the inactivation of MAP during HTST treatment.

## Dairy Outbreaks

Dairy products have frequently been implicated in food poisoning outbreaks. There are two main reasons for that; one is the fact that the gross chemical composition for most of the dairy products favors the growth and survival of microorganisms (i.e., high protein content, presence of lactose, available moisture—water activity, $a_w$), and second, the sheer volumes of dairy products consumed by almost all population groups making dairy products a very good vehicle for microorganisms. Fortunately, processing of milk and milk products (i.e., pasteurization, ultra high temperature, drying, evaporating, filtrating of bacteria, fermenting/acidification) minimizes risks either by posing hurdles for bacteria to grow (low pH), decreasing available water (lowering water activity), or by destroying bacteria (temperature).

Nonetheless, milk and other dairy products (i.e., cheese) could also be consumed raw (according to the provisions of legislation EC 853/2004); therefore strict hygienic conditions must apply; at farm, during milk transportation, during manufacture at the factory and at retail level in order to avoid food poisoning outbreaks. Table 3 below summarizes recent alert cases as reported by the Rapid Alert System for Food and Feed (RASFF), European Union concerning milk and milk products and certain common pathogens.

Langer et al. (2012) published a study (results are shown in Table 4) where they describe the outbreaks related to dairy products both raw and pasteurized in the USA between the years 1993–2006. They reported that the causative agent was identified for all 73 outbreaks involving raw dairy products; all were caused by bacteria. One outbreak was caused by *Campylobacter* spp. and shigatoxin producing *Escherichia coli*. Among the remaining 72 outbreaks, 39 (54%) were caused by *Campylobacter* spp., 16 (22%) by *Salmonella* spp., 9 (13%) by shiga-toxin producing *E. coli*, 3 (4%) by *Brucella* spp., 3 (4%) by *Listeria* spp., and 2 (3%) by *Shigella* spp. Among the 30 outbreaks involving pasteurized dairy products for which the

**Table 3.** Milk and milk products involved in recent alert cases (RASFF, EU).

| Pathogen | Milk and Milk products | No. of cases 2013–03/2014 | Severity | Notes |
|---|---|---|---|---|
| *Listeria monocytogenes* | Cheese (Raw or pasteurized) | 25 | alert | Most cases involved sheep or goat raw milk cheeses from France |
| *Salmonella enteritidis* | Cheese | 1 | alert | |
| *Salmonella* spp. | Pasteurised cow milk cheese, Raw milk cheese | 2 | alert | |
| *E. coli* (shigatoxin producing) | Raw milk cheeses | 5 | alert | Country of origin: France |

**Table 4.** Characteristics of disease outbreaks after consumption of dairy products 1993–2006, United States.

| Product | Total Outbreaks | Associated illnesses | Associated hospitalizations | Associated deaths |
|---|---|---|---|---|
| Raw | | | | |
| Fluid milk | 46 | 930 | 71 | 0 |
| Cheese | 27 | 641 | 131 | 2 |
| Total | 73 | 1,571 | 202 | 2 |
| Pasteurized | | | | |
| Fluid milk | 10 | 2,098 | 20 | 0 |
| Cheese | 38 | 744 | 17 | 1 |
| Total | 48 | 2,842 | 37 | 1 |
| All dairy | 121 | 4,413 | 239 | 3 |

After: Langer et al. (2012)

causative agent was reported, 13 (44%) were caused by norovirus, 6 (20%) by *Salmonella* spp., 4 (13%) by *Campylobacter* spp., 3 (10%) by *Staphylococcus aureus*, and 1 (3%) each by *Clostridium perfringens, Bacillus cereus, Listeria* spp., and *Shigella* spp.

A total of 48 reported outbreaks involved pasteurized dairy products. The source of contamination was reported for 7 (14%) of these outbreaks, of which at least 4 (57%) probably resulted from post-pasteurization contamination by an infected food handler.

More recent reports (Food safety News 2012) state that in the USA between the years 2010–2012; 24 raw dairy outbreaks were reported with 309 illnesses and no deaths (22 fluid raw milk, 2 aged raw milk cheeses); 2 pasteurized dairy outbreaks with 39 illnesses and no deaths; 1 pasteurized Mexican-style cheese sporadic illness and no deaths; 2 queso fresco Mexican-style cheese outbreak with 67 illnesses and no deaths and finally 3 sporadic illnesses and hospitalizations from illegal Mexican-style cheese with no deaths. The major pathogens involved in most of the cases stated above were: *L. monocytogenes, E. coli O157:H7, C. jejuni,* and *S. aureus.* A comprehensive study carried out by de Buyser et al. (2001) reported the implication of milk and milk products in food-borne disease in France and other industrialsed countries. Four etiologic agents were considered: *Salmonella* spp., *S. aureus, L. monocytogenes,* and pathogenic *E. coli* and the results showed that in France, out of 69 reported outbreaks for which the implication of milk and milk products was considered as confirmed because the etiologic agent was isolated from food; 48% of the food vehicles were from raw milk, but for 51%, the type of milk was not specified. A majority of these outbreaks were traced to cheeses made from raw or unspecified milk. *S. aureus* was by far the most frequent pathogen associated with these outbreaks, followed by *Salmonella* spp. In England and Wales, the situation seems different because the most frequently reported outbreaks were salmonellosis associated with the consumption of raw milk.

## Microbiological Sampling Plans

Foodstuffs should not contain micro-organisms or their toxins or metabolites in quantities that present an unacceptable risk for human health. The safety of foodstuffs is mainly ensured by a preventive approach, such as implementation of good hygiene practice (a prerequisite programme) and application of procedures based on hazard analysis and critical control point (HACCP) principles. Testing against the criteria as set down in Regulation (EC) No 2073/2005 should be undertaken when validating or verifying the correct functioning of systems in place. In addition, food business operators (FBOs) should determine shelf-life by a strict testing programme to ensure that the criteria are met over the entire intended shelf-life of the product.

Article 4 of Regulation (EC) No 852/2004 places an obligation on FBOs to comply with microbiological criteria for foodstuffs. This should include testing against the values set for the criteria through the taking of samples, the conduct of analyses and the implementation of corrective actions, in accordance with food law and the instructions given by the competent authority. The producer or manufacturer of a food product has to decide whether the product is ready to be consumed as such, without the need to cook or otherwise process it in order to ensure its safety and compliance with the microbiological criteria.

In ensuring compliance with the relevant microbiological criteria set out in Annex I to Regulation (EC) No 2073/2005, the FBOs at each stage of food production, processing and distribution, including retail, must take measures, as part of their procedures based on HACCP principles together with the implementation of good hygiene practice, to ensure the following:

a) that the supply, handling and processing of raw materials and foodstuffs under their control are carried out in such a way that the process hygiene criteria are met,
b) that the food safety criteria applicable throughout the shelf-life of the products can be met under reasonably foreseeable conditions of distribution, storage and use.

The Commission Regulation (EC) No 2073/2005 and its amendments; Commission Regulation (EC) 1441/2007 and 365/2010 on microbiological criteria for foodstuffs for Milk and Milk products state the following:

Food category: Pasteurized milk and other pasteurized liquid dairy products[#] | Micro-organisms: Enterobacteriaceae | n = 5 | c = 0 | m = < 1/ml, M = 5/ml | Analytical reference method: ISO 21528–2 | Stage where the criterion applies: End of the manufacturing process | Action in case of unsatisfactory results: Check on the efficiency of heat-treatment and prevention of recontamination as well as the quality of raw materials.

Food category: Cheeses made from milk or whey that has undergone heat treatment | Micro-organisms: *E. coli** | n = 5 | c = 2 | m= 100 cfu/g | M=1000 cfu/g | Analytical reference method ISO 16649–1 or 2 | Stage where the criterion applies: At the time during the manufacturing process when the

[#] The criterion does not apply to products intended for further processing in the food industry.
* *E. coli* is used here as an indicator for the level of hygiene.

*E. coli* count is expected to be highest** | Action in case of unsatisfactory results: Improvements in production hygiene and selection of raw materials.

Food category: Cheeses made from raw milk | Micro-organisms: Coagulase-positive staphylococci | n = 5 | c = 2 | m = $10^4$ cfu/g | M = $10^5$ cfu/g | Analytical reference method: EN/ISO 6888–2 | Stage where the criterion applies: At the time during the manufacturing process when the number of staphylococci is expected to be highest | Action in case of unsatisfactory results: Improvements in production hygiene and selection of raw materials. If values > $10^5$ cfu/g are detected, the cheese batch has to be tested for staphylococcal enterotoxins.

Food category: Cheeses made from milk that has undergone a lower heat treatment than pasteurization⁺ and ripened cheeses made from milk or whey that has undergone pasteurization or a stronger heat treatment ⁺ | Micro-organisms : Coagulase-positive staphylococci | n = 5 | c = 2 | m = 100 cfu/g | M = 1000 cfu/g | Analytical reference method: EN/ISO 6888–1 or 2

Food category: Unripened soft cheeses (fresh cheeses) made from milk or whey that has undergone pasteurization or a stronger heat treatment⁺ | Micro-organisms: Coagulase-positive staphylococci | n = 5 | c = 2 | m = 10 cfu/g | M = 100 cfu/g | Analytical reference method: EN/ISO 6888–1 or 2 | Stage where the criterion applies: End of the manufacturing process | Action in case of unsatisfactory results: Improvements in production hygiene. If values > $10^5$ cfu/g are detected, the cheese batch has to be tested for staphylococcal enterotoxins.

Food category: Butter and cream made from raw milk or milk that has undergone a lower heat treatment than pasteurization | Micro-organisms: *E. coli** | n = 5 | c = 2 | n = 10 cfu/g | n = 100 cfu/g | Analytical reference method: ISO 16649–1 or 2 | Stage where the criterion applies: End of the manufacturing process | Action in case of unsatisfactory results: Improvements in production hygiene and selection of raw materials |.

Food category: Milk powder and whey powder# | Micro-organisms: Enterobacteriaceae | n = 5 | c = 0 | m,M = 10 cfu/g## | Analytical reference method: ISO 21528–1 | Stage where the criterion applies: End of the manufacturing process | Action in case of unsatisfactory results: Check

** For cheeses which are not able to support the growth of *E. coli*, the *E. coli* count is usually the highest at the beginning of the ripening period, and for cheeses which are able to support the growth of *E. coli*, it is normally at the end of the ripening period.

⁺ Excluding cheeses where the manufacturer can demonstrate, to the satisfaction of the competent authorities, that the product does not pose a risk of staphylococcal enterotoxins.

# The criterion does not apply to products intended for further processing in the food industry.

## m = M.

on the efficiency of heat treatment and prevention of recontamination | Micro-organisms : Coagulase-positive staphylococci | n = 5 | c = 2 | m = 10 cfu/g | M = 100 cfu/g | Analytical reference method: EN/ISO 6888–1 or 2 | Stage where the criterion applies: End of the manufacturing process | Action in case of unsatisfactory results: Improvements in production hygiene. If values > $10^5$ cfu/g are detected, the batch has to be tested for staphylococcal enterotoxins. |.

Food category: Ice cream[++] and frozen dairy desserts | Micro-organisms: Enterobacteriaceae | n = 5 | c = 2 | m = 10 cfu/g | M = 100 cfu/g | Analytical reference method: ISO 21528– 2 | Stage where the criterion applies: End of the manufacturing process | Action in case of unsatisfactory results: Improvements in production hygiene |.

Food category: Dried infant formulae and dried dietary foods for special medical purposes intended for infants below six months of age | Micro-organisms: Enterobacteriaceae | n = 10 | c = 0 | m, M[+] = Absence in 10 g | Analytical reference method: ISO 21528– 1 | Stage where the criterion applies: End of the manufacturing process | Action in case of unsatisfactory results: Improvements in production hygiene to minimise contamination[$] |.

Food Category: Dried follow-on formulae | Micro-organisms: Enterobacteriaceae | n = 5 | c = 0 | m, M[+] = Absence in 10 g | Analytical reference method: ISO 21528– 1 | Stage where the criterion applies: End of the manufacturing process | Action in case of unsatisfactory results: Improvements in production hygiene to minimise contamination |.

Food Category: Dried infant formulae and dried dietary foods for special medical purposes intended for infants below six months of age | Micro-organisms: Presumptive *Bacillus cereus* | n = 5 | c = 1 | m = 50 cfu/g, M = 500 cfu/g | Analytical reference method: EN/ISO 7932 | Stage where the criterion applies: End of the manufacturing process | Action in case of unsatisfactory results: Improvements in production hygiene. Prevention of recontamination. Selection of raw material |.

n = number of units comprising the sample; c = number of sample units giving values between m and M.

[++] Only ice creams containing milk ingredients.

[+] m=M.

[$] Parallel testing for Enterobacteriaceae and *E. sakazakii* shall be conducted, unless a correlation between these micro-organisms has been established at an individual plant level. If Enterobacteriaceae are detected in any of the product samples tested in such a plant, the batch has to be tested for *E. sakazakii*. It shall be the responsibility of the manufacturer to demonstrate to the satisfaction of the competent authority whether such a correlation exists between Enterobacteriaceae and *E. sakazakii*.

Interpreting the Legislation

As an example: in the food category *cheeses made from raw milk* the sample is made up of 5 units *(n)*, the *maximum* number of sample units giving values between *m* and *M* is 2(c). The batch is rejected if 1 unit gives values higher than M and if c is surpassed.

It is recomended that the reader, with regards to EU legislation, should seek for current updates in the European Community's website or webpages such as Food Safety Ireland (www.fsai.ie).

### *Food Safety Management Systems and the Dairy Industry*

The implementation of Food Safety Management Systems (FSMS) in the dairy industry is of paramount importance if we are to produce dairy products that are safe for consumption. The ISO/TS 22002–1:2009 specifies detailed requirements to be specifically considered in relation to ISO 22000:2005, i.e. a) construction and layout of buildings and associated utilities; b) layout of premises, including workspace and employee facilities; c) supplies of air, water, energy, and other utilities; d) supporting services, including waste and sewage disposal; e) suitability of equipment and its accessibility for cleaning, maintenance and preventive maintenance; f) management of purchased materials; g) measures for the prevention of cross-contamination; h) cleaning and sanitizing; i) pest control; j) personnel hygiene. These aspects together with the required on-farm practices are discussed in Chapter 7.

## Conclusions

Dairy pathogens are of paramount importance as they can cause major food scares to the consumers. The fact that dairy products are widely consumed, daily, in many forms by the majority of the world's population including special groups (i.e., infants, the elderly, and the immune-compromised) makes their strict control inherent. We should also be aware of emerging dairy-related pathogens (i.e., *Mycobacterium avium* subsp. *paratuberculosis*) and assess their impact to human health, apart from the obvious food poisoning. Surveys from underdeveloped countries as well as from the developed part of the world should continue to yield results in order to enhance our understanding on dairy pathogens that up to now may not have been given the attention they required. Finally, the acquired knowledge on prevalence and impact of dairy pathogens in the food chain would enable the scientific community to make informed decisions regarding issues such as the direct selling (or not) of raw milk, from animal species.

## References Cited

Ackers, M.L., S. Schoenfeld, J. Markman, M.G. Smith, M.A. Nicholson, W. DeWitt, D.N. Cameron, P.M. Griffin and L. Slutsker. 2000. An outbreak of Yersinia enterocolitica O:8 infections associated with pasteurized milk. J. Infect Dis. 181(5): 1834–1837.

AES Chemunex. Culture Media for the Food Industry, Rapid Microbiology Company, France.

Ahmed, A.M. and T. Shimamoto. 2014. Isolation and molecular characterization of Salmonella enterica, *Escherichia coli* O157: H7 and *Shigella* spp. from meat and dairy products in Egypt. International Journal of Food Microbiology 168-169: 57–62.

Andersson, M.A., E.L. Jaaskelainen, R. Shaheen, T. Pirhonen, L.M. Wijnands and M.S. Salkinoja-Salonen. 2004. Sperm bioassay for rapid detection of cereulide-producing Bacillus cereus in food and related environments. Int. J. Food Microbiol. 94: 175–183.

Argudín, M.A., M.C. Mendoza and M.R. Rodicio. 2010. Food Poisoning and *Staphylococcus aureus* enterotoxins. Toxins 2(7): 1751–1773.

Arnesen, L.P., A. Fagerlund and P.E. Granum. 2008. From soil to gut: *Bacillus cereus* and its food poisoning toxins, FEMS Microbiol. Rev. 32: 579–606.

Arsalam, A., Z. Anwar, I. Ahmad, Z. Shad and S. Ahmed. 2013. *Cronobacter sakazakii*: an emerging contaminant in pediatric infant milk formula. Int. Res. J. Pharm. 4(4): 17–22.

Asperger, H. and P. Zangerl. 2003. *Staphylococcus aureus*. pp. 2563–2569. *In*: H. Roginski, J. Fuquay and P. Fox [eds.]. Encyclopedia of Dairy Science. San Diego: Academic Press.

Belessi, C.I.A., S. Papanikolaou, E.H. Drosinos and P.N. Skandamis. 2008. Survival and Acid Resistance of Listeria innocua in Feta Cheese and Yogurt, in the Presence or Absence of Fungi. J. Food Prot. 4: 742–749.

Bernardino-Varo, L., E.I. Quiñones-Ramírez, F.J. Fernandez and C. Vazquez-Salinas. 2013. Prevalence of Yersinia enterocolitica in raw cow's milk collected from stables of Mexico City. J. Food Prot. 75(4): 694–698.

Botsaris, G., I. Slana, M. Liapi, C. Dodd, C. Economides, C. Rees and I. Pavlik. 2010. Rapid detection methods for viable Mycobacterium avium subspecies paratuberculosis in milk and cheese. Int. J. Food Microbiol. 141: S87–S90.

Bottone, E.J. 1997. *Yersinia enterocolitica*: the charisma continues. Clin. Microbiol. Rev. 10(2): 257–276.

Bottone, E.J. 2010. *Bacillus cereus*, a Volatile Human Pathogen. Clin. Microbiol. Rev. 23(2): 382.

Ceuppens, S., N. Boon and M. Uyttendaele. 2013. Diversity of *Bacillus cereus* group strains is reflected in their broad range of pathogenicity and diverse ecological lifestyles. FEMS Microbiol. Ecol. 84: 433–450.

Commission Regulation (EC) No. 853/2004 of the European Parliament and of the Council laying down specific hygiene rules for the hygiene of foodstuffs. OJ L139.

Commission Regulation (EC) No 2073/2005 of 15 November 2005 on microbiological criteria for foodstuffs. OJ L338.

Commission Regulation (EC) No 1441/2007 of 5 December 2007 amending Regulation (EC) No 2073/2005 on microbiological criteria for foodstuffs. OJ L322/12.

Commission Regulation (EU) No. 365/2010 amending Regulation (EC) No. 2073/2005 on microbiological criteria for foodstuffs as regards Enterobacteriaceae in pasteurised milk and other pasteurised liquid dairy products and Listeria monocytogenes in food grade salt. OJ L107. 9–11.

Conedera, G., P. Dalvit, M. Martini, G. Galierod, M. Gramaglia, E. Goffredo, G. Loffredo, S. Morabito, D. Ottaviani, F. Paterlini, G. Pezzotti, M. Pisanu, P. Semprini and A. Caprioli. 2004. Verocytotoxin-producing *Escherichia coli* O157 in minced beef and dairy products in Italy. Int. J. Food Microbiol. 96(1): 67–73.

De Buyser, M.L., B. Dufour, M. Mair and V. Lafarge. 2001. Implication of milk and milk products in food-borne diseases in France and in different industrialised countries. Int. J. Food Microbiol. 67(1-2): 1–17.

Dumen, E. 2010. *Cronobacter sakazakii* (*Enterobacter sakazakii*): Only An Infant Problem? Kafkas Univ. Vet. Fak. Derg. 16: S171–S178.

Ehling-Schulz, M., M. Fricker and S. Scherer. 2004. *Bacillus cereus*, the causative agent of an emetic type of food-borne illness. Mol. Nutr. Food Res. 48: 479–487.

El-Gazzar, F.E. and E.H. Marth. 1992. Salmonellae, salmonellosis, and dairy foods: a review. J. Dairy Sci. 75: 2327–2343.

El Sharoud, W.M. 2009. Prevalence and survival of Campylobacter in Egyptian dairy products. Food Res. Int. 42(5-6): 622–626.

European Food Safety Authority. 2003. Opinion of the Scientific Committee on Veterinary Measures relating to public health on Staphulococcal enterotoxins in milk products, particularly cheeses.

Farmer, J.J., M.A. Asbury, F.W. Hickman and D.J. Brenner. 1980. *Enterobacter sakazakii*, new species of Enterobacteriaceae isolated from clinical specimens. Int. J. Syst. Bacteriol. 30: 569–584.

Farrokh, C., K. Jordan, F. Auvray, K. Glass, H. Oppegaard, S. Raynaud, D. Thevenot, R. Condron, K. De Reu, A. Govaris, K. Heggum, M. Heyndrickx, J. Hummerjohann, D. Lindsay, S. Miszczycha, S. Moussiegt, K. Verstraete and O. Cerf. 2013. Review of Shiga-toxin-producing *Escherichia coli* (STEC) and their significance in dairy production. Int. J. Food Microbiol. 162(2): 190–212.

Food and Drug Administration. 2002. Isolation and enumeration of *Enterobacter sakazakii* from dehydrated powdered infant formula.

Food Safety news. Dairy-Related Outbreaks, Illnesses, Recalls: 2010 to Present. http://www.foodsafetynews.com/2012/04/dairy-related-outbreaks-illnesses-recalls-2010-to-present. Date accessed 17/3/2014.

Fredriksson-Ahomaa, M. and H. Korkeala. 2003. Low occurrence of pathogenic *Yersinia enterocolitica* in clinical, food, and environmental samples: a methodological problem. Clin. Microbiol. Rev. 16: 220–229.

Gasanov, U., D. Hughes and P.M. Hansbro. 2005. Methods for the isolation and identification of *Listeria* spp. and *Listeria monocytogenes*: a review. FEMS Microbiology Reviews 29: 851–875.

Glickman, M.S. and W.R. Jacobs. 2001. Microbial Pathogenesis Review of *Mycobacterium tuberculosis*: dawn of discipline. Cell. 104: 477–485.

Gougouli, M., A.S. Angelidis and K. Koutsoumanis. 2008. A study on the kinetic behavior of *Listeria monocytogenes* in ice cream stored under static and dynamic chilling and freezing conditions. J. Dairy Sci. 91(2): 523–530.

Grant, I.R. 2006. *Mycobacterium avium* subsp. *paratuberculosis* in foods: Current evidence and potential consequences. Int. J. Dairy Technol. 59: 112–117.

Granum, P.E. and T. Lund. 1997. *Bacillus cereus* and its food poisoning toxins. FEMS Microbiol. Lett. 157: 223–228.

Gurtler, J.B., J.L. Kornacki and L.R. Beuchat. 2005. *Enterobacter sakazakii*: A coliform of increased concern to infant health. Int. J. Food Microbiol. 104: 1–34.

Hammer, P., C. Kiesner and H.G. Walte. 2014. Short communication: Effect of homogenization on heat inactivation of Mycobacterium avium subspecies paratuberculosis in milk. J. Dairy Sci. 97(4): 2045–2048.

Hariharan, H., G.A. Murphey and I. Kempf. 2004. Campylobacter jejuni: Public health hazards and potential control methods in poultry: a review Vet. Med. – Czech 49(11): 441–446.

Hauge, S. 1955. Food Poisoning caused by aerobic spore-forming bacilli. J. Applied Bacter. 18(3): 591–595.

Heuvelink, A.E., C. Van Heerwaarden, A. Zwartkruis-Nahuis, J.J Tilburg, M.H. Bos, F.G. Heilmann, A. Hofhuis, T. Hoekstra and E. de Boer. 2009. Two outbreaks of campylobacteriosis associated with the consumption of raw cows' milk. Int. J. Food Microbiol. 134(1-2): 70–74.

Hitchins, A.D. 2003. Detection and Enumeration of *Listeria monocytogenes* in Foods. Bacteriological Analytical Manual Online. U.S. department of Health and Human

Services, U.S. Food and Drug Administration Center for Food Safety & Applied Nutrition. Chapter 10.

Holt, J.G. 1994. Bergey's manual of determination bacteriology, 9th Edition.

Huang, D.B., A. Mohanty, H.L. DuPont, P.C. Okhuysen and T. Chiang. 2006. A review of An emerging enteric pathogen: enteroaggregative *Escherichia coli*. J. Med. Microbiol. 55: 1303–1311.

International Dairy Federation [IDF] Technical Bulletin. 2012. Shiga-toxin-producing *Escherichia coli* (STEC), IDF.

International Organization for Standardization. 1996. Microbiology of food and animal feeding stuffs—Horizontal method for the detection and enumeration of *Listeria monocytogenes* —Part 1: Detection method. International Standard ISO 11290–1, Geneva, Switzerland.

International Organization for Standardization. 1998. Microbiology of food and animal feeding stuffs—Horizontal method for the detection and enumeration of *Listeria monocytogenes*—Part 2: Enumeration method. International Standard ISO 11290–2, Geneva, Switzerland.

International Organization for Standardization. 2002. ISO 6579: (Annex D) Detection of *Salmonella* spp. In animal faeces and in samples of the primary production stage—Horizontal method for the detection of *Salmonella* spp., International Organization for Standardization, 1, rue de Varembé, Case postale 56 CH-1211 Geneva 20, Switzerland.

International Organization for Standardization. 1995. Microbiology of Food and Animal Feeding Stuffs—Horizontal Method for Detection of Thermotolerant Campylobacter, ISO 10270:1995(E). Geneva, Switzerland.

International Organization for Standardization. 2006. Microbiology of Food and Animal Feeding Stuffs—Horizontal Method for Detection and Enumeration of *Campylobacter* spp., ISO 10272–1:2006(E). Geneva, Switzerland.

International Organization for Standardization. 2006. Milk and milk products—Detection of Enterobacter sakazakii ISO/TS 22964:2006. Geneva, Switzerland.

International Organization for Standardization. 2001. Microbiology of food and animal feeding stuffs—Horizontal method for the detection of *Escherichia coli* O157. ISO 16654:2001. Geneva, Switzerland.

International Organization for Standardization. 2004. Microbiology of food and animal feeding stuffs —Horizontal methods for the detection and enumeration of Enterobacteriaceae—Part 2: Colony-count method. ISO 21528-2:2004. Geneva, Switzerland.

International Organization for Standardization. 2001. Microbiology of food and animal feeding stuffs —Horizontal method for the enumeration of beta-glucuronidase-positive *Escherichia coli*—Part 1: Colony-count technique at 44 degrees C using membranes and 5-bromo-4-chloro-3-indolyl beta-D-glucuronide. ISO 16649-1:2001. Geneva, Switzerland.

International Organization for Standardization. 1999. Microbiology of food and animal feeding stuffs —Horizontal method for the enumeration of coagulase-positive staphylococci (*Staphylococcus aureus* and other species)—Part 2: Technique using rabbit plasma fibrinogen agar medium. ISO 6888-2:1999. Geneva, Switzerland.

International Organization for Standardization. 2004. Microbiology of food and animal feeding stuffs —Horizontal method for the enumeration of presumptive Bacillus cereus—Colony-count technique at 30 degrees C. ISO 7932:2004. Geneva, Switzerland.

International Organization for Standardization. 2009. Prerequisite programmes on food manufacturing safety—Part 1: Food ISO/TS 22002-1:2009. Geneva, Switzerland.

International Organization for Standardization. 2005. Food safety management systems—Requirements for any organization in the food chain. ISO 22000:2005. Geneva, Switzerland.

Iversen, C. and S. Forsythe. 2003. Risk profile of *Enterobacter sakazakii*, an emergent pathogen associated with infant milk formula. Trends Food Sci. Technol. 14: 443–454.

Iversen, C., N. Mullane, B. McCardell, B.D. Tall, A. Lehner, S. Fanning, R. and H. Joosten. 2008. Cronobacter gen. nov., a new genus to accommodate the biogroups of *Enterobacter sakazakii*, and proposal of *Cronobacter sakazakii* gen. nov., comb. nov., *Cronobacter malonaticus* sp. nov., *Cronobacter turicensis* sp. nov., *Cronobacter muytjensii* sp. nov., *Cronobacter dublinensis* sp. nov., *Cronobacter genomospecies* 1, and of three subspecies, *Cronobacter dublinensis* subsp. *dublinensis* subsp. nov., *Cronobacter dublinensis* subsp. *lausannensis* subsp. nov. and *Cronobacter dublinensis* subsp. *lactaridi* subsp. nov. Int. J. Syst. Evol. Microbiol. 58: 1442–1447.

Jafari, A., M.M. Aslani and S. Bouzari. 2012. *Escherichia coli*: a brief review of diarrheagenic pathotypes and their role in diarrheal diseases in Iran. Iran. J. Microbiol. 4(3): 102–117.

Janssen, R., K.A. Krogfelt, S.A. Cawthraw, W. van Pelt, J.A. Wagenaar and R.J. Owen. 2008. Host-pathogen interactions in Campylobacter infections: the host perspective. Clin. Microbiol. Rev. 21(3): 505–518.

Jemmi, T. and S. Stephan. 2006. *Listeria monocytogenes*: food-borne pathogen and hygiene indicator. Rev. Sci. Tech. 25(2): 571–580.

Jennison, A.V. and N.K. Verma. 2004. Shigella flexneri infection: pathogenesis and vaccine development. FEMS Microbiol. Rev. 28(1): 43–58.

Jeyaletchumi, P., R. Tunung, S.P. Margaret, R. Son, M.G. Farinazleen and Y.K. Cheah. 2010. Detection of *Listeria monocytogenes* in foods. Int. Food Res. J. 17: 1–11.

Kaper, J.B. 1998. Enterohemorrhagic *Escherichia coli.* Current Opinion in Microbiology 1: 103–108.

Kasalica, A., V. Vuković, A. Vranješ and N. Memiši. 2011. *Listeria monocytogenes* in milk and dairy products. Biotech Animal Husbandry 27(3): 1067–1082.

Langer, A.J., T. Ayers, J. Grass, M. Lynch, F.J. Angulo and B.E. Mahon. 2012. Nonpasteurized dairy products, disease outbreaks, and State laws-United States, 1993–2006. Emerg. Inf. Dis. 18: 385–391.

Le Loir, Y., F. Baron and M. Gautier. 2003. *Staphylococcus aureus* and food poisoning. Genet. Mol. Res. 2: 63–76.

Levin, R.E. 2007. *Campylobacter jejuni*: A Review of its characteristics, pathogenicity, ecology, distribution, subspecies characterization and molecular methods of detection. Food Biotech. 21(4): 271–347.

Liapi, M., G. Botsaris, I. Slana, M. Moravkova, V. Babak, M. Avraam, A. Di Provvido, S. Georgiadou and I. Pavlik. 2013. Short communication: *Mycobacterium avium* subsp. *paratuberculosis* Sheep Strains Isolated from Cyprus Sheep and Goats. Transb. Emerg. Diseas.

Liu, D. 2006. Identification, subtyping and virulence determination of *Listeria monocytogenes*, an important food borne pathogen. J. Med. Microbiol. 55: 645–659.

Longenberger, A.H., A.J. Palumbo, A.K. Chu, M.E. Moll, A. Weltman and S.M. Ostroff. 2013. Campylobacter jejuni infections associated with unpasteurized milk—multiple States. Clinical infectious diseases: an official publication of the Infectious Diseases Society of America 57(2): 263–266.

Lund, T., M.L. De Buyser and P.E. Granum. 2000. A new cytotoxin from *Bacillus cereus* that may cause necrotic enteritis. Mol. Microbiol. 38(2): 254–261.

Marchetti, G., M. Ricchi, A. Serraino, F. Giacometti, E. Bonfante and N. Arrigoni. 2013. Prevalence of *Mycobacterium avium* subsp. *paratuberculosis* in milk and dairy cattle in Southern Italy: preliminary results. Ital. J. Food Saf. 2:e35: 124–127.

Miszczycha, S.D., J. Thévenot, S. Denis, C. Callon, V. Livrelli, M. Alric, M.-C. Montel, S. Blanquet-Diot and D. Thevenot-Sergentet. 2014. Survival of *Escherichia coli* O26: H11 exceeds that of *Escherichia coli* O157: H7 as assessed by simulated human digestion of contaminated raw milk cheeses. Int. J. Food Microbiol. 172: 40–48.

Nataro, J.P. 2005. Enteroaggregative *Escherichia coli* pathogenesis. Curr. Opin. Gastroenterol. 21(1): 4–8.

Niyogi, S.K. 2005. Shigellosis. The J. Microbiol. 43(2): 133–143.

Oliver, S.P., B.M. Jayarao and R.A. Almeida. 2005. Foodborne pathogens in milk and the dairy farm environment: food safety and public health implications. Foodborne Pathog. Dis. 2(2): 115–129.

Ooi, S.T. and B. Lorber. 2005. Gastroenteritis Due to *Listeria monocytogenes.* Clin. Infect. Dis. 40(9): 1327–1332.

Papageorgiou, D.K. and E.H. Marth. 1989. Fate of Listeria monocytogenes during the manufacture and ripening of blue cheese. J. Food Prot. 52: 459–465.

Phalipon, A. and P.J. Sansonetti. 2007. Shigella's ways of manipulating the host intestinal innate and adaptive immune system: a tool box for survival? Immun. Cell Biol. 85(2): 119–129.

Pui, C.F., W.C. Wong. L.C. Chai, R. Tunung, P. Jeyaletchumi, M.S. Noor Hidayah, A. Ubong, M.G. Farinazleen, Y.K. Cheah and R. Son. 2011. Salmonella: A Foodborne Pathogen. Int. Food Res. J. 18(2): 465–473.

Ryan, C.A., M.K. Nickels, N.T. Hargrett-Bean, M.E. Potter, T. Endo, L. Mayer, C.W. Langkop, C. Gibson, R.C. McDonald, R.T. Kenney, N.D. Puhr, P.J. McDonnell, R.J. Martin, M.L. Cohen and P.A. Blake. 1987. Massive Outbreak of Antimicrobial-Resistant Salmonellosis Traced to Pasteurized Milk JAMA. 258(22): 3269–3274.

Ryser, E.T. and E.H. Marth. 2004. Listeria, listeriosis and food safety. Third Edition. Marcel Dekker Inc., New York, USA.

Sabina, Y., A. Rahman, R.C. Ray and D. Montet. 2011. Yersinia enterocolitica: Mode of Transmission, Molecular Insights of Virulence, and Pathogenesis of Infection. J. Pathog. 429069.

[Scharlau] Standardized procedures for microbiological examinations of food products. Culture Medium Handbook, Edition 9.

Shaker, R., T. Osaili, W. Al-Omary, Z. Jaradat and M. Al-Zuby. 2007. Isolation of *Enterobacter sakazakii* and other *Enterobacter* sp. from food and food production environments. Food Control 18: 1241–1245.

Silva, J., D. Leite, M. Fernandes, C. Mena, P.A. Gibbs and P. Teixeira. 2011. *Campylobacter* spp. as a foodborne pathogen: a review. Front Microbiol. 2: 200.

Singh, P. and A. Prakash. 2010. Prevalence of coagulase positive pathogenic Staphylococcus aureus in milk and milk products collected from unorganized sector of Agra. Acta Argiculturae Slovenica 96(1): 37–41.

Tacket, C.O., J.P. Narain, R. Sattin, J.P. Lofgren, C. Konigsberg, R.C. Rendtorff, A. Rausa, B.R. Davis and M.L. Cohen. 1984. A Multistate Outbreak of Infections caused by *Yersinia enterocolitica* Transmitted by pasteurized milk. JAMA 251(4): 483–486.

Teunis, P.F.M., K. Takumi and K. Shinagawa. 2004. Dose response for infection by *Escherichia coli* O157:H7 from outbreak data. Risk Analysis. 24: 401–408.

Tindall, B.J., P.A. Grimont, G.M. Garrity and J.P. Euzeby. 2005. Nomenclature and taxonomy of the genus *Salmonella*. Int. J. Syst. Evol. Microbiol. 55: 521–524.

United States Department of Agriculture (USDA). 2013. Isolation and Identification of Listeria monocytogenes from Red Meat, Poultry and Egg Products, and Environmental Samples, 9th Rev. 1–20.

Yan, Q.Q., O. Condell, K. Power, F. Butler, B.D. Tall and S. Fanning. 2012. Cronobacter species (formerly known as *Enterobacter sakazakii*) in powdered infant formula: a review of our current understanding of the biology of this bacterium. J. of Appl. Microbiol. 113: 1–15.

Young, K.T., L.M. Davis and V.J. DiRita. 2007. *Campylobacter jejuni*: molecular biology and pathogenesis. Nat. Rev. Microbiol. 5: 665–679.

Yousef, A.E. and E.H. Marth. 1990. Fate of *Listeria monocytogenes* During the Manufacture and Ripening of Parmesan Cheese. J. Dairy Sc. 73(12): 3351–3356.

Zaika, L.L. and J.G. Phillips. 2005. Model for the combined effects of temperature, pH and sodium chloride concentration on survival of Shigella flexneri strain 5348 under aerobic conditions. Int. J. Food Microbiol. 101: 179–187.

CHAPTER 4

# Dairy Starter Cultures

*Thomas Bintsis*[1,*] and *Antonis Athanasoulas*[2]

## INTRODUCTION

Fermented milk products, such as yogurt and cheese, appeared in human diet about 8000–10000 years ago. Up to the 20th century, milk fermentation remained an unregulated process, and, the discovery and characterization of lactic acid bacteria (LAB) have changed the views on milk fermentation. In the early 1960s, commercial starter culture companies developed the production technology to freeze-dry liquid cultures and produce concentrated frozen starter cultures for the direct inoculation of bulk starter tanks at the dairy. In the beginning of 1980s Chr. Hansen released the first Direct-Vat-Set (DVS) culture comprising defined single strains. Today, several commercial starter companies offer an extensive range of frozen and freeze-dried concentrated cultures for direct inoculation, eliminating the need for use of bulk starters, and thus, propagation problems. In fact, these fermentation businesses, together with probiotic products, represent a total global market value of over 100 billion Euros (de Vos 2011).

While the traditional method for the manufacture of fermented dairy products was the "inoculation" of the milk with a sample of a previous day product, i.e., back-slopping. This method had certain drawbacks and is not used anymore, except for some home-made products. There was a significant microbial variability both in the milk and the natural starter, hence a great fluctuation in the quality of the product. Thus, the substitution

[1] PhD, Dairy Science, 25 Kappadokias St. 55134 Thessaloniki, Greece.
[2] Starter Cultures Technical Manager, TYRAS SA 5th klm Trikala-Pili, 42100, Trikala, Greece.
[*] Corresponding author: tbintsis@gmail.com

of the back-slopping with a selected starter culture was conceived a necessity very early. Nowadays, since the production of dairy products is automated and requires large quantities of milk, and total control of the process, the use of commercial starter cultures has become an integral part of a successful production of any fermented dairy product.

Bacterial starter cultures are defined as "prepared cultures that contain one or several strains of microorganisms at high counts (in general more than $10^8$ cfu/ml or $10^8$ cfu/g of viable bacteria) being added to bring about a desirable enzymatic reaction (e.g., fermentation of lactose resulting in acid production, degradation of lactic acid to propionic acid or other metabolic activities directly related to specific product properties" (International Organization for Standardization/International Dairy Federation 2010).

The main function of the starter is to produce lactic acid, that is, the acidification of milk. Besides, the acidifying bacteria contribute to the flavour, texture and nutritional value of the fermented foods, through production of aroma components, production or degradation of exopolysaccharides, lipids and proteins, and the production of nutritional components such as vitamins. In addition, they contribute to the inhibition of adventitious organisms and pathogens.

The process for the production of starter cultures has been reviewed by Høier et al. (2010) and Tamime (2002), and a typical process may include the following steps:

a) handling of inoculation material,
b) preparation of media,
c) propagation of cultures in fermenters under pH control,
d) concentration,
e) freezing,
f) drying, and
g) packaging and storage.

## Classification of Starter Cultures

Starter cultures used by the dairy industry can be divided into two broad groups: mesophilic and thermophilic. Mesophilic starters have an optimum temperature for growth at 30°C and are used in the production of most cheese varieties (soft and semi-hard), while thermophilic have an optimum temperature at 37°C and are used in the production of yogurt and hard and semi-hard cheeses with high cooking temperatures (Table 1).

The mesophilic cultures are divided into LD cultures and O cultures. LD cultures contain citrate-fermenting bacteria (L = *Leuconostoc* species and D = *Lc. lactis* subsp. *lactis* biovar. *diacetylactis*), which produce aroma and $CO_2$ from citrate. The O cultures contain only acid-producing strains,

**Table 1.** Types of starter cultures and species of lactic acid bacteria used in some dairy products.

| Type of culture | Species | Product |
|---|---|---|
| Mesophiles | | |
| Type O | *Lactococcus lactis* subsp. *lactis*<br>*Lactococcus lactis* subsp. *cremoris* | Cheddar, White-brined cheeses, Cottage, Raclette, Edam, Compté, St. Paulin. |
| Type LD | *Lactococcus lactis* subsp. *lactis*<br>*Lactococcus lactis* subsp. *cremoris*<br>*Lactococcus lactis* subsp. *lactis* biovar. *diacetylactis*<br>*Leuconostoc mesenteroides* subsp. *cremoris* | Gouda, Tilsitter, Blue cheeses, Camembert and other mould ripened cheeses, Sour cream, Butter. |
| Thermophiles | | |
| | *Streptococcus thermoplillus* | Fermented milks, Mozzarella, Emmental, Compté, Asiago, Sarde. |
| | *Streptococcus thermoplillus*<br>*Lactobacillus* delbreuckii subsp. *bulgaricus* | Yogurt, Fermented milks, Mozzarella, Feta, White-brined cheeses. |
| | *Lactobacillus helveticus*<br>*Lactobacillus* delbreuckii subsp. *lactis* | Emmental, Compté, Beaufort, Asiago, Sarde. |
| Mixed | | |
| | *Lactococcus lactis* subsp. *lactis*<br>*Lactococcus lactis* subsp. *cremoris*<br>*Streptococcus thermoplillus* | Cheddar, Colby, Chester, Leicester, Gouda, Manchego. |
| | *Lactococcus lactis* subsp. *lactis*<br>*Lactococcus lactis* subsp. *cremoris*<br>*Streptococcus thermoplillus*<br>*Lactobacillus* delbreuckii subsp. *bulgaricus* | White-brined cheeses |
| | *Lactobacillus casei* subsp. *casei* | Yakult |
| | *Lactobacillus acidophilus* | Acidophilus milk |

and produce no gas. The L cultures and D cultures also exist, but are only used to a minor degree in the cheese industry. Traditional, mesophilic O cultures are used in many cheeses, where the main focus is on a rapid and consistent acidification of the milk. The LD cultures are used in most continental semi-hard and soft cheeses (Table 1).

Starters are also subdivided into defined- and mixed-strain cultures. Defined-strain cultures are pure cultures with known physiological characteristics and technological properties. These consist of 2–6 strains, used in rotation as paired single strains or as multiple strains and enable industrial-scale production of high quality products. Mixed-strain cultures contain unknown numbers of strains of the same species and may also contain bacteria from different species or genera of LAB (Sheehan 2007). For a detailed classification of starter cultures see Tamime (2002) and Parente and Cogan (2004).

An authoritative list of microorganisms with a documented use in food was established as a result of a joint project between the International Dairy Federation (IDF) and the European Food and Feed Cultures Association (EFFCA) (Bourdichon et al. 2012a). The "2002 IDF Inventory" listed 82 bacterial species and 31 species of yeast and molds whereas the 2012 IDF-EFFCA updated inventory contains 195 bacterial species and 69 species of yeasts and molds (Bourdichon et al. 2012b).

### *Adjunct Cheese Cultures*

Adjunct cultures are defined as any cultures that are deliberately added at some point of the manufacture of cheese, but whose primary role is not acid production (Chamba and Irlinger 2004). The application of adjuncts could be seen as an attempt by the cheese-maker to add back to the product, in a controlled manner, some of the biodiversity removed by pasteurization, improved hygiene and the defined strain starter system. Adjuncts can be added in the milk or at a later stage of the cheese-making, e.g., in the salting brines (Bintsis et al. 2002).

During cheese ripening, the starter bacteria and most other organisms in the curd die; they autolyse and release intracellular enzymes as ripening progresses (Khalid and Marth 1990a). On the other hand, the nonstarter lactic acid bacteria (NSLAB) persist and can grow from levels of $10^2$ to $10^4$ cfu/g to ~$10^8$ cfu/g after manufacture, where they can have a significant impact on flavour, either positively or negatively (Khalid and Marth 1990, Cogan et al. 2007). The role of the secondary micro-flora is evident in certain cheese varieties, e.g., "eye" formation due to fermentation of lactate by propionibacteria in Swiss-type cheeses (i.e., Emmental, Gruyère de Comte) and neutralization of the pH of the surface of a cheese due to metabolism of lactate by yeasts and moulds in mould- and smear-ripened cheeses (i.e., Camembert and Limburger) (Dezmazeaud and Cogan 1996). *Penicillium camemberti* cultures are commercially available for the manufacture of mould-ripened soft cheeses (i.e., Brie and Camembert), and *Penicillium roqueforti* for blue-veined cheeses (i.e., Danish Blue, Stilton, Gorgonzola and Roquefort). In addition, *Geotrichum candidum* are used in mould-ripened cheeses since its deacidification activity stimulates the growth of moulds on the cheese surface, and yeasts such as *Debaryomyces hansenii, Kluyveromyces lactis, Candida utilis, Saccharomyces cerevisiae* and *Rhodosporium infirmominiatum* are used in bacterial smear cheeses to stimulate the growth of smear bacteria, due to their deacidification potential, and for their debittering effect, due to their aminopeptidase activity (Chamba and Irlinger 2004). For the manufacture of bacterial smear-ripened cheeses, cultures of *Brevibacterium linens* and *Brevibacterium casei* are available, since they belong to the typical surface bacterial flora of such cheeses and contribute to the formation of aromatic sulphur compounds (Bockelmann 1999).

In addition, it has been suggested that addition of the adjunct cultures may be an indirect solution for controlling the growth of non-desirable bacteria in cheese (Crow et al. 2001, Di Cagno et al. 2003). The term "protective cultures" has been applied to microbial food cultures exhibiting a metabolic activity contributing to inhibiting or controlling the growth of undesired microorganisms in food (European Food and Feed Cultures Association 2011). These undesired microorganisms could be pathogenic or toxigenic bacteria and fungi but spoilage causing species may also be included.

The technological properties of a number of *Lactobacillus* strains isolated from cheese have been studied, including milk acidification kinetics and proteolytic properties, as well as tolerance to salts and phage resistance (Briggiler-Marco et al. 2007). For example, *Lactobacillus casei* I90 and *Lactobacillus rhamnosus* I91 with low acidifying activity were found to be appropriate for use as adjunct cultures as they did not alter the composition of the cheese products and improved their sensory attributes. Interestingly, the contribution of *Lactobacillus paracasei* subsp. *paracasei* and *Lactobacillus plantarum* to proteolysis of Cheddar cheese appears to be mainly at the level of free amino acids (i.e., due to exopeptidase activity) (Lynch et al. 1999). Thus, depending on the biochemical activities of the selected strains, together with the rate of autolysis and/or release of enzymes, the water-soluble peptides can be further degraded to small peptides and free amino acids. Khalid and Marth (1990) emphasized the role of the proteolytic and lipolytic activities in the development of cheese flavour and Peterson and Marshall (1990) concluded that the high proteolytic and peptidolytic activities of lactobacilli have an influence on the extent of proteolysis, but the growth of heterofermentative strains have been associated with undesirable flavours in Cheddar cheese.

In order to become an appropriate adjunct culture, it has been suggested that a microorganism needs to be able to:

1. reach and maintain high levels of cell density during ripening,
2. cause no defect in the product, and
3. impact positively on the overall quality of the cheese.

Probiotic cultures have been extensively used during the last 10 years in the dairy industry. Probiotic bacteria are "live microorganisms which when administered in adequate amounts confer a health benefit on the host" (Food and Agriculture Organization/World Health Organization 2001). Recently, the addition of probiotic bacteria to cheese has been increasing applied (Ross et al. 2002, Karimi et al. 2011, Kasimoglu et al. 2004, Souza et al. 2008, Rodgers 2008, Ong et al. 2006, Gardiner et al. 2002, Cruz et al. 2009, Araujo et al. 2010).

EU Regulation EC No. 1924/2006 on nutrition and health claims made on foods has led to increased focus on the clinical documentation available for probiotic strains (EU 2006). It is generally assumed that in order to provide a beneficial health effect, the probiotic bacteria must be viable at the time of consumption and remain viable throughout the gastrointestinal tract. Some of the health benefits from probiotic bacteria include: (a) improving intestinal tract health, (b) enhancing the immune system, (c) synthesising and enhancing the bioavailability of nutrients, (d) reducing symptoms of lactose intolerance, (e) decreasing the prevalence of allergy in susceptible individuals, and (f) reducing risk of certain cancers (Parvez et al. 2006).

## Genetics

LAB used for starter cultures in dairy products belong to a number of bacterial genera including *Lactococcus*, *Streptococcus*, *Lactobacillus*, *Pediococcus*, and *Leuconostoc*, all members of the *Firmicutes*. Moreover, some probiotic cultures are mostly members of the genus *Bifidobacterium*, which also produce lactic acid as one of their major fermentation end-products, however, from the taxonomical point of view, they are members of the *Actinobacteria*. LAB have either homofermentative metabolism, which mainly produce lactic acid, or heterofermentative metabolism, which, apart from lactic acid, yield a variety of fermentation products such as acetic acid, ethanol, carbon dioxide and formic acid (Hutkins 2001).

The genetics of the LAB used as starter cultures in the dairy industry have been reviewed (Broadbent 2001, Callanan and Ross 2004, Klaenhammer et al. 2002, Morelli et al. 2004, Mills et al. 2010) and an overview is presented below. In addition, complete genome sequences of a number of LAB have been published, and an updated review of the available nucleic acid databases is published every year in the Nucleic Acid Research (Galperin and Fernandez-Suarez 2012).

### *Lactococcus* spp.

Lactococci are mesophilic LAB that were first isolated from green plants (Klaenhammer et al. 2002). These bacteria, previously designated as the lactic streptococci (*Streptococcus lactis* subsp. *lactis* or *S. lactis* subsp. *cremoris*) were placed in this new taxon in 1987 by Schleifer (Klaenhammer et al. 2002). Lactococci are selected for use as starters based on their metabolic stability, their resistance to bacteriophage, and their ability to produce unique compounds—often from amino acid catabolism. *Lc. lactis* subsp. *lactis* form one of the main constituents in starter cultures (Table 1) where their most important role lies in their ability to produce acid in milk and to convert milk fat and protein into flavour compounds.

To date, the complete genome sequences of three lactococcal strains have been published (Mayo et al. 2008); *Lc. lactis* subsp. *lactis* IL1403 (Bolotin et al. 2001), *Lc. lactis* subsp. *cremoris* MG1363 and NZ9000 (Wegmann et al. 2007, Linares et al. 2010) and *Lc. lactis* subsp. *cremoris* SK11 (Makarova et al. 2006).

There are noticeable differences between strains, e.g., the chromosome of *Lc. lactis* subsp. *lactis* MG1363 is 160 kb larger than that of *Lc. lactis* subsp. *lactis* IL1403 and has an average Guanine + Cytosine (G+C) content of 35.8%, and thus, encodes more proteins (Table 2).

*Lc. lactis* subsp. *cremoris* strains are preferred over *Lc. lactis* subsp. *lactis* strains because of their superior contribution to product flavor via unique metabolic mechanisms (Salama et al. 1991). With the knowledge of the complete genome sequences, *Lc. lactis* subsp. *cremoris* was found to contain greater genome sizes than *Lc. lactis* subsp. *lactis* IL1403 (approximately 2.37 Mb), with *Lc. lactis* subsp. *cremoris* MG1363 containing the largest genome size of approximately 2.53 Mb, followed by *Lc. lactis* subsp. *cremoris* SK11 with a genome size of 2.44 Mb (Mills et al. 2010). Approximately 85% DNA sequence identity was observed between the coding domains of *Lc. lactis* subsp. *lactis* IL1403 and *Lc. lactis* subsp. *cremoris* MG1363, whereas 97.7% identity was observed between *Lc. lactis* subsp. *cremoris* MG1363 and *Lc. lactis* subsp. *cremoris* SK11 (Wegmann et al. 2007). Interestingly, a complete set of competence genes was observed on the *Lc. lactis* subsp. *lactis* IL1403 genome, indicating that the strain may have the ability to undergo natural DNA transformation (Mills et al. 2010).

Many of the traits in lactococci which render these microorganisms suitable for dairy fermentations are in fact encoded on plasmids (Mills et al. 2006, McKay 1985). Traits such as lactose utilization, casein breakdown, bacteriophage resistance, bacteriocin production, antibiotic resistance,

**Table 2.** Comparison of characteristics of sequences for *Lc. lactis* subsp. *lactis* MG1363 and *Lc. lactis* subsp. *lactis* IL1403.

| | ***Lc. lactis* subsp. *lactis* MG1363** | ***Lc. lactis* subsp. *lactis* IL1403** |
|---|---|---|
| Size (Mb) | 2.53 | 2.37 |
| No of ORFs | ~2500 | ~2310 |
| Total GC% | 35.8 | 35.4 |
| GC% ORFs | 36.7 | 36.1 |
| No phage genes | ~200 | 293 |
| Phage DNA (kb) | 134 | 293 |
| IS elements | 92 | 52 |

After: Kok et al. (2005)

resistance to and transport of metal ions, and exopolysaccharide (EPS) production have all been associated with extra-chromosomal plasmid DNA. Plasmids isolated from lactococci range in size from 3 to 130 kb, have a G+C content of 30–40% and vary in function and distribution, with most strains carrying between 4 and 7 per cell (Davidson et al. 1996). Plasmids are commonly exchanged between strains via conjugation (McKay 1985, Dunny and McKay 1999) and with the chromosome by insertion sequence (IS) elements (Hughes 2000). Presumably, these exchanges and rearrangements mediate rapid strain adaptation and evolution but also add to the instability of important metabolic functions (Klaenhammer et al. 2002).

### *Streptococcus thermophilus*

*Streptococcus thermophilus* is the second most commercially important starter culture. *S. thermophilus* is used, along with *Lactobacillus* spp., as a starter culture for the manufacture of several important fermented dairy foods, including fermented milks, yogurt, Feta and Mozzarella cheeses (Table 1). Although research on the physiology of *S. thermophilus* has revealed important information on some of these properties, including sugar and protein metabolism, polysaccharide production, and flavor generation, only recently has the genetic basis for many of these traits been determined.

To date, three complete genome sequences have been published for *S. thermophilus* (Mayo et al. 2008); *S. thermophilus* CNRZ1066, *S. thermophilus* LMD-9 and *S. thermophilus* LMG18311. The genome of *S. thermophilus* is 1.8 Mb, making it among the smallest genomes of all LAB. Although a moderate thermophile, it is phylogenetically related to the more mesophilic lactococci and has a comparable low G+C ratio between 36.8 and 39% (Table 3). Moreover, *S. thermophilus* is related to human pathogenic strains of streptococci such as *Streptococcus pneumoniae*, *Streptococcus pyogenes* and *Streptococcus agalactiae* (Hols et al. 2005). However, the most important pathogenic determinants are either absent or present as pseudogenes, unless they encode basic cellular functions (Bolotin et al. 2004). *S. thermophilus* has therefore diverged from its pathogenic relatives to occupy the well-defined ecological niche of milk (Mills et al. 2010). Pastink et al. (2009) compared, using a genome-scale metabolic model of *S. thermophilus* LMG18311 with those of *Lc. lactis* subsp. *lactis* and reported the minimal amino acid auxotrophy (only histidine and methionine or cysteine) of *S. thermophilus* and the broad range of volatiles produced by the strain compared to lactococci. The unique pathway for acetaldehyde production, which is responsible for yogurt flavour, was also identified in *S. thermophilus*.

Unlike *Lactococcus* spp., plasmids are thought to play a relatively insignificant role in *S. thermophilus*, reported to be found in about 20–59% of

**Table 3.** Genomes of lactic acid bacteria used as starter cultures in dairy industry.

| Genus | Species | Strain | Size (Mb) | %GC |
|---|---|---|---|---|
| *Lactococcus* | *lactis* subsp. *Lactis* | IL1403 | 2.3 | 35.4 |
| | *lactis* subsp. *cremoris* | MG1363 | 2.6 | 37.1 |
| | *lactis* subsp. *cremoris* | SK11 | 2.3 | 30.9 |
| *Streptococcus* | *thermophilus* | CNRZ1066 | 1.8 | 39.0 |
| | *thermophilus* | LMD-9 | 1.8 | 36.8 |
| | *thermophilus* | LMG18311 | 1.9 | 39.0 |
| *Lactobacillus* | *Rhamnosus* | HN001 | 2.4 | 46.4 |
| | *delbrueckii* subsp. *bulgaricus* | ATCC11842 | 1.8 | 50.0 |
| | *gasseri* | ATCC33323 | 1.8 | 35.1 |
| | *acidophilus* | NCFM | 2.0 | 34.7 |
| | *johnsonii* | NCC533 | 2.0 | 34.6 |
| | *plantarum* | WCFS1 | 3.3 | 44.5 |
| | *helveticus* | CNRZ32 | 2.4 | 37.1 |
| *Pediococcus* | *Pentosaceus* | ATCC25745 | 2.0 | 37.0 |
| *Leuconostoc* | *Mesenteroides* | ATCC8293 | 2.0 | 37.4 |
| *Propionibacterium* | *Freudenreichii* | ATCC6207 | 2.6 | 67.4 |
| *Bifidobacterium* | *Longum* | NCC2705 | 2.3 | 60.1 |

After: Klaenhammer et al. (2002)

strains examined (Madera et al. 2003, Turgeon et al. 2004, Hols et al. 2005). Streptococcal plasmids are generally small, ranging in size from 2.1 to 10 kb and encode few industrially useful phenotypic traits, which include low molecular weight, heat-stress proteins and specificity subunits of bacteriophage-resistant restriction modification (R/M) systems (O' Sullivan et al. 1999) (Solow and Somkuti 2000, El Demerdash et al. 2006).

## *Lactobacillus* spp.

The genus *Lactobacillus* encompasses a large number of different species that display a relatively large degree of diversity. Similar to *S. thermophilus*, the lactobacilli also belong to the thermophilic group of LAB starter cultures. Lactobacilli commonly used for dairy fermentations include *L. delbreuckii* subsp. *bulgaricus*, *L. delbreuckii* subsp. *lactis*, *L. helveticus* and *L. acidophilus* (Thunell and Sandine 1985). To date a number of complete genome sequences are available for lactobacilli (Mayo et al. 2008), including the starter strains *L. delbreuckii* subsp. *bulgaricus* ATCC11842 (van de Guchte et al. 2006), *L. delbreuckii* subsp. *bulgaricus* ATCC BAA-365 with genome sizes of ~1.8 Mb (Makarova et al. 2006) and *L. helveticus* DPC4571 with a genome size of ~2.0 Mb (Callanan et al. 2008).

*L. plantarum* has one of the largest genomes known among LAB (Chevalier et al. 1994, Daniel 1995). The circular chromosome of *L. plantarum* WCFS1 consists of ~3.3 Mb with an average G+C content of 44.5%, and is among the largest of lactic acid bacteria. In addition, the strain harbours three plasmids of ~36 kb (G+C 40.8%), 2.3 kb (G+C 34.3%), and ~1.9 kb (G+C 39.5%), respectively (Klaenhammer et al. 2002).

*Lactobacillus rhamnosus* is one of the few species of *Lactobacillus* that have been used as probiotic organisms in functional foods. A strain of *L. rhamnosus*, designated HN001, has been identified that has both flavour enhancing and probiotic attributes, therefore, it can be used as an adjunct during cheese manufacture to reduce adventitious microflora, accelerate cheese ripening, and improve cheese flavour (Klaenhammer et al. 2002).

*Lactobacillus johnsonii* strains have been mainly isolated from the feces of humans and animals (Johnson et al. 1980, Fujisawa et al. 1992), suggesting that these bacteria constitute part of the natural intestinal flora. *L. johnsonii* La1 (formerly *L. acidophilus* La1) has been extensively studied for its probiotic properties and is commercialized in the LC1 fermented milk products (Klaenhammer et al. 2002). La1 shows immunomodulatory properties (Link-Amster et al. 1994, Schiffrin et al. 1995) and antimicrobial properties (Bernet-Camard et al. 1997, Felley et al. 2001, Pérez et al. 2001).

*L. helveticus* is quite closely related (< 10% sequence divergence) to *L. amylovorus, L. acidophilus, L. delbrueckii, L. acetotolerans, L. gasseri,* and *L. amylophilus* (Schleifer and Ludwig 1995). The genome size of *L. helveticus* has been determined to be 2.4 Mb by PFGE (Klaenhammer et al. 2002). Approximately 40 chromosomal genes and four plasmids have been sequenced from *L. helveticus. L. helveticus* is a component of "thermophilic" starter cultures used in the manufacture of a number of fermented dairy products (Hassan and Frank 2001) and grows on a relatively restricted number of carbohydrates that includes lactose and galactose and typically requires riboflavin, pantothenic acid and pyridoxal for growth (Hammes and Vogel 1995).

The *L. delbrueckii* species contains three subspecies, *L. delbrueckii* subsp. *delbrueckii, L. delbrueckii* subsp. *lactis,* and *L. delbrueckii* subsp. *bulgaricus. L. delbrueckii* subsp. *bulgaricus* grows on a relatively restricted number of carbohydrates and typically requires pantothenic acid and niacin (Hammes and Vogel 1995).

Phylogenetically, *L. delbrueckii* subsp. *bulgaricus* is closely related to *L. amylovorus, L. acidophilus, L. helveticus, L. acetotolerans, L. gasseri,* and *L. amylophilus* (Schleifer and Ludwig 1995). The G+C ratio of *L. delbrueckii* subsp. *bulgaricus* (49–51%) is somewhat higher than that found among other species (34–46%) within this phylogenetic tree (Hammes and Vogel 1995). The genome size of *L. delbrueckii* subsp. *bulgaricus* has been determined to be 1.8 Mb by PFGA (Klaenhammer et al. 2002).

Very few chromosomal genes (< 15) have been sequenced from *L. delbrueckii* subsp. *bulgaricus,* however the complete sequence of a small cryptic plasmid and the partial sequence of a bacteriophage are known.

Gene transfer systems for *L. delbrueckii* subsp. *bulgaricus* include two conjugation-based gene transfer systems (Thompson et al. 1999) and electrotransformation (Serror et al. 2002).

## *Pediococcus* spp.

Phylogenetically, *Pediococcus* and *Lactobacillus* are related and form a super-cluster; all species of *Pediococcus* fall within the *Lactobacillus casei–Pediococcus* sub-cluster. However, morphologically, they are distinct since they form tetrads via cell division in two perpendicular directions in a single plane. *Pediococcus* can be described as the only acidophilic, homofermentative, LAB that divide alternatively in two perpendicular directions to form tetrads (Simpson and Taguchi 1995).

*Pediococcus pentosaceus* can be isolated from a variety of plant materials as well as bacterial-ripened cheeses and is a typical component of the NSLAB of most cheese varieties during ripening (Beresford et al. 2001) and has been suggested as an acid producing starter culture in the dairy fermentations (Caldwell et al. 1996, 1998).

Strains of *P. pentosaceus* have been reported to contain between three and five resident plasmids (Graham and McKay 1985). Plasmid-linked traits include the ability to ferment raffinose, melibiose, and sucrose, as well as, the production of bacteriocins (Daeschel and Klaenhammer 1985, Gonzalez and Kunka 1986). Plasmids can be conjugally transferred between *Pediococcus* and *Enterococcus*, *Streptococcus*, or *Lactococcus* and electrotransformation has been utilized to introduce plasmids into pediococci, including *P. pentosaceus* (Klaenhammer et al. 2002).

## *Leuconostoc* spp.

*Leuconostoc mesenteroides* is a facultative anaerobe requiring complex growth factors and amino acids (Garvie 1986). Most strains in liquid culture appear as cocci, occurring singly or in pairs and short chains; however, morphology can vary with growth conditions; cells grown in glucose or on solid media may have an elongated or rod-shaped morphology. Cells are Gram-positive, asporogenous and non-motile. Although *L. mesenteroides* is commonly found on fruits and vegetables, it has been extensively used as an industrial dairy starter culture (Table 1). Under microaerophilic conditions, has a heterofermentative reaction. Glucose and other hexose sugars are converted to equimolar amount of D-lactate, ethanol and $CO_2$ via a combination of the hexose monophosphate and pentose phosphate pathways (Garvie 1986). Other metabolic pathways include conversion of citrate to diacetyl and acetoin and production of dextrans and levan from sucrose (Cogan et al. 1981). Viscous polysaccharides produced by *L. mesenteroides* are widely recognized as causing product losses and processing problems in the production of sucrose from sugar cane and sugar beets (Klaenhammer et al. 2002).

### *Brevibacterium* spp.

The *Brevibacterium* genus is a heterogeneous mixture of coryneform organisms that have particular application to industrial production of vitamins, amino acids for fine chemical production, and are commonly used in cheese production (Rattray and Fox 1999). This genus contains 9 species from diverse habitats, such as soil, poultry, fish, human skin, and food. While *Brevibacterium linens* is phenotypically similar to *Arthrobacter globiformis,* cellular pigmentation, cell wall composition, DNA/DNA hybridization and 5S RNA analysis show that *Brevibacterium* is distinctly different (Park et al. 1987). PFGE analysis indicates that diversity within the species is related to polymorphisms in the 16S rRNA genes with genome sizes that range from 3.2 and 3.9 Mb (Lima and Correia 2000).

*B. linens* is a non-motile, non-spore forming, Gram-positive coryneform that tolerates high salt concentrations (8–20%) and is capable of growing in a broad pH range (5.5–9.5), with an optimum of pH 7.0, without being a fast acid-producer. *B. linens* has been found to produce extracellular protease (Rattray et al. 1997), high levels of volatile compounds from amino acid catabolism (Ferchichi et al. 1985, Ummadi and Weimer 2001) and bacteriocins (Valdes-Stauber and Scherer 1996).

### *Propionibacterium* spp.

Propionibacteria are high G+C Gram-positive bacteria belonging to the class of *Actinobacteria* that prefer anaerobic growth conditions and have a peculiar physiology (Axelsson 2004). They produce propionate as their major fermentation product. Propionate fermentation yields more energy and, consequently, biomass than any other anaerobic microbial fermentation. Furthermore, propionibacteria utilize polyphosphate and pyrophosphate instead of ATP for several energy-dependent reactions and their metabolism is tuned to synthesize high levels of porphyrins, in particular $B_{12}$ (Mills et al. 2010).

Propionibacteria have long been employed in the production process of Swiss-type cheeses for which they are indispensable for the typical "eye" formation and production of characteristic taste components (Guinee and O'Callanhan 2010).

### *Bifidobacterium* spp.

Species of the genus *Bifidobacterium* are Gram positive bacteria, strictly anaerobic, fermentative rods, often Y-shaped or clubbed at the end and contain DNA with a relatively high G+C content (Table 2). Bifidobacteria have been shown to be the predominant species in the gastrointestinal tract of infants, and represent the third most numerous species encountered in the gastrointestinal of adult humans, considerably out-numbering other microbial groups. The role of these bacteria in human health has stimulated significant interest in the dairy industry, and has highlighted the position of these bacteria in the development of functional foods, used as adjunct cultures.

Despite growing consumer interest, key aspects regarding *Bifidobacterium* species, such as metabolic activities (particularly relating to catabolism of prebiotics) and physiology are still poorly understood. The determination of the complete genome of the *B. breve* strain NCIM 8807 (Klaenhammer et al. 2002), later designated as *B. breve* UCC 2003 (Sheehan et al. 2007), was undertaken as a first step towards the molecular analysis of a probiotic *Bifidobacterium* species. The genus is comprised of 31 characterized species, 11 of which have been detected in human feces (Tannock 1999). *B. longum* is often the dominant species detected in humans and is one of the few species to regularly harbor plasmids. It is a leading member of the probiotic bacteria, and numerous studies have suggested its potential health benefits (Ballongue 2004). These include modulation of the host immune system, resistance to infectious diseases, and control of inflammatory bowel disease and prevention of colorectal cancer (Crittenden 2004). The potential health benefits attributed to the *B. longum* species clearly illustrate that this species possesses many very interesting characteristics. It is anticipated that identification and functional analysis of the genetic determinants involved in these activities will strengthen the evidence for the involvement of *B. longum* in these significant health benefits. It should be noted that the selection of suitable strains for probiotic purposes is very difficult as inherent characteristics of strains of *B. longum* that are necessary for its survival and competition in the human large intestine are currently very poorly understood (O'Sullivan 1999).

## Identification and Typing Methods

Traditionally, LAB species have been identified on the basis of cell morphology, analysis of fermentation products and associated enzyme activities, and the ability to utilize various carbohydrate substrates (Axelson 1998). The application of these approaches in the classification and identification of LAB has been the subject of several reviews (Lane et al. 1985, Axelson 1998, Hammes and Vogel 1995, Tannock 1999). In general, phenotypic methods suffer from a lack of reproducibility generated by conditions of culture related to different laboratories, and are often unable to distinguish between closely related strains (Albuquerque et al. 2009, Huys et al. 2006, Schleifer et al. 1995).

### *Genus to Species Identification*

The use of nucleic acid probes, that is fragments of a single-stranded nucleic acid that will specifically bind (hybridize) to complementary regions of a target nucleic acid has been extensively used for the identification of LAB (Ben Amor et al. 2007). Analysis of nucleic acids provides a basis for identification methods that are reproducible from one laboratory to another (Bintsis et al. 2008).

The application of 16S or 23S rRNA targeted probes is one of the most reliable approaches. Today, with the availability of rapid and automatic DNA sequencing technology, direct sequencing of the 16S rRNA gene has emerged as the most powerful and relatively easy one-step method for classification of bacteria (Axelsson 2004). Bacterial rRNA subunit sequences, namely 16S rRNA genes, intergenic spacer 16S–23S rRNA and 23S rRNA genes, have been widely used over the years to study bacterial phylogeny (Rudi et al. 2007, Myers et al. 2006, Eom et al. 2007). With automated sequencing systems and convenient direct PCR sequencing methods, it has become an easy task to determine the 16S rRNA sequence from any bacterium in a short time. The low variability of 16S rRNA genes from closely related species, the intra-genomic variability among the different chromosomal 16S rRNA gene copies (Coenye and Vandamme 2003, Morandi et al. 2005), and the failure to identify a collective conserved region for the design of universal primers (Baker et al. 2003) are the main reported limitations.

#### *Genotypic typing: strain identification*

Typing methods refer to methods capable of discriminating microorganisms at or near the strain level. These methods are generally applied after the previous identification of the target microorganisms in which clonal discrimination resolution is required. They have been routinely used in epidemiological studies (Schleifer et al. 1995).

The most powerful of these are genetic-based molecular methods known as DNA fingerprinting techniques, e.g., pulsed-field gel electrophoresis (PFGE) of rare-cutting restriction fragments, ribotyping, randomly amplified polymorphic DNA (RAPD), and amplified fragment length polymorphism (AFLP), which have been applied extensively for the infraspecific identification and genotyping of LAB and bifidobacteria isolated from fermented food products as well as from the human gastrointestinal tract (McCartney 2002). Basically, these methods rely on the detection of DNA polymorphisms between species or strains and differ in their dynamic range of taxonomic discriminatory power, reproducibility, ease of interpretation, and standardization.

Ribotyping is a variation of the conventional PFGE analysis and the probes used in ribotyping vary from partial sequences of the rDNA genes or the intergenic spacer regions to the whole rDNA operon (O' Sullivan 1999). Ribotyping has been used to characterize strains of *Lactobacillus* spp. and *Bifidobacterium* spp. from commercial products as well as from human fecal samples (Giraffa et al. 2000, Zhong et al. 1998). However, ribotyping provides high discriminatory power at the species and subspecies level rather than on the strain level. PFGE was shown to be more discriminatory in typing closely related *L. casei* and *L. rhamnosus* as well as *L. johnsonii* strains than either ribotyping or RAPD analysis (Tynkkynen et al. 1999, Ventura et al. 2002).

Randomly amplified polymorphic DNA arbitrary amplification, also known as RAPD, has been widely reported as a rapid, sensitive, and inexpensive method for genetic typing of different strains of LAB and bifidobacteria. RAPD profiling has been applied to distinguish between strains of *Bifidobacterium* spp. and between strains of the *L. acidophilus* group and related strains (Tynkkynen et al. 1999, Roy et al. 2000, Torriani et al. 1999). Several factors have been reported to influence the reproducibility and discriminatory power of the RAPD fingerprints, i.e., annealing temperature, DNA template purity and concentration, and primer combinations.

## Metabolism of Starter Cultures and Flavor Development

The three main pathways which are involved in the development of flavor in fermented dairy products are glycolysis (fermentation of lactose), lipolysis (degradation of fat) and proteolysis (degradation of caseins) (Law 1999, Smit et al. 2005, Tamime and Robinson 1999). Lactate is the main product generated from the metabolism of lactose and a fraction of the intermediate pyruvate can alternatively be converted to diacetyl, acetoin, acetaldehyde or acetic acid (some of which can be important for typical yogurt flavours). The contribution of LAB to lipolysis is relatively little, but proteolysis is the key biochemical pathway for the development of flavour in fermented dairy products (Souza et al. 2001). Degradation of caseins by the activities of rennet enzymes and the cell-envelope proteinase and peptidases yields small peptides and free amino acids, the latter of which can be further converted to various alcohols, aldehydes, acids, esters and sulphur compounds for specific flavour development (Tamime and Robinson 1999, Smit et al. 2005).

### *Lactose Metabolism*

LAB need a sugar for energy production and subsequent growth. In dairy products, the sugar is lactose—a disaccharide composed of glucose and galactose. Transport of lactose into the cell is carried out using a translocation system involving the phospoenol pyruvate phosphotransferase (PTS) in lactococci, while *S. thermophilus* use a proton motive force (PMF) (Cogan and Hill 1993). Lactose fermentation by LAB has been reviewed by Broome et al. (2002), Cogan and Hill (1993), Cocaign-Bousquet et al. (1996), Poolman (2002), and Tamime and Robinson 1999.

After transporting into the cell, lactose is fermented with one of the four pathways as shown in Fig. 1. For example, in lactococci the tagarose pathway is followed and lactose transport and the enzymes for the pathway are plasmid encoded (Crow et al. 1983). Galactose is only metabolized by *L. helveticus* and some strains of *L. delbrueckii* subsp. *lactis* (Gal+) and probably leuconstoc via the Leloir pathway. Glucose-6-P is metabolized by the glucolytic pathway in the lactobacilli and by the phosphoketolase pathway in leuconstoc. L-lactate is generally the sole product of fermentation, but when LAB are grown on galactose, maltose or low levels of glucose other products are formed, form pyruvate metabolism (Fig. 2).

Citrate is present at a low concentration in milk and is metabolizes by *Leuconostoc* spp. and some strains of *Lc. lactis* subsp. *lactis* (citrate-utilizing, $Cit^{+}$) to $CO_2$ which is responsible for "eye" formation in some cheeses (Parente and Cogan 2004). In addition, other important aroma compounds are produced in fermented milks, cheese and butter (Fig. 3). $Cit^{+}$ strains of *Lc. lactis* subsp. *lactis* contain a plasmid which encodes the transport of citrate. Citrate metabolism has been reviewed by Hugenholtz (1993).

The presence of a citrate permease is essential for the metabolism of citrate. The citrate permeases of both *Lc. lactis* subsp. *lactis* and *Leuconostoc* spp. were found to be pH dependant and their highest acidity was between pH 5.0 and 6.0. The citrate inside the cell is converted to oxaloacetate, by the enzyme citrate lyase, and then oxaloacetate is decarboxylated to pyruvate. In lactococci, pyruvate is then converted to acetate, diacetyl, acetoin, 2, 3-butanediol and $CO_2$. The enzyme pyruvate formate lyase is able to convert pyruvate to formate, acetate, acetaldehyde and ethanol under anaerobic conditions and at high pH (> 7.0). Under aerobic conditions and at pH 5.5 to 6.5, pyruvate can be converted to acetate, acetaldehyde, ethanol and the minor products acetoin, diacetyl and 2, 3-butanediol via the multienzyme pyruvate dehydrogenase complex (Fig. 3). In *Leuconostoc* spp., the pyruvate produced from citrate is converted to lactate, although at low pH and in the absence of glucose (or lactose) *Leuconostoc* spp. will produce diacetyl and acetoin. Acetate is also formed via the heterofermentative metabolism of lactose during co-metabolism with citrate (Broome et al. 2002).

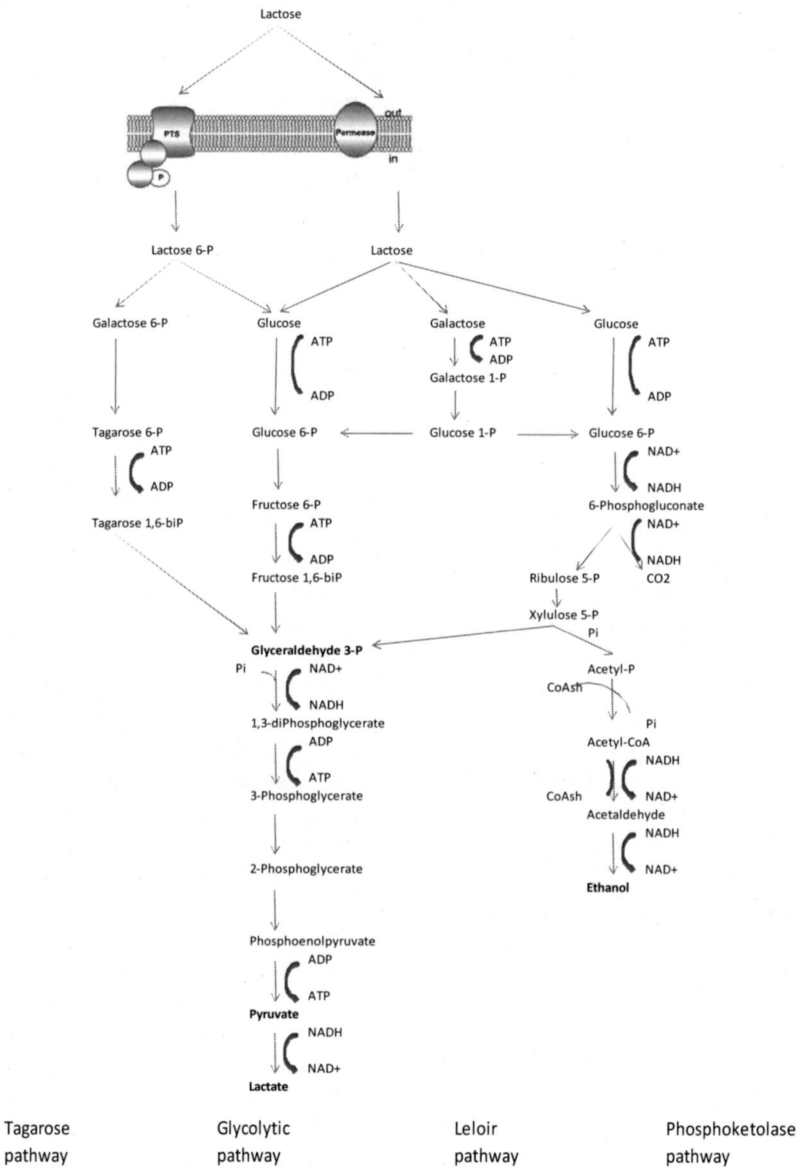


**Figure 1.** Lactose metabolism pathways in lactic acid bacteria (After: Cogan and Hill 1993, Hutkins 2001, Tamime and Robinson 1999).

Over the last 15–20 years numerous attempts have been made to change metabolite production in LAB, via metabolic engineering, from lactic acid to production of other flavour compounds—usually by removing lactate

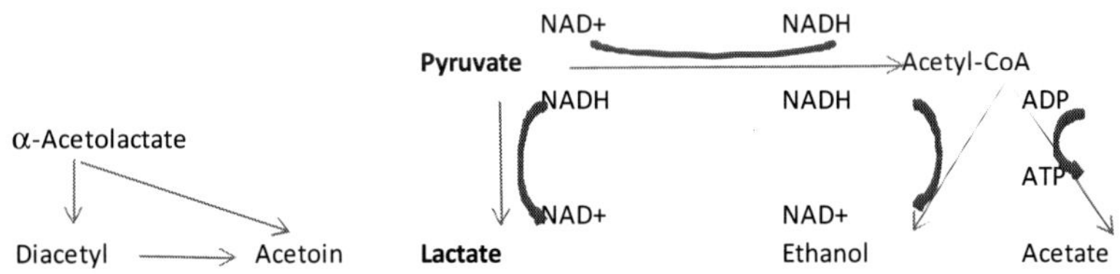


**Figure 2.** Pyruvate metabolism in lactic acid bacteria (After: Cogan and Hill 1993, Hutkins 2001, Tamime and Robinson 1999).

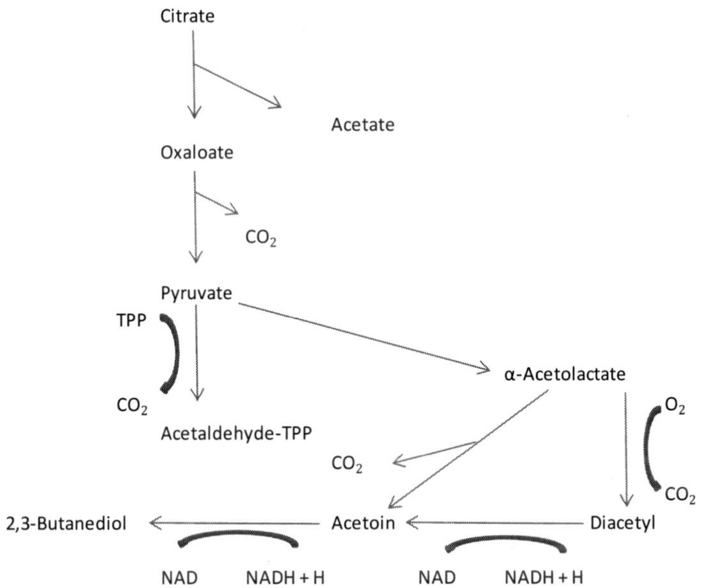


**Figure 3.** Citrate metabolism in lactic acid bacteria (After: Cogan and Hill 1993, Hutkins 2001, Tamime and Robinson 1999).

dehydrogenase (LDH), the enzyme directly responsible for reduction of pyruvate to lactate (Hugenholtz 2008, Hugenholtz et al. 2000). The relative simplicity of *Lc. lactis* subsp. *lactis* sugar metabolism via the pyruvate pathway together with the availability of the complete genome sequence (Bolotin et al. 2001) makes this bacterium a top model for the study of LAB. The metabolic re-routing of sugar metabolism has been extensively reviewed (Hugenholtz and Kleerebezem 1999, Hugenholtz and Smid 2002, Kleerebezem and Hugenholtz 2003). Initial metabolic engineering of *Lc. lactis* has focussed primarily on the rerouting of pyruvate metabolism. Sugar metabolism was diverted towards the production of α-acetolactate, the precursor of diacetyl, by either disruption of lactate dehydrogenase (LDH) or by the nisin inducible expression system (NICE), through the overproduction of NADH oxidase. By combining the latter strategy with disruption of the gene encoding α-acetolactate decarboxylase, high diacetyl production from glucose and lactose was achieved.

### *Protein Metabolism*

LAB are fastidious microorganisms and are unable to synthesize many amino acids, vitamins and nucleic acid bases (Parente and Cogan 2004). Depending on the species and the strain, LAB require from 6 to 14 different amino acids (Chopin 1993, Kunji et al. 1996). Since free amino acids in milk are limited and amino acids are present as protein components, the growth of LAB requires the hydrolysis of milk proteins.

The hydrolysis of peptides to free amino acids and the subsequent utilization of these amino acids is a central metabolic activity in LAB (Christensen et al. 1999), and proteolysis has been identified as the key process influencing the rate of flavour and texture development in most cheese varieties and have been reviewed (Souza et al. 2001, Upadhyay et al. 2004, Law 1999) and the catabolism of amino acids has been reviewed by Curtin and McSweeney (2004). The degradation of milk proteins to peptides is catalysed by proteolytic enzymes present in LAB (Christensen et al. 1999, Thomas and Mills 1981), and peptides are then further hydrolysed by exopeptidases and endopeptidases to small peptides and amino acids (Fox and Wallace 1997). The main peptidases derived from lactic acid bacteria used as dairy starter cultures are shown in Table 4.

There are several reports in the literature on the biochemical characterization of peptidolytic and aminopeptidase activities from a number of LAB (Atlan et al. 1990, Bockelmann et al. 1991, Bintsis et al. 2003, Booth et al. 1990, El Abboudi et al. 1992, Khalid and Marth 1990b). However, since the genomes of many LAB have been sequenced, a thorough comparative analysis of their proteolytic systems has been advanced at a genome scale (Liu et al. 2010, Varmanen et al. 2000, Vesanto et al. 1995). It has been found that the proteolytic activity of the LAB is chromosome linked and the number of plasmids have been detected in most of the strains is very limited. Thus, the gene encoding the cell surface endopeptidase from *L. delbrueckii* subsp. *bulgaricus* has been sequenced and a comparison of DNA sequences for the cell surface endopeptidases of *L. delbrueckii* subsp. *bulgaricus* and lactococci showed little genetic homology (Gilbert et al. 1996). In addition, the LAB genomes in the *L. acidophilus* group encode a relatively higher number and variety of proteolytic system components. Some enzymes are only found in a few LAB strains, such as the cell-wall bound proteinase (PrtP). PrtP was only found on the chromosome of *L. acidophilus*, *L. johnsonii*, *L. bulgaricus*, *L. casei*, *L. rhamnosus* and *S. thermophilus* strain LMD9, as well as on the plasmid of *Lc. lactis*

**Table 4.** Some peptidases derived from lactic acid bacteria used as dairy starter cultures .

| Peptidase | Substrate hydrolyzed |
|---|---|
| Aminopeptidases | |
| PepA | Glu/Asp —↓— (X)n |
| PepC | X —↓— (X)n |
| PepL | Leu —↓— X |
| PepN | X —↓— (X)n |
| PepP | X —↓— Pro — (X)n |
| PepX | X — Pro —↓— (X)n |
| Dipeptidases | |
| PepV | X —↓— X |
| PepD | X —↓— X |
| Tripeptidase T (PepT) | X —↓— X — X |
| Proiminopeptidase (PepI) | Pro —↓— X — (X)n |
| Prolidase (PepQ) | X —↓— Pro |
| Prolinase (PepR) | Pro —↓— X |
| Endopeptidases | |
| PepF | (X)n — X — X —↓— X — (X)n |
| PepO | (X)n — X —↓— X — (X)n |
| PepE | (X)n — X —↓— X — (X)n |
| PepG | (X)n — X —↓— X — (X)n |

After: Hutkins (2001), Christensen et al. (1999)

subsp. *cremoris* SK11 (Liu et al. 2010). Members of both the PepE/PepG (endopeptidases) and PepI/PepR/PepL (proline peptidases) superfamilies are absent in lactococci and streptococci. On the other hand, many of the peptidases, such as aminopeptidases PepC, PepN, and PepM, and proline peptidases PepX and PepQ are present in all LAB genomes, usually with one gene per genome, and seem to be essential for bacterial growth or survival (Liu et al. 2010).

The formation of amino acids from the proteolysis of cheese caseins is very important for the development of flavour during the maturation process. Free amino acids contribute directly to the cheese flavour (Seth and Robinson 1988) and, additionally, they serve as substrates for other flavour-generating compounds (Fox and Wallace 1997). In addition, undesirable bitter peptides may lead to the formation of off-flavours, and the ability of the micro-flora to hydrolyse bitter peptides is equally important (Arora and Lee 1990). The conversion of free amino acids in more specific flavour components is the next, and much more complicated, process in overall flavour generation. The principal compounds formed are amines, other amino acids, α-keto acids and sulphur compounds. In lactococci, transamination in which the α-amino group of the amino acid is transferred to a keto acid acceptor appears to be the first step in the degradation of aromatic and branched chain amino acids. One of the key amino acids in the formation of flavour compounds is methionine. This amino acid is transaminated using α-ketoglutarate as the amino group acceptor to form 4-methylthio-2-oxobutyrate which appears to be degraded either nonenzymically or enzymically to form methanethiol. Methanethiol in turn is the precursor to a number of potential flavour compounds such as dimethyl sulphide, dimethyl disulphide and S-methylthioesters (Broome et al. 2002).

Overproduction of specific amino acid-converting enzymes in starter bacteria, such as cystathione-β-lyase involved in methionine- and cysteine-metabolism has had little effect on the ripening in cheese (Fernandez et al. 2002), while increased production of α-ketoglutarate and overproduction of aminotransferases gave a clear stimulation of the flavour formation by LAB (Rijnen et al. 1999, 2000). The limited impact of metabolic engineering in this area is mainly due to the lack of descriptive, and predictive, models for overall amino acid metabolism in the LAB, which could guide the different metabolic engineering strategies. Genomics techniques, however, have been very productive in this area with the development of different high throughput analyses for amino acid-converting enzymes and DNA microarrays for analysis of the presence and expression of relevant genes in different *Lc. lactis* strains (Hugenholtz 2008).

### *Lipid Metabolism*

The enzymatic metabolism of milk fat is limited during the manufacture of fermented dairy products. The degradation of milk fat releases free fatty acids and glycerol, monoacylglycerides or diacylglycerides. However, certain free fatty acids are essential flavor compounds in certain cheeses (e.g., caprine milk cheeses). In addition, they react with alcohols or free sulphydryl groups to form esters and thioesters, respectively, or act as precursors of a number of other flavour compounds, such as lactones (Fox and Wallace 1997). Esterase activity has been detected in various lactobacilli (Khalid et al. 1990), and esters contribute to the characteristic flavor of Swiss-type cheeses (Fox and Wallace 1997) and Feta cheese (Bintsis et al. 2003).

## Exopolysaccharide Production

Extracellular polysaccharides (EPSs) are produced by a variety of bacteria and are present as capsular polysaccharides (CPS and LPS) bound to the cell surface, or are released into the growth medium (Hassan 2008). These polymers can consist of a single type of sugar (homopolysaccharides) or are regular, repeating units consisting of different sugars (heteropolysaccharides). EPSs play a major role in the production of yogurt, cheese, fermented cream and milk-based desserts (Jolly et al. 2002) where they contribute to texture, mouth-feel, taste perception and stability of the final products (Hassan 2008). Capsular EPSs form capsules around the cell wall and are not secreted into the medium. Unattached EPSs are secreted outside the cell wall of the producer and are responsible for the "ropy" phenotype observed in fermented milks. The genetic machinery responsible for EPSs production in lactococci is generally encoded on large operons containing more than 10 genes. As the producing strains are food-grade, the EPSs are also considered food-grade. In addition, it has been suggested that these EPSs or fermented milks containing these EPSs are active as prebiotics (Gibson and Robertfroid 1995), cholesterol-lowering (Nakajima et al. 1992) and immunomodulants (Hosono et al. 1997). EPS-producing strains of *S. thermophilus* and *L. delbreuckii* subsp. *bulgaricus* have been shown to enhance the texture and viscosity of yogurt and to reduce syneresis (Hassan et al. 2003, Doleyres et al. 2005). Broadbent et al. (2003) extensively reviewed the topic of EPS production in *S. thermophilus*, providing a detailed overview of its biochemistry, genetics and applications. With regard to the genetic machinery encoding EPS production in *S. thermophilus*, it has been suggested that a dozen or more unique EPSs gene clusters may occur (Rallu et al. 2002). These clusters range in size from 15 to 20 kb and are chromosomally encoded.

## Bacteriophage Resistance

Bacteriophages (phages) are viruses infecting bacteria and multiply in the bacterial host cells resulting in cell lysis (i.e., virulent phages). The word originates from the Greek word "φάγω", which means "eat", so bacteriophages are the viruses that attack bacteria and they are the main reason for slow or faulty fermentations during dairy-products making. They cause no harm to human beings, animals or plants and cannot cause any human diseases. With a size in the range of 0.1–0.2 µm, phage are considerably smaller than bacteria. Therefore they can be transferred through air and can thus be present in all production areas (Chr. Hansen 2002). The general morphology of a bacteriophage consists of a head and protruding tail. The hexagonal head is a protein envelope. Inside this envelope is the genetic information (DNA or RNA) of the phage. Under the head, the collar and the tail are located through which the DNA is injected into the bacterium. At the end of the tail, a baseplate and tail fibres are located. They will read and recognise the host specific receptors on the bacterial cell wall (Chr. Hansen 2002).

Over the years different methods have been proposed to classify bacteriophages, but they were not accepted universally. However, a recent approach to bacteriophage taxonomy, which is accepted universally, has identified three groups known as bacteriophage families, namely the *Myoviridae*, *Podoviridae* and *Siphoviridae* (Tamime et al. 1999). Based on morphological distinctions, there are three main types of bacteriophages: morphotypes A, B, and C, with the B or *Siphoviridae* group being the most common type that infect lactic acid bacteria. The group B phages have long tails with heads that have either a small isometric (B1 sub-group), prolate (B2), or elongated (B3) shape. The lactococcal phages generally belong to the B1 or B2 group, whereas the phages that infect *S. thermophilus* are all in the B1 group (Hutkins 2001).

While total loss of the final product as a result of phage infection is infrequent nowadays, phage infection continues to result in quality defects that affect the flavour, texture and even safety of dairy products and so continues to receive attention in both research and manufacturing communities. Phage resistance mechanisms are probably best characterized in lactococci. Indeed, within lactococci there are four main cellular defences which interfere with different stages of the phage lytic cycle, namely, adsorption inhibition (Ads), injection blocking, R/M and abortive infection (Abi). The Ads phenotype prevents the attachment of the phage particle to the cell surface and can often be induced through the generation of bacteriophage insensitive mutants (BIMs) (Mills et al. 2007). In *Lactococcus* spp., this has been attributed to nonspecific point mutations in chromosomal genes encoding cell receptor sites by masking of receptors through, for example, polysaccharide production (Forde and Fitzgerald 1999). Potentially, one of the most exciting breakthroughs in phage resistance research recently is the study of *S. thermophilus* BIMs. Interestingly, within streptococci the CRISPR loci have recently been shown to play a role in BIM formation. These loci consist of highly conserved repeats of approximately 21–48 base pairs, which are separated by variable sequences of constant and similar length, called spacers of approximately 20–58 base pairs (Horvath et al. 2008). It has been proposed that these spacers function as small interfering RNAs resulting in termination of the phage lytic cycle. A recent comparative study of the CRISPR loci in LAB genomes resulted in the identification of multiple CRISPR families within *Bifidobacterium* and *Lactobacillus* as well as *Streptococcus*, and similar CRISPR loci were found in distant organisms, suggesting that these loci have evolved independently in select lineages, partly due to selective pressure from phage predation (Horvath et al. 2009). Exploitation of the CRISPR system offers many opportunities for strain improvement such as strain typing, engineered defence against phage, selective silencing of endogenous genes (Sorek et al. 2008), as well as development of intelligent starter rotation strategies through exploitation of the diversity introduced into an otherwise homogeneous population (Mills et al. 2009).

Constant exposure to phage has resulted in the evolution of a diverse range of defence mechanisms in lactococci, including a large range of different Abi systems. This mechanism comes into play after injection of phage DNA and includes a broad range of defences which can interfere with genome replication, transcription, translation and packaging or assembly of phage particles. This interruption of phage development leads to the release of few or no phage and to the death of the infected cell, thus further propagation of phage is prevented and the bacterial population survives (Chopin et al. 2005). To date, of the 21 Abis' identified in *Lactococcus* spp. most are plasmid encoded and range from AbiA to AbiV. The phenotype is most often conferred by a single gene, although there are a few exceptions where the involvement of two genes has been proposed (AbiE, G, L and T). Protein homology has rarely been observed between lactococcal Abis and no homology with known proteins has been found, preventing any prediction being made on their mode of action (Chopin et al. 2005).

Bacteriophages enter the dairy industry through raw milk, and thus, it is a difficult task to eliminate entry of phages into the dairy plant, because raw milk continually enters the facility. Phages are disseminated throughout the dairy plant by aerosol and human carriers (Hassan and Frank 2001). They multiply at incredible rates; in the time it takes for one "non-infected" bacterium to produce four new bacteria by two generations of fission, one bacteriophage has grown to a total of 22,500 phages (Tetrapak 1995). So it is obvious that it is absolutely impossible not to have phages in a dairy industry. Instead, in everyday life we have to learn how to coexit with them.

Currently, although the commercial starter cultures are phage insensitive when launched in the market, it is usually possible to detect bacteriophage after a period of time due to the rapid evolution of the phage. During cheese production, problems with phage attack against *Lc. lactis* spp. are most common, followed by problems with phage attacking *S. thermophilus*. Phages attacking *Lactobacillus* spp. and *Leuconostoc* spp. in starter cultures represent only a minor problem during dairy fermentations (Høier et al. 2010). Practically, it is impossible to keep bacteriophages out of industry production but there are a lot of preventive measures (i.e., use of commercial phage insensitive starters and avoid use of bulk starters, use of sterile air in the headspace of the incubation tanks, use of defined-strain starter cultures with different phage-host spectra within a carefully designed rotation scheme) that can be applied in order to control the phage proliferation.

## Bacteriocin Production

Bacteriocins are polypeptides synthesized ribosomally by bacteria and can have a bacteriocidal or bacteriostatic effect on other bacteria (McAuliffe et al. 2001, Ross et al. 2002). In general, bacteriocins lead to cell death by inhibiting cell wall biosynthesis or by disrupting the membrane through pore formation (Twomey et al. 2002). Bacteriocins are therefore important in food fermentations where they can prevent food spoilage or the inhibition of food pathogens. Although different classification schemes have been proposed (Klaenhammer 1993, Nes et al. 1996, Kemperman et al. 2003), the bacteriocins of LAB have been classified into two major classes. Bacteriocins are classified into either lanthionine-containing bacteriocins (class I) (lantibiotics) or the non-lanthionine-containing bacteriocins (class II). The class II family is relatively diverse in terms of sequence variations as well as in modes of action, and has therefore been divided further into subgroups (Diep et al. 2009). Lantibiotics contain post-translationally modified amino acids such as lanthionine, β-methyllanthionine and the dehydrated residues dehydroalanine and dehydrobutyrine. The best known lantibiotic is nisin, which has gained widespread application in the food industry and is used as a food additive in at least 50 countries, particularly in processed cheese, dairy products and canned foods (Delves-Broughton et al. 1996). Another useful bacteriocin in terms of starter culture improvement is lacticin 3147, encoded on the 60.2 kb plasmid, pMRC01, and this plasmid has been transferred to over 30 different lactococcal hosts (Ryan et al. 1996, Coakley et al. 1997). Lacticin 3147 has been shown to be effective in many food systems for the control of food spoilage or pathogenic bacteria (McAuliffe et al. 1999, Ross et al. 1999, Morgan et al. 2001, O'Sullivan et al. 2006). Recently, *Streptococcus macedonicus* ACA-DC 198, was found to produce a lantibiotic named macedovicin, which inhibits a broad spectrum of lactic acid bacteria, several food spoilage species (e.g., *Clostridium* spp.) and oral streptococci (Georgalaki et al. 2013); *Streptococcus macedonicus* ACA-DC 198 has been isolated from Greek Kasseri cheese and produce the lacticin 481 lantibiotic macedocin (Georgalaki et al. 2002).

Bacteriocin-producing LAB have also been applied to improve the flavour and quality of fermented foods through two strategies; by using bacteriocin-producing LAB to control adventitious microbial populations (Ryan et al. 1996) and secondly by using bacteriocin-producing LAB as cell lysis-inducing agents to increase the rate of proteolysis in cheese (Garde et al. 2002, Oumer et al. 2001). Many strategies have been developed for the detection of antibacterial production by microorganisms including phenotypic methods using indicator strains (Morgan et al. 2000) and liquid chromatography/mass spectrometry techniques (Zendo et al. 2008). The detection and identification of bacteriocins in producing strains has also been greatly aided by PCR amplification of putative bacteriocin genes using specific primers designed to known bacteriocins (Trmcic et al. 2008) and by *in silico* searching amongst bacterial DNA sequences (Draper et al. 2009).

## Safety Aspects of Genetic Engineering

A complete understanding of the metabolic potential of a LAB, established through knowledge of gene function, pathway reconstruction and prediction of phenotype through metabolic models, is a powerful tool in developing metabolic engineering strategies (de Vos and Hugenholtz 2004). Both directed and uncontrolled (i.e., occurring during random mutagenesis) genetic alterations result in a change of the genetic code of the microorganism that may affect the transcription and translation processes and, consequently, may influence metabolic processes in the cell. The nature of the DNA modification could dictate the necessity for proceeding to a safety assessment procedure. Interestingly, self-cloning genetic modifications are considered as GMOs, whereas, uncontrolled genetic modifications (e.g., mutations via IS) together with conjugation and transduction are not.

Spontaneous mutations may occur in LAB by natural events such as IS elements (de Visser et al. 2004), radiation, erroneous DNA replication or transcription, and other factors. For example, in *L. bulgaricus* a spontaneous IS element-mediated deletion of the *lacZ* gene altered lactose metabolism resulting in limited fermentation capacity, and thus, manufacture of a high quality (Mollet and Delley 1990). The level of such mutations depends upon the growth conditions, and by the screening of natural isolates of LAB, strains with improved fermentation characteristics can be selected (Sybesma et al. 2006). The frequency of mutations can be further increased by exposing LAB to mutagenic conditions such as UV light (Bintsis et al. 2000).

For controlled genetic modification, a variety of techniques have been developed to generate genetic modified starters, such as cloning systems, chromosome modification systems and expression systems (de Vos 2001). The most popular transformation system for generating directed genetic alterations in LAB is electroporation with self-replicating vectors. Alternative systems are conjugation and transduction (Gasson 1990).

Targeted gene replacement or removal and inactivation of genes can also be applied via (non-replicating) vectors using the natural event of crossing-over during cell division and DNA replication (Leenhouts et al. 1996). Compared to the use of replicating vectors in genetic modified starters that result in new or enhanced cellular behaviour, the deletion of genes after double cross-over events (by using a non replicating plasmid) does not result in the addition of any DNA to the genetic content of the cell (Sybesma et al. 2006).

A specific aspect related to the application of vectors in industrial strain improvement is the use of selection markers. These markers should be carefully selected and must be food-grade. For example, antibiotic resistance markers cannot be used and strains carrying transferable antibiotic resistance genes, such as enterococci should not be used as starters.

Certain targeted modifications of the genetic content of DNA may occur via conjugation and transduction (Gasson 1990). These processes are considered natural events, and thus, bacteria that are modified by using these transfer systems are not considered as GMOs.

Starter cultures are substances used in foods and in the United States are regulated according to the Food Drug and Cosmetic Act, in which the status of Generally Recognized As Safe (GRAS) was introduced (Food and Drug Administration 2012). Accordingly, a GRAS substance is generally recognized, among qualified experts, as having been adequately shown to be safe under the conditions of its intended use. Starter cultures are an integral part of traditional fermented foods and, as a significant number of people have consumed these foods for many centuries before 1958, the fermenting microorganisms of these products can be said to be GRAS (Bourdichon et al. 2012c).

In the European Union, starter cultures are considered ingredients and must satisfy the legal requirements of food legislation. The regulation EC 258/97 amended in regulation EC 1829/2003 lay down the framework for use of genetically modified food and feed (EC 2003). According to Sybesma et al. (2006), the legislation regarding the use of GMOs is not yet completely clear on a number of important scientific matters. The legislation predominantly focuses on the methodology rather than on the end product and hence, holds too strongly to the definition of GMOs. Consequently, organisms in which the genetic material has been altered by recombinant DNA techniques in a way that does occur naturally, for instance by point mutations or small deletions, are considered to be GMOs.

Since LAB are widely used as starter cultures for fermentation in the dairy, meat and other food industries and have a long history of use by man for food production and preservation, they are considered as food-grade microorganisms.

The European Food Safety Authority (EFSA) has recently introduced the "Qualified Presumption of Safety" (QPS) for a premarket safety assessment of microorganisms used in food and feed production (European Food Safety Authority 2006). QPS is applicable to food and feed additives, food enzymes and plant protection products. The QPS system was proposed to harmonize approaches to the safety assessment of microorganisms across the various EFSA scientific panels. The QPS approach is meant to be a fast track for species for which there is a sufficient body of knowledge that all strains within a species are assumed to be safe.

In the QPS approach, safety assessment will depend upon the body of knowledge available for a given microorganism. For some bacterial groups (e.g., most lactobacilli, including those dominating natural cultures such as *L. helveticus* and *L. delbrueckii*, or *S. thermophilus*) the identification at genus/species level and the lack of acquired antibiotic resistance will be sufficient to assure a safety status (European Food Safety Authority 2006), while identification and characterization at strain level should be required for some microbial groups, such as for example enterococci (European Food Safety Authority 2010).

The QPS list covers only selected groups of microorganisms which have been referred to EFSA for a formal assessment of safety (European Food Safety Authority 2006, Leuschner et al. 2010). Seventy-nine species of microorganisms have so far been submitted to EFSA for a safety assessment (Pedersen et al. 2005); the list is updated annually (European Food Safety Authority 2010). It should be noted that the absence of a particular organism from the QPS list does not necessarily imply a risk associated with its use. Individual strains may be safe, but this cannot be ascertained from the existing knowledge of the taxonomic unit to which it belongs. Another reason for a species not being on the list could be that EFSA has not been asked to assess the safety of any strains of the species.

## Specifications

Commercial starter cultures are supplied either in a frozen or freeze-dried form and a food safety management system based on the principles of HACCP is applied in the manufacturing procedure in order to assure the stability and safety of the product (Tamime 2002). The required specifications are presented in Table 5.

**Table 5.** Specifications for starter cultures.

| Type of criterion | Contaminant[a] | Unit | Liquid and frozen | Dry |
|---|---|---|---|---|
| Process hygiene | Non-lactic acid bacteria[b] | CFU/g | < 500 | < 500 |
| | Yeasts and moulds | CFU/g | < 1 | < 10 |
| | *Enterobacteriaceae* | CFU/g | < 1 | < 10 |
| | Coagulase-positive staphylococci | CFU/g | < 1 | < 10 |
| Food safety | *Salmonella* spp. | Absence in 1 g | Absence | Absence |
| | *Listeria monocytogenes* | Absence in 1 g | Absence | Absence |

[a] Contaminants can be tested in process environment and in process or products samples. The set-up of environmental samples compared to process or product samples shall be based on HACCP principles and justified against the specifications shown in the Table above.

[b] This criterion is only relevant as a contaminant in cultures containing only lactic acid bacteria.

After: International Organization for Standardization and International Dairy Federation (2010)

The safety of starter cultures has been advanced through molecular methods for their identification and taxonomy and modern methods are used to assure safety during manufacture (Hansen 2002).

Labeling shall be in accordance with national legislation, where applicable. The following items should appear on the starter culture product label:

a) name of product,
b) type of product (e.g., mesophilic culture) or bacterial composition in accordance with international scientific nomenclature,
c) type of products (e.g., freeze-dried, concentrated),
d) net contents which may be indicated in one of the following units: grams, milliliters, units,
e) name and address of the manufacturer, packer, distributor, importer, exporter or vendor,
f) country of manufacture (optional),
g) code and lot identification,
h) expiry date (month and year), and
i) storage conditions.

In addition, technical data should accompany the product, including: a) application areas of use, b) instructions for use (inoculation rate, incubation temperature, etc.), c) composition (types of bacteria, type of culture, etc.), and d) certificate of analysis, certificate of compliance or similar.

## Conclusions

Starter cultures are a part of a rapidly developing field, and the results from research in the last 15 years, since the discovery of the complete genome sequence of *Lc. lactis* subsp. *lactis* IL1403 by Bolotin et al. (2001), have led to the development of commercial starters with desirable properties. Advances in the genetics, molecular biology, physiology, and biochemistry of LAB have provided new insights and applications for these bacteria in the dairy industry. On the other hand, the constant risk of bacteriophage attack justifies the continuous need to search for new strains for improved processes.

Dairy industry is now capable of producing safe and nutritious products with different flavours, sometimes with special health-promoting properties, which satisfy the demands of all consumer and market niches, and resemble the characteristics of the traditional products. In addition, the use of selected strains of given species with known metabolic properties and high technological performances has improved the total quality control of the manufacturing process. Interestingly, the potential intended and unintended effects and related risks of using genetic modified starter cultures can be predicted in an accurate way and verification is feasible. This combined with profound safety assessments, such as QPS, ensures safety for consumers and safety for the environment.

These developments could lead to novel functional foods, produced from processes with well-characterized metabolic pathways, illustrating new possibilities for the application of these food-grade microorganisms.

## References Cited

Albuquerque, P., M.V. Mendes, C.L. Santos, P. Moradas-Ferreira and F. Tavares. 2009. DNA signature-based approaches for bacterial detection and identification. Sci. of the Total Environm. 407: 3641–3651.

Araujo, E.A., A.F. Carvalho, E.S. Leandro, M.M. Furtado and C.A. Moraes. 2010. Development of a symbiotic cottage cheese added with *Lactobacillus delbrueckii* UFV H2b20 and inulin. J. of Functional Foods 2: 85–89.

Arora, G. and B.H. Lee. 1990. Comparative studies of peptidases of *Lactobacillus casei* subspecies. Journal of Dairy Science 73: 274–279.

Atlan, D., P. Laloi and R. Portalier. 1990. X-prolyldipeptidyl-aminopeptidase of *Lactobacillus delbrueckii* subsp. *bulgaricus*: characterization of the enzyme and isolation of deficient mutants. Appl. Environ. Microbiol. 56: 2174–2179.

Axelsson, L.T. 1998. Lactic acid bacteria: Classification and physiology. pp. 1–72. *In:* S. Salminen and A. von Wright [eds.]. Lactic Acid Bacteria, Marcel Dekker, New York.

Axelsson, L.T. 2004. Lactic acid bacteria: Classification and physiology. pp. 1–65. *In:* S. Salminen, A. von Wright and A. Ouwehand [eds.]. Lactic Acid Bacteria, Marcel Dekker, New York.

Baker, G.C., J.J. Smith and D.A. Cowan. 2003. Review and re-analysis of domain-specific 16S primers. J. Microbiol. Meth. 55: 541–555.

Ballongue, J. 2004. Bifidobacteria and Probiotic Action. pp. 66–124. *In:* S. Salminen, A. von Wright and A. Ouwehand [eds.]. Lactic Acid Bacteria, Marcel Dekker, New York.

Ben Amor, K., E.E. Vaughan and W.M. de Vos. 2007. Advanced Molecular Tools for the Identification of Lactic Acid Bacteria. The Journal of Nutrition, Supplement: Effect of Probiotics and Prebiotics 741S–747S.

Beresford, T.P., N.A. Fitzsimons, N.L. Brennan and T.M. Cogan. 2001. Recent advances in cheese microbiology. Int. Dairy J. 11: 259–274.

Bernet-Camard, M.-F., V. Liıvin, D. Brassart, J.-R. Neeser, A.L. Servin and S. Hudault. 1997. The human *Lactobacillus acidophilus* strain La1 secretes a non bacteriocin antibacterial substance active *in vitro* and *in vivo*. Appl. Environ. Microbiol. 63: 2747–2753.

Bintsis, T., A. Vafopoulou-Mastrojiannaki, E. Litopoulou-Tzanetaki and R.K. Robinson. 2003. Protease, peptidase and esterase activities by lactobacilli and yeast isolates from Feta cheese brine. Journal of Applied Microbiology 95: 68–77.

Bintsis, T., A.S. Angelidis and L. Psoni. 2008. Some modern laboratory practices—analysis of dairy products. pp. 183–261. *In:* T. Britz and R.K. Robinson [eds.]. Advanced Dairy Technology, Blackwell Publishing Ltd., Oxford.

Bintsis, T., E. Litopoulou-Tzanetaki and R.K. Robinson. 2000. Review: Existing and potential applications of ultraviolet light in the food industry—a critical review. J. Sci. Food Agric. 80: 637–645.

Bintsis, T., E.-Z. Chissimelli, A. Papadimitriou and R.K. Robinson. 2002. Brine microbiology and the chemical and sensory properties of mature feta-type cheese. Australian Dairy Foods 8: 30–32.

Bockelmann, W., M. Fobker and M. Teuber. 1991. Purification and characterisation of the X-prolyldipeptidyl-aminopeptidase from *Lactobacillus delbrueckii* ssp. *bulgaricus* and *Lactobacillus acidophilus*. Int. Dairy J. 1: 51–66.

Bockelmann, W. 1999. Secondary cheese cultures. pp. 132–162. *In:* B.A. Law [ed.]. Technology of cheesemaking. Sheffild Academic Press, Sheffild, UK.

Bolotin, A., B. Quinquis, P. Renault, A. Sorokin, S.D. Ehrlich, S. Kulakauskas, A. Lapidus, E. Goltsman, M. Mazur, G.D. Pusch, M. Fonstein, R. Overbeek, N. Kyprides, B. Purnelle, D. Prozzi, K. Nqui, D. Masuy, F. Hancy, S. Burteau, M. Boutry, J. Delcour and P. Hols. 2004. Complete sequence and comparative genome analysis of the dairy bacterium *Streptococcus thermophilus*. Nature Biotechnology 22: 1554–1558.

Bolotin, A., P. Wincker, S. Mauger, O. Jaillon, K. Malarme, J. Weissenbach, S.D. Ehrlich and A. Sorokin. 2001. The complete genome sequence of the lactic acid bacterium *Lactococcus lactis* ssp. *lactis* IL1403. Genome Research 11: 731–753.

Booth, M., D. Fhaolain, P.V. Jennings and G. O'Cuinn. 1990. Purification and characterization of a postproline dipeptidyl-aminopeptidase from *Streptococcus cremoris* AM2. J. Dairy Res. 57: 89–99.

Bourdichon, F., P. Boyaval, Casaregola, J. Dupont, C. Forrokh, J.C. Frisvad, W.P. Hammes, J. Harnett, G. Huys, J.L. Jany, S. Laululd, A. Ouwehand, Y. Seto, A. Zgoda and E.B. Hansen. 2012a. The 2012 Inventory of Microbial Species with technological beneficial role in fermented food products. Bulletin of the IDF 455: 22–61.

Bourdichon, F., B. Berger, S. Casaregola, C. Forrokh, J.C. Frisvad, M.L. Gerds, W.P. Hammes, J. Harnett, G. Huys, S. Laululd, A. Ouwehand, I.B. Powell, J.B. Prajapati, Y. Seto, E. Ter Schure, A. Van Boven, V. Vankerckhoven, A. Zgoda and E.B. Hansen. 2012b. A Safety Assessment of Microbial Food Cultures with History of Use in Fermented Dairy Products. Bulletin of the IDF 455: 2–12.

Bourdichon, F., S. Casaregola, C. Forrokh, J.V. Frisvad, M.L. Gerds, W.P. Hammes, J. Harnett, G. Huys, S. Laululd, A. Ouwehand, I.B. Powell, J.B. Prajapati, Y. Seto, E.T. Schure, A. van Boven, V. Vankerckhoven, A. Zgoda, S. Tuijelaars and E.B. Hansen. 2012c. Food fermentations: Microorganisms with technological beneficial use. International Journal of Food Microbiology 154: 87–97.

Briggiler-Marco, M., M.L. Capra, A. Quiberoni, G. Vinderola, J.A. Reinheimer and E. Hynes. 2007. Nonstarter Lactobacillus strains as adjunct cultures for cheese making: *in vitro* characterization and performance in two model cheeses. Journal of Dairy Science 90: 4532–4542.

Broadbent, J.R., D.J. McMahon, D.L. Welker, C.J. Oberg and S. Moineau. 2003. Biochemistry, genetics, and applications of exopolysaccharide production in *Streptococcus thermophilus*: a review. Journal of Dairy Science 86: 407–423.

Broadbent, J.R. 2001. Genetics of lactic acid bacteria. *In:* J.L. Steele and E.H. Marth [eds.]. Applied Dairy Microbiology, 2nd ed. Marcel Dekker, New York.

Broome, M.C., I.B. Powell and G.K.Y. Limsowtin. 2002. Starter cultures: Specific Properties. pp. 269–275. *In:* H. Roginski, J.W. Fuguay and P.F. Fox [eds.]. Encyclopedia of Dairy Science, Vol. I, Elsevier Science Ltd., London.

Caldwell, S., D.J. McMahon, C.J. Oberg and J.R. Broadbent. 1996. Development and characterization of lactose-positive *Pediococcus* species for milk fermentation. Appl. Environ. Microbiol. 62: 936–941.

Caldwell, S., R.W. Hutkins, D.J. McMahon, C.J. Oberg and J.R. Broadbent. 1998. Lactose and galactose uptake by genetically engineered *Pediococcus* species. Appl. Microbiol. Biotechnol. 49: 315–320.

Callanan, M., P. Kaleta and J. O'Callaghan. 2008. Genome sequence of *Lactobacillus helveticus*, an organism distinguished by selective gene loss and insertion sequence element expansion. Journal of Bacteriology 190: 727–735.

Callanan, M.J. and R.P. Ross. 2004. Starter cultures: genetics. pp. 149–161. *In:* P.F. Fox, P.L.H. McSweeney, T.M. Cogan and T.P. Guinee [eds.]. Vol 1, 4th Ed. Elsevier Academic Press, London, UK.

Chamba, J.-F. and F. Irlinger. 2004. Secondary and adjunct cultures. pp. 191–206. *In:* P.F. Fox, P.L.H. McSweeney, T.M. Cogan and T.P. Guinee [eds.]. Vol 1, 4th Ed. Elsevier Academic Press, London, UK.

Chevallier, B., J.C. Hubert and B. Kammerer. 1994. Determination of chromosome size and number of rrn loci in Lactobacillus plantarum by pulsed-field gel electrophoresis. FEMS Microbiol. Lett. 120: 51–56.

Chopin, A. 1993. Organization and regulation of genes for amino acid biosynthesis in lactic acid bacteria. FEMS Microbiol. Rev. 12: 21–38.

Chopin, M.C., A. Chopin and E. Bidnenko. 2005. Phage abortive infection in lactococci: variations on a theme. Current Opinion in Microbiology 8: 473–479.

Chr. Hansen. 2002. Bacteriophage: A latent problem in the cheese industry. 3rd Edition, Horsholm, Denmark, pp. 2–7.

Christensen, J.E., E.G. Dudley, J.A. Pederson and J.L. Steele. 1999. Peptidases and amino acid catabolism in lactic acid bacteria. Antonie van Leeuwenhoek 76: 217–246.

Coakley, M., G. Fitzgerald and R.P. Ross. 1997. Application and evaluation of the phage resistance- and bacteriocin encoding plasmid pMRC01 for the improvement of dairy starter cultures. Applied and Environmental Microbiology 63: 1434–1440.

Cocaign-Bousquet, M., C. Garrigues, L. Novak, P. Loubiere and N.D. Lindley. 1996. Physiology of pyruvate metabolism in *Lactococccus lactis*. Antonie van Leewenhoek 70: 253–267.

Coenye, T. and P. Vandamme. 2003. Intragenomic heterogeneity between multiple 16S ribosomal RNA operons in sequenced bacterial genomes. FEMS Microbiol. Lett. 228: 45–49.

Cogan, T.M. and C. Hill. 1993. Cheese starter cultures. pp. 193–255. *In:* P.F. Fox [ed.]. Cheese: Chemistry, Physics and Microbiology, Vol. 1, 2nd Ed. Chapman & Hall, London, UK.

Cogan, T.M., M. O'Dowd and D. Mellerick. 1981. Effects of sugar on acetoin production from citrate by *Leuconostoc lactis*. Appl. Environ. Microbiol. 41: 1–8.

Cogan, T.M., T.P. Beresford, J. Steele, J. Broadbent, N.P. Shah and Z. Ustunol. 2007. Invited review: advances in starter cultures and cultured foods. Journal of Dairy Science 90: 4005–4021.

Crittenden, R. 2004. An update on probiotic bifidobacteria. pp. 125–158. *In:* S. Salminen, A. von Wright and A. Ouwehand [eds.]. Lactic Acid Bacteria, Marcel Dekker, New York.

Crow, V.L., B. Curry and M. Hayes. 2001. The ecology of non-starter lactic acid bacteria (NSLAB) and their use as adjuncts in New Zealand Cheddar. International Dairy Journal 11: 275–283.

Crow, V.L., G.P. Davey, L.E. Pearce and T.D. Thomas. 1983. Plasmid linkage of the D-tagatose 6-phosphate pathway in *Streptococcus lactis:* Effect on lactose and galactose metabolism. J. Bacteriol. 153: 76–83.

Cruz, A.G., F.C.A. Buriti, C.H.B. Souza, J.A.F. Faria and S.M.I. Saad. 2009. Probiotic cheese: health benefits, technological and stability aspects. Trends in Food Science & Technology 20: 344–354.

Curtin, A.C. and P.L.H. McSweeney. 2004. Catabolism of amino acids in cheese during ripening. pp. 435–454. *In:* P.F. Fox, P.L.H. McSweeney, T.M. Cogan and T.P. Guinee [eds.]. Vol 1, 4th Ed. Elsevier Academic Press, London, UK.

Daeschel, M.A. and T.R. Klaenhammer. 1985. Association of a 13.6-megadalton plasmid in *Pediococcus pentosaceus* with bacteriocin activity. Appl. Environ. Microbiol. 50: 1528–1541.

Daniel, P. 1995. Sizing the *Lactobacillus plantarum* genome and other lactic bacteria species by transverse alternating field electrophoresis. Curr. Microbiol. 30: 243–246.

Davidson, B., N. Kordis, M. Dobos and A. Hillier. 1996. Genomic organization of lactic acid bacteria. Antonie van Leeuwenhoek 70: 161–183.

de Visser, J.A., A.D. Akkermans, R.F. Hoekstra and W.M. de Vos. 2004. Insertion-sequence mediated mutations isolated during adaptation to growth and starvation in *Lactococcus lactis*. Genetics 168: 1145–1157.

de Vos, W.M. 2001. Advances in genomics for microbial food fermentations and safety. Curr. Opin. Biotechnol. 12: 493–498.

de Vos, W.M. 2011. Systems solutions by lactic acid bacteria: from paradigms to practice. Microbial. Cell Factories 10(Suppl. 1): S2.

de Vos, W.M. and J. Hugenholtz. 2004. Engineering metabolic highways in Lactococci and other lactic acid bacteria. Trends in Biotechnology 22: 72–79.

Delves-Broughton, J., P. Blackburn, R.J. Evans and J. Hugenholtz. 1996. Applications of the bacteriocin, nisin. Antonie Van Leeuwenhoek 69: 193–202.

Dezmazeaud, M. and T.M. Cogan. 1996. Role of cultures in cheese ripening. pp. 207–232. *In*: T.M. Cogan and J.-P. Accolas [eds.]. Dairy Starter Cultures. VCH Publishers Ltd., Cambridge, UK.

Di Cagno, R., M. De Angelis, V. Upadhyay, P.L.H. McSweeney, F. Minervini, G. Gallo and M. Gobbetti. 2003. Effect of proteinases of starter bacteria on the growth and proteolytic activity of *Lactobacillus plantarum* DPC2741. International Dairy Journal 13: 145–157.

Diep, D.B., D. Straume, M. Kjos, C. Torres and I.F. Nes. 2009. An overview of the mosaic bacteriocin pln loci from *Lactobacillus plantarum*. Peptides 30: 1562–1574.

Doleyres, Y., L. Schaub and C. Lacroix. 2005. Comparison of the functionality of exopolysaccharides produced *in situ* or added as bioingredients on yogurt properties. Journal of Dairy Science 88: 4146–4156.

Draper, L.A., K. Grainger, L.H. Deegan, P.D. Cotter, C. Hill and R.P. Ross. 2009. Cross-immunity and immune mimicry as mechanisms of resistance to the lantibiotic lacticin 3147. Molecular Microbiology 71: 1043–1054.

Dunny, G. and L.L. McKay. 1999. Group II introns and expression of conjugative transfer functions in lactic acid bacteria. Antonie van Leeuwenhoek 76: 77–88.

EC European Council. 2003. Regulation (EC) No 1829/2003 of the European Parliament and of the Council of 22 September 2003 on genetically modified food and feed. Official Journal of the European Union. L268: 1–23.

[EC] European Council 2006. Regulation (EC)No 1924/2006 of the European Parliament and of the Council of 20 December 2006 on nutrition and health claims made on foods. Official Journal of the European Union. L12: 3–18.

El Abboudi, M., M. El Soda, S. Pandian, R.E. Simard and N.F. Olson. 1992. Purification of X-prolyl dipeptidyl aminopeptidase from *Lactobacillus casei* subspecies, Int. J. Food Microbiol. 15: 87–98.

El Demerdash, H.A.M., J. Oxmann, K.J. Heller and A. Geis. 2006. Yoghurt fermentation at elevated temperatures by strains of *Streptococcus thermophilus* expressing a small heat-shock protein: application of a two-plasmid system for constructing food-grade strains of *Streptococcus thermophilus*. Biotechnol. Journal 1: 398–404.

Eom, H.S., B.H. Hwang, D.H. Kim, I.B. Lee, Y.H. Kim and H.J. Cha. 2007. Multiple detection of food-borne pathogenic bacteria using a novel 16S rDNA-based oligonucleotide signature chip. Biosens. Bioelectron 22: 845–853.

European Food & Feed Cultures Association. 2011. Protective Cultures. http://www.effca.org/sites/effca.drupalgardens.com/files/Protective-cultures-effca-statement_0.pdf.

European Food Safety Authority. 2010. Scientific Opinion on the maintenance of the list of QPS biological agents intentionally added to food and feed (2010 update). EFSA Journal 8, 1944: 1–56.

European Food Safety Authority. 2007. Introduction of a Qualified Presumption of Safety (QPS) approach for assessment of selected microorganisms referred to EFSA. The EFSA Journal 587: 1–16.

European Food Safety Authority. 2006. Qualified Presumption of Safety of Micro-organisms in Food and Feed. EFSA Scientific Colloquium Summary Report. Brussels, Belgium.

Felley, C.P., I. Corthısy-Theulaz, J.-L. Blanco Rivero, P. Sipponen, M. Kaufmann, P. Bauerfeind, P.H. Wiesel, D. Brassart, A. Pfeifer, A.L. Blum and P. Michetti. 2001. Favourable effect of an acidified milk (LC-1) on *Helicobacter pylori* gastritis in man. Eur. J. Gastroenterol. Hepatol. 13: 25–29.

Ferchichi, M., D. Hemme, M. Nardi and N. Pamboukdjian. 1985. Production of methanethiol from methionine by *Brevibacterium linens* CNRZ 918. J. Gen. Microbiol. 131: 715.

Fernandez, M., M. Kleerebezem, O.P. Kuipers, R.J. Siezen and R. Kranenburg. 2002. Regulation of the metC-cysK operon, involved in sulphur metabolism in *Lactococcus lactis*. Journal of Bacteriology 184: 82–90.

Food and Agriculture Organization and World Health Organization. 2001. Health and Nutritional Properties of Probiotics in Food including Powder Milk with Live Lactic Acid Bacteria. http://www.who.int/foodsafety/publications/fs_management/en/probiotics.pdf.

Forde, A. and G.F. Fitzgerald. 1999. Bacteriophage defence systems in lactic acid bacteria. Antonie van Leeuwenhoek 76: 89–113.

Fox, P.F. and J.M. Wallace. 1997. Formation of flavour compounds in cheese. Advances in Applied Microbiology 45: 17–85.

Fujisawa, T., Y. Benno, T. Yaeshima and T. Mitsuoka. 1992. Taxonomic study of the *Lactobacillus acidophilus* group, with recognition of *Lactobacillus gallinarum* sp. nov. and *Lactobacillus johnsonii* sp. nov. and synonymy of *Lactobacillus acidophilus* group A3 with the type strain of *Lactobacillus amylovorus*. Int. J. System. Bacteriol. 42: 487–491.

Galperin, M.Y. and X.M. Fernandez. 2012. The Nucleic Acids Research Database Issue and the online Molecular Biology Database Collection. Nucleic Acids Research 40, Database Issue: D1–D8.

Garde, S., J. Tomillo, P. Gaya, M. Medina and M. Nunez. 2002. Proteolysis in Hispanico cheese manufactured using a mesophilic starter, a thermophilic adjunct culture and bacteriocin-producing *Lactococcus lactis* subsp. *lactis* INIA 415 adjunct culture. J. Agric. Food Chem. 50: 3479–3485.

Gardiner, G.E., P. Bouchier, E. O'sullivan, J. Kelly, K. Collins, G. Fitzgerald, R.P. Ross and C. Stanton. 2002. A spray-dried culture for porbiotic Cheddar chesse manufacture. International Dairy Journal 12: 749–756.

Garvie, E.I. 1986. Genus *Leuconostoc*. pp. 1071–1075. *In:* P.H.A. Sneath, N.S. Mair, M.E. Sharpe and J.G. Holt [eds.]. Bergey's Manual of Systematic Bacteriology, vol 2, 9th ed. Williams and Wilkins, Baltimore, US.

Gasson, M.J. 1990. *In vivo* genetic systems in lactic acid bacteria. FEMS Microbiol. Rev. 7: 43–60.

Gasson, M.J., J. Godon, C.J. Pillidge, T.J. Eaton, K. Jury and C.J. Shearman. 1995. Characterization and exploitation of conjugation in *Lactococcus lactis*. International Dairy Journal 5: 757–762.

Georgalaki, M.D., K. Papadimitriou, R. Anastasiou, B. Pot, G. Van Driessche, B. Devreese and E. Tsakalidou. 2013. Macedovicin, the second food-grade lantibiotic produced by Streptococcus macedonicus ACA-DC 198. Food Microb. 33: 124–130.

Georgalaki, M.D., E. Van Den Berghe, D. Kritikos, B. Devreese, J. Van Beeumen, G. Kalantzopoulos, L. De Vuyst and E. Tsakalidou. 2002. Macedocin, a food-grade lantibiotic produced by Streptococcus macedonicus ACA-DC 198. Appl. Environ. Microbiol. 68: 5891–5903.

Gibson, G.R. and M.B. Robertfroid. 1995. Dietary modulation of the human colonic microbiota: introducing the concept of prebiotics. Journal of Nutrition 124: 1401–1412.

Gilbert, C., D. Atlan, B. Blanc, R. Portailer, J.E. Germond, L. Lapierre and B. Mollet. 1996. A new cell surface proteinase: sequencing and analysis of the prtB gene from *Lactobacillus delbruekii* subsp. *bulgaricus*. J. Bacteriol. 178: 3059–3065.

Giraffa, G., M. Gatti, L. Rossetti, L. Senini and E. Neviani. 2000. Molecular diversity within *Lactobacillus helveticus* as revealed by genotypic characterization. Appl. Environ. Microbiol. 66: 1259–1265.

Gonzalez, C.F. and B.S. Kunka. 1986. Evidence for plasmid linkage of raffinose utilization and associated α-galactosidase and sucrose hydrolase activity in *Pediococcus pentosaceus*. Appl. Environ. Microbiol. 51: 105–109.

Graham, D.C. and L.L. McKay. 1985. Plasmid DNA in strains of *Pediococcus cerevisiae* and *Pediococcus pentosaceus*. Appl. Environ. Microbiol. 50: 532–534.

Guinee, T.P. and D.J. O'Callaghan. 2010. Control and prediction of quality characteristics in the manufacture and ripening of Cheese. pp. 260–329. *In:* B.A. Law and A.Y. Tamime [eds.]. Technology of Cheesemaking 2nd Edition, John Wiley & Sons Ltd., Chichester, UK.

Hammes, W. P. and R. F. Vogel. 1995. The genus *Lactobacillus*. pp. 19–54. *In*: B.J.B. Wood and W.H. Holzapfel [eds.]. The lactic acid bacteria. Vol 2. The genera of lactic acid bacteria. Blackie Academic and Professional, London.

Hansen, E.B. 2002. Commercial bacterial starter cultures for fermented foods of the future. International Journal of Food Microbiology 78: 119–131.

Hassan, A.N. 2008. ADSA Foundation Scholar Award: possibilities and challenges of exopolysaccharide-producing lactic cultures in dairy foods. Journal of Dairy Science 91: 1282–1298.

Hassan, A.N. and J.F. Frank. 2001. Starter cultures and their use. pp 151–206. *In:* E.H. Marth and J.L. Steele [eds.]. Applied Dairy Microbiology, 2nd edition. Marcel Dekker, Inc, New York.

Hassan, A.N., R. Ipsen, T. Janzen and K.B. Qvist. 2003. Microstructure and rheology of yogurt made with cultures differing only in their ability to produce exopolysaccharides. Journal of Dairy Science 86: 1632–1638.

Høier, E., T. Janzen, F. Rattray, K. Sørensen, M.W. Børsting, E. Brockmann and E. Johansen. 2010. The production, application and action of Lactic cheese starter cultures. pp. 166–192. *In*: B.A. Law and A.Y. Tamime [eds.]. Technology of Cheesemaking, 2nd ed. John Wiley & Sons Ltd., Oxford.

Hols, P., F. Hancy, L. Fontaine, B. Grossiord, D. Prozzi, N. Leblond-Bourget, B. Decaris, A. Bolotin, C. Delorme, S.D. Ehrlich, E. Guedon, V. Monnet, P. Renault and M. Kleerebezem. 2005. New insights in the molecular biology and physiology of *Streptococcus thermophilus* revealed by comparative genomics. FEMS Microbiology Reviews 29: 435–463.

Hols, P., M. Kleerebezem, A.N. Schanck, T. Ferain, J. Hugenholtz, J. Delcour and W.M. de Vos. 1999. Conversion of *Lactococcus lactis* from homolactic to homoalanine fermentation through metabolic engineering. Nature Biotechnology 17: 588–592.

Horvath, P., A.C. Coute-Monvoisin, D.A. Romero, P. Boyaval, C. Fremaux and R. Barrangou. 2009. Comparative analysis of CRISPR loci in lactic acid bacteria genomes. International Journal of Food Microbiolology 131: 62–70.

Horvath, P., D.A. Romero, A.C. Coute-Monvoisin, M. Richards, H. Deveau, S. Moineau, P. Boyaval, C. Fremaux and R. Barrangou. 2008. Diversity, activity, and evolution of CRISPR loci in *Streptococcus thermophilus*. Journal of Bacteriology 190: 1401–1412.

Hosono, A., J. Lee, A. Ametani, M. Natsume, M. Hirayama, T. Adachi and S. Kaminogawa. 1997. Characterization of a water-soluble polysaccharide fraction with immunopotentiating activity from *Bifidobacterium adolescentis* M101-4. Bioscience, Biotechnology and Biochemistry 61: 312–316.

Hugenholtz, J. 1993. Citrate metabolism in lactic acid bacteria. FEMS Microbiol. Rev. 12: 165–178.

Hugenholtz, J. 2008. The lactic acid bacterium as a cell factory for food ingredient production. International Dairy Journal 18: 466–475.

Hugenholtz, J. and M. Kleerebezem. 1999. Metabolic engineering of lactic acid bacteria: overview of the approaches and results of pathway rerouting involved in food fermentations. Current Opinion in Biotechnology 10: 492–497.

Hugenholtz, J. and E.K. Smid. 2002. Nutraceutical production with food-grade microorganisms. Current Opinion in Biotechnology 13: 497–507.

Hugenholtz, J., M. Kleerebezem, M. Starrenburg, J. Delcour, W. de Vos and P. Hols. 2000. Lactococcus lactis as a cell factory for high-level diacetyl production. Applied and Environmental Microbiology 66: 4112–4114.

Hughes, D. 2000. Evaluating genome dynamics: The constraints on rearrangements within bacterial genomes. Genome Biol. 1: Reviews 0006.1–0006.8.

Hutkins, R.W. 2001. Metabolism of Starter cultures. pp. 207–242. *In:* E.H. Marth and J.L. Steele [eds.]. Applied Dairy Microbiology. Marcel Dekker, Inc., New York.

Huys, G., M. Vancanney, K. D'Haene, V. Vankerckhoven, H. Goossens and J. Swings. 2006. Accuracy of species identity of commercial bacterial cultures intended for probiotic or nutritional use. Research in Microbiology 157: 803–810.

ISO 27205:2005 and IDF 149:2010—Fermented milk products—Bacterial starter cultures—Standard of identity. International Organization for Standardization, Geneva.

Johnson, J.L., C.F. Phelps, C.S. Cummins, J. London and F. Gasser. 1980. Taxonomy of the *Lactobacillus acidophilus* group. Int. J. System. Bacteriol. 30: 53–68.

Jolly, L., S.J. Vincent, P. Duboc and J.R. Neeser. 2002. Exploiting expolysaccharides from lactic acid bacteria. Antonie van Leeuwenhoek 82: 367–374.

Karimi, R., A.M. Mortazavian and A.G. Da Cruz. 2011. Viability of probiotic microorganisms in cheese during production and storage: a review. Dairy Science and Technology: 91: 283–308.

Kasimoglu, A., M. Goncuoglu and S. Akgun. 2004. Probiotic White cheese with *Lactobacillus acidophilus*. International Dairy Journal 14: 1067–1073.

Kemperman, R., A. Kuipers, H. Karsens, A. Nauta, O. Kuipers and J. Kok. 2003. Identification and characterization of two novel clostridial bacteriocins, circularin A and closticin 574. Applied and Environmental Microbiology 69: 1589–1597.

Khalid, N.M. and E.M. Marth. 1990a. Lactobacilli—their enzymes and role in ripening and spoilage of cheese: a review. Journal of Dairy Science 73: 2669–2684.

Khalid, N.M. and E.M. Marth. 1990b. Purification and characterization of prolyl dipeptidyl aminopeptidase from *Lactobacillus helveticus* CNRZ 32. Appl. Environ. Microbiol. 56: 381–388.

Khalid, N.M., M. El-Soda and E.H. Marth 1990. Esterases of *Lactobacillus helveticus* and *Lactobacillus delbrueckii* ssp. *bulgaricus*. Journal of Dairy Science 73: 2711–2719.

Klaenhammer, T., E. Altermann, F. Arigoni, A. Bolotin, F. Breidt, J. Broadbent, R. Cano, S. Chaillou, J. Deutscher, M. Gasson, M. van de Guchte, J. Guzzo, A. Hartke, T. Hawkins, P. Hols, R. Hutkins, M. Kleerebezem, J. Kok, O. Kuipers, M. Lubbers, E. Maguin, L. McKay, D. Mills, A. Nauta, R. Overbeek, H. Pel, D. Pridmore, M. Saier, D. van Sinderen, A. Sorokin, J. Steele, D. O'Sullivan, W. de Vos, B. Weimer, M. Zagorec and R. Siezen. 2002. Discovering lactic acid bacteria by genomics. Antonie van Leeuwenhoek 82: 29–58.

Klaenhammer, T.R. 1993. Genetics of bacteriocins produced by lactic acid bacteria. FEMS Microbiology Reviews 12: 39–85.

Kleerebezem, M., P. Hols and J. Hugenholtz. 2000. Lactic acid bacteria as a cell factory: rerouting of carbon metabolism in *Lactococcus lactis* by metabolic engineering. Enzyme Microb. Technol. 26: 840–848.

Kleerebezem, M. and J. Hugenholtz. 2003. Metabolic pathway engineering in lactic acid bacteria. Current Opinion in Biotechnology 14: 232–237.

Kok, J., G. Buist, A. Zomer, S.A.F.T. vn Hijum and O.P. Kuipers. 2005. Comparative and functional genomics of lactococci. FEMS Microb. Rev. 29: 411–433.

Kunji, E.R.S., I. Mierau, A. Hagting, B. Poolman and W.N. Konings. 1996. The proteolytic system of lactic acid bacteria. Antonie van Leewenhoek 70: 187–221.

Lane, D.J., B. Pace, G.J. Olsen, D.A. Stahl, M.L. Sogin and N.R. Pace. 1985. Rapid determination of 16S ribosomal RNA sequences for phylogenetic analyses. Proceedings of the National Academy of Sciences of the United States of America 82: 6955–6959.

Law, B.A. 1999. Cheese ripening and cheese flavour technology. pp. 163–192. *In:* B.A. Law [ed.] Technology of Cheesemaking. Sheffield Academic Press Ltd., Sheffield, UK.

Leenhouts, L., G. Buist, A. Bolhuis, A.A. ten Berge, J. Kiel, I. Mierau, M. Dabrowska, G. Venema and J. Kok. 1996. A general system for generating unlabelled gene replacements in bacterial chromosomes. Molecular and General Genetics 253: 217–224.

Leuschner, R.-G.K., T. Robinson, M. Hugas, P.S. Cocconcelli, F. Richard-Forget, G. Klein, T.R. Licht, C. Nguyen-The, A. Querol, M. Richardson, J.E. Suarez, J.M. Vlak and A. von Wright. 2010. Qualified presumption of safety (QPS): a generic risk assessment approach for biological agents notified to the European Food Safety Authority (EFSA). Trends Food Sci. Technol. 21: 425–435.

Lima, P.T. and A.M. Correia. 2000. Genetic fingerprinting of *Brevibacterium linens* by pulsed-field gel electrophoresis and ribotyping. Curr. Microbiol. 41: 50–55.

Linares, D.M., J. Kok and B. Poolman. 2010. Genome Sequences of *Lactococcus lactis* MG1363 (Revised) and NZ9000 and Comparative Physiological Studies. Journal of Bacteriology 192: 5806–5812.

Link-Amster, H., F. Rochat, K.Y. Saudan, O. Mignot and J.-M. Aeschlimann. 1994. Modulation of a specific humoral immune response and changes in intestinal flora mediated through fermented milk intake. FEMS Immunol. Med. Microbiol. 10: 55–64.

Liu, M., J.R. Bayjanov, B. Renckens, A. Nauta and R.J. Siezen. 2010. The proteolytic system of lactic acid bacteria revisited: a genomic comparison. BMC Genomics 11: 36–51.

Lynch, C.M., D.D. Muir, J.M. Banks, P.L.H. McSweeney and P.F. Fox. 1999. Influence of adjunct cultures of *Lactobacillus paracasei* ssp. *paracasei* or *Lactobacillus plantarum* on Cheddar cheese ripening. Journal of Dairy Science 82: 1618–1628.

Madera, C., P. Garcia, T. Janzen, A. Rodriguez and J.E. Suarez. 2003. Characterization of technologically proficient wild *Lactococcus lactis* strains resistant to phage infection. Int. Journal of Food Microb. 86: 213–222.

Makarova, K., A. Slesarev, Y. Wolf, A. Sorokin, B. Mirkin, E. Koonin, A. Pavlov, N. Pavlova, V. Karamychev, N. Polouchine, V. Shakhova, I. Grigoriev, Y. Lou, D. Rohksar, S. Lucas, K. Huang, D.M. Goodstein, T. Hawkins, V. Plengvidhya, D. Welker, J. Hughes, Y. Goh, A. Benson, K. Baldwin, J.H. Lee, I. Díaz-Muñiz, B. Dosti, V. Smeianov, W. Wechter, R. Barabote, G. Lorca, E. Altermann, R. Barrangou, B. Ganesan, Y. Xie, H. Rawsthorne, D. Tamir, C. Parker, F. Breidt, J. Broadbent, R. Hutkins, D. O'Sullivan, J. Steele, G. Unlu, M. Saier, T. Klaenhammer, P. Richardson, S. Kozyavkin, B. Weimer and D. Mills. 2006. Comparative genomics of the lactic acid bacteria. Proc. Natl. Acad. Sci. USA 103: 15611–15616.

Mayo, B., D. van Sinderen and M. Ventura. 2008. Genome analysis of food grade lactic acid-producing bacteria: from basics to applications. Current Genomics 9: 169–183.

McAuliffe, O., C. Hill and R.P. Ross. 1999. Inhibition of *Listeria monocytogenes* in cottage cheese manufactured with a lacticin 3147-producing starter culture. Journal of Applied Microbiology 86: 251–256.

McAuliffe, O., R.P. Ross and C. Hill. 2001. Lantibiotics: structure, biosynthesis and mode of action. FEMS Microbiol. Rev. 25: 285–308.

McCartney, A. 2002. Application of molecular biological methods for studying probiotics and the gut flora. Br. J. Nutr. 88: 29–37.

McKay. L.L. 1985. Roles of plasmids in starter cultures. pp. 159–174. *In:* S.E. Gilliland [ed.]. Bacterial Starter Cultures for Food. CRC Press, Boca Raton, US.

Mills, S., A. Coffey, O.E. McAuliffe, W.C. Meijer, B. Hafkamp and R.P. Ross. 2007. Efficient method for generation of bacteriophage insensitive mutants of *Streptococcus thermophilus* yoghurt and mozzarella strains. Journal of Microbiological Methods 70: 159–164.

Mills, S., C. Griffin, A. Coffey, W.C. Meijer, B. Hafkamp and R.P. Ross. 2009. CRISPR analysis of bacteriophage insensitive mutants (BIMs) of industrial *Streptococcus thermophilus*—implications for starter design. Journal of Applied Microbiology 108: 945–955.

Mills, S., O. O'Sullivan, C. Hill, G. Fitzgerald and R.P. Ross. 2010. The changing face of dairy starter culture research: From genomics to economics. International Journal of the Dairy Technology 63: 149–170.

Mills, S., O.E. McAuliffe, A. Coffey, G.F. Fitzgerald and R.P. Ross. 2006. Plasmids of lactococci—genetic accessories or genetic necessities? FEMS Microbiology Reviews 30: 243–273.

Mollet, B. and M. Delley. 1990. Spontaneous deletion formation within the beta-galactosidase of *Lactobacillus bulgaricus*. J. of Bacteriol. 172: 5670–5676.

Morandi, A., O. Zhaxybayeva, J.P. Gogarten and J. Graf. 2005. Evolutionary and diagnostic implications of intragenomic heterogeneity in the 16S rRNA gene in *Aeromonas* strains. J. Bacteriol. 187: 6561–6564.

Morelli, L., F.K. Vogensen and A. von Wright. 2004. Genetics of Lactic Acid Bacteria. pp. 249–293. *In*: S. Salminen, A. von Wright and A. Ouwehand [eds.]. Lactic Acid Bacteria—Microbiological and Functional Aspects, 3rd Edition. Marcel Dekker Inc., New York.

Morgan, S.M., M. Galvin, R.P. Ross and C. Hill. 2001. Evaluation of a spray-dried lacticin 3147 powder for the control of *Listeria monocytogenes* and *Bacillus cereus* in a range of food systems. Letters in Applied Microbiology 33: 387–391.

Morgan, S.M., R. Hickey, R.P. Ross and C. Hill. 2000. Efficient method for the detection of microbially-produced antibacterial substances from food systems. Journal of Applied Microbiology 89: 56–62.

Myers, K.M., J. Gaba and S.F. Al-Khaldi. 2006. Molecular identification of *Yersnia enterocolitica* isolated from pasteurized whole milk using DNA microarray chip hybridization. Mol. Cell Probes 20: 71–80.

Nakajima, H., Y. Suzuki, H. Kaizu and T. Hirota. 1992. Cholesterol-lowering activity of ropy fermented milk. Journal of Food Science 57: 1327–1329.

Nes, I.F., D.B. Diep, L.S. Havarstein, M.B. Brurberg, V. Eijsink and H. Holo. 1996. Biosynthesis of bacteriocins in lactic acid bacteria. Antonie van Leeuwenhoek 70: 113–128.

O'Sullivan, D.J. 1999. Methods for analysis of the intestinal microflora. *In:* G.W. Tannock [ed.]. Probiotics: A Critical Review. Horizon Scientific Press Norfolk, UK.

O'Sullivan, L., E.B. O'Connor, R.P. Ross and C. Hill. 2006. Evaluation of live-culture-producing lacticin 3147 as a treatment for the control of *Listeria monocytogenes* on the surface of smear-ripened cheese. Journal of Applied Microbiology 100: 125–143.

O'Sullivan, O., J. O'Callaghan, A. Sangrador-Vegas, O. McAuliffe, L. Slattery, P. Kaleta, M. Callanan, G.F. Fitzgerald, R.P. Ross and T. Beresford. 2009. Comparative genomics of lactic acid bacteria reveals a niche-specific gene set. BMC Microbiology 9: 50–58.

Ong, L., A. Henriksson and N.P. Shah. 2006. Development of probiotic Cheddar cheese containing *Lactobacillus acidophilus*, *Lb. casei*, *Lb. paracasei* and *Bifidobacterium* spp. and the influence of these bacteria on proteolytic patterns and production of organic acid. International Dairy Journal 16: 446–456.

Oumer, B.A., E. Fernandez-Garcia, R. Marciaca, S. Garde, M. Medina and M. Nunez. 2001. Proteolysis and formation of volatile compounds in cheese manufactured with a bacteriocin-producing adjunct. J. Dairy Res. 68: 119–127.

Parente, E. and T.M. Cogan. 2004. Starter Cultures: General Aspects. pp. 123–147. *In:* P.F. Fox, P.L.H. McSweeney, T.M. Cogan and T.P. Guinee, Vol. 1, 4th Ed. Elsevier Academic Press, London, UK.

Park, Y.H., H. Hori, K. Suzuki, S. Osawa and K. Komagata. 1987. Phylogenetic analysis of the coryneform bacteria by 5S rRNA sequences. J. Bacteriol. 169: 1801–1806.

Parvez, S., K.A. Malik, S.A. Kang and H.Y. Kim. 2006. Probiotics and their fermented food products are beneficial for health. Journal of Applied Microbiology 100: 1171–1185.

Pastink, M.I., B. Teusink, P. Hols, S. Visser, W.M. de Vos and J. Hugenholtz 2009. Genome-Scale Model of *Streptococcus thermophilus* LMG18311 for Metabolic Comparison of Lactic Acid Bacteria. Appl. and Environm. Microb. 75: 3627–3633.

Pedersen, M.B., S.L. Iversen, K.K. Sorensen and E. Johansen. 2005. The long and winding road from the research laboratory to industrial applications of lactic acid bacteria. FEMS Microbiology Reviews 29: 611–624.

Pérez, P.F., J. Minnaard, M. Rouvet, C. Knabenhans, D. Brassart, G.L. De Antoni and E.J. Schiffrin. 2001. *Inhibition of Giardia intestinalis by extracellular factors from lactobacilli: an in vitro* study. Appl. Environ. Microbiol. 67: 5037–*5042*.

Peterson, S.D. and R.T. Marshall 1990. Nonstarter Lactobacilli in Cheddar Cheese: A Review. Journal of Dairy Science 73: 1395–1410.

Poolman, B. 2002. Transporters and their role in LAB cell physiology. Antonie van Leewenhoek 82: 147–164.

Rallu, F., P. Taillez, D. Ehrlich and P. Renault. 2002. Common scheme of evolution between eps clusters of the food bacteria *Streptococcus thermophilus* and cps clusters of the pathogenic streptococci. 6th American Society of Microbiology Conference on Streptococcal Genetics, 112. Asheville, NC, USA.

Rattray, F.P., P.F. Fox and A. Healy. 1997. Specificity of an extracellular proteinase from *Brevibacterium linens* ATCC 9174 on bovine beta-casein. Appl. Environ. Microbiol. 63: 2468–2471.

Rattray, F.P. and P.F. Fox. 1999. Aspects of enzymology and biochemical properties of *Brevibacterium linens* relevant to cheese ripening: a review. J. Dairy Sci. 82: 891–909.

Rijnen, L., P. Courtin, J.C. Gripon and M. Yvon. 2000. Expression of heterologous glutamate dehydrogenase gene in *Lactococcus lactis* highly improves the conversion of amino acids to aroma compounds. Applied and Environmental Microbiology 66: 1354–1359.

Rijnen, L., A. Delacroix Buchet, D. Demaiziere, J.L. Le Quere, J.C. Gripon and M. Yvon. 1999. Inactivation of lactococcal aromatic aminotransferase prevents the formation of floral aroma compounds from aromatic amino acids in semi-hard cheese. International Dairy Journal 9: 877–885.

Rodgers, S. 2008. Novel applications of live bacteria in food services: probiotics and protective cultures. Trends Food Sci. Tech. 19: 188–197.

Ross, R.P., M. Galvin, O. McAuliffe, S.M. Morgan, M.P. Ryan, D.P. Twomey, W.J. Meaney and C. Hill. 1999. Developing applications for lactococcal bacteriocins. Antonie van Leeuwenhoek 76: 337–346.

Ross, R.P., S. Morgan and C. Hill. 2002. Preservation and fermentation: past, present and future. International Journal of Food Microbiology 79: 3–16.

Roy, D., P. Ward, D. Vincent and F. Mondou. 2000. Molecular identification of potentially probiotic lactobacilli. Curr. Microbiol. 40: 40–46.

Rudi, K., M. Zimonja, B. Kvenshagen, J. Rugtveit, T. Midtvedt and M. Eggesbo. 2007. Alignment-independent comparisons of human gastrointestinal tract microbial communities in a multidimensional 16S rRNA gene evolutionary space. Appl. Environ. Microbiol. 73: 2727–2734.

Ryan, M.P., M.C. Rea, C. Hill and R.P. Ross. 1996. An application in cheddar cheese manufacture for a strain of *Lactococcus lactis* producing a novel broad-spectrum bacteriocin, lacticin 3147. Applied and Environmental Microbiology 62: 612–619.

Salama, M., W.E. Sandine and S. Giovannoni. 1991. Development and application of oligonucleotide probes for identification of *Lactococcus lactis* subsp. *cremoris*. Appl. Environ. Microbiol. 57: 1313–1318.

Schiffrin, E.J., F. Rochat, H. Link-Amster, J.-M. Aeschlimann and A. Donnet-Hughes. 1995. Immunomodulation of human blood cells following the ingestion of lactic acid bacteria. J. Dairy Sci. 78: 491–497.

Schleifer, K.H., J. Kraus, C. Dvorak, R. Kilpper-Balz, M.D. Collins and W. Fisher. 1985. Transfer of *Streptococcus lactis* and related streptococci to the genus *Lactococcus*. Systematic and Applied Microbiology 6: 183–195.

Schleifer, K.-H., M. Ehrmann, C. Beimfohr, E. Brockmann, W. Ludwig and R. Amann. 1995. Application of molecular methods for the classification and identification of Lactic Acid Bacteria. Int. Dairy Journal 5: 1081–1094.

Schleifer, K.H. and W. Ludwig. 1995. Phylogenetic relationships of lactic acid bacteria. pp. 7–18. *In:* B.J.B. Wood and W.H. Holzapfel [eds.]. The Genera of Lactic Acid Bacteria. Chapman & Hall, London.

Serror, P., T. Sasaki, S.D. Ehrlich and E. Maguin. 2002. Electrotransformation of *Lactobacillus delbrueckii* subsp. *bulgaricus* and *L. delbrueckii* subsp. *lactis* with various plasmids. Appl. Environ. Microbiol. 68: 46–52.

Seth, R.J. and R.K. Robinson. 1988. Factors contributing to the flavour characteristics of mould-ripened cheese. pp 23–47. *In:* R.K. Robinson [ed.]. Developments in Food Microbiology, Vol. 4. Elsevier Applied Science, London, UK.

Sheehan, J.J. 2007. What are starters and what starter types are used for cheesemaking? pp. 36–37. *In*: P.L.H. McSweeney [ed.]. Cheese problems solved. Woodhead Publishing Ltd., Boca Raton, US.

Simpson, W.J. and H. Taguchi. 1995. The genus *Pediococcus*, with notes on the genera *Tetratogenococcus* and *Aerococcus*. pp 125–172. *In:* B.J.B. Wood and W.H. Holzapfel [eds]. The Genera of Lactic Acid Bacteria. Chapman & Hall, London.

Smit, G., B.A. Smit and W.J. Engels. 2005. Flavour formation by lactic acid bacteria and biochemical flavour profiling of cheese products. FEMS Microbiology Reviews 29: 591–610.

Solow, B.T. and G.A. Somkuti. 2000. Molecular properties of *Streptococcus thermophilus* plasmid pER35 encoding a restriction modification system. Curr. Microbiol. 42: 122–128.

Sorek, R., V. Kunin and P. Hugenholtz. 2008. CRISPR—a widespread system that provides acquired resistance against phages in bacteria and *archaea*. Nature Reviews Microbiology 6: 181–186.

Souza, C.H.B., F.C.A. Buriti, J.H. Behrens and S.M.I. Saad. 2008. Sensory evaluation of probiotic Minas fresh cheese with *Lactobacillus acidophilus* added solely or in co-culture with a thermophilic starter culture. International Journal of Food Science and Technology 43: 871–877.

Souza, M.J., Y. Ardo and P.L.H. McSweeney. 2001. Advances in the study of proteolysis in cheese. Int. Dairy J. 11: 327–345.

Sybesma, W., J. Hugenholtz, W.M. de Vos and E.J. Smid. 2006. Safe Use of Genetically Modified Lactic Acid Bacteria in Food. Electronic Journal of Biotechnology 9: 1–25.

Tamime, A.Y. 2002. Microbiology of Starter Cultures, pp. 261–366. *In:* R.K. Robinson [ed.]. Dairy Microbiology Handbook. 3rd Edition John Wiley & Sons Inc., New York.

Tamime, A.Y. and R.K. Robinson. 1999. Yoghurt Science and Technology. 2nd Edition. Woodhead Publishing Ltd. Cambridge.

Tannock, G.W. 1999. Probiotics: A Critical Review. Horizon Scientific Press, Wymondham, U.K.

Tetrapak. 1995. Dairy Processing Handbook, Lund, Sweden 64: 233–240.

Thomas, T.D. and O.E. Mills 1981. Proteolytic enzymes of starter bacteria. Netherlands Milk and Dairy Journal 35: 255–273.

Thompson, J.K., K.J. McConville, C. McReynolds and M.A. Collins. 1999. Potential of conjugal transfer as a strategy for the introduction of recombinant genetic material into strains of *Lactobacillus helveticus*. Appl. Environ. Microbiol. 65: 1910–1914.

Thunell, R.K. and W.E. Sandine. 1985. Types of starter cultures. pp. 127–144. *In:* S.E. Gilliland [ed.]. Bacterial Starter Cultures for Foods. CRC Press Inc. Boca Raton, US.

Torriani, S., G. Zapparoli and F. Dellaglio. 1999. Use of PCR-based methods for rapid differentiation of *Lactobacillus delbrueckii* subsp. *bulgaricus* and *L. delbrueckii* subsp. *lactis*. Appl. Environ. Microbiol. 65: 4351–4356.

Trmcic, A., T. Obermajer, I. Rogelj and B. Bogovic Matijasic. 2008. Short communication: culture-independent detection of lactic acid bacteria bacteriocin genes in two traditional slovenian raw milk cheeses and their microbial consortia. J. Dairy Sci. 91: 4535–4541.

Turgeon, N., M. Frenette and S. Moineau. 2004. Characterization of a theta-replicating plasmid from *Streptococcus thermophilus*. Plasmid 51: 24–36.

Twomey, D., R.P. Ross, M. Ryan, B. Meaney and C. Hill. 2002. Lantibiotics produced by lactic acid bacteria: structure, function and applications. Antonie Van Leeuwenhoek 82: 165–185.

Tynkkynen, S., R. Satokari, M. Saarela, T. Mattila-Sandholm and M. Saxelin. 1999. Comparison of ribotyping, randomly amplified polymorphic DNA Analysis, and pulsed-field gel electrophoresis in typing of *Lactobacillus rhamnosus* and *L. casei* strains. Appl. Environ. Microbiol. 65: 3908–3914.

Ummadi, M. and B.C. Weimer. 2001. Tryptophan metabolism in *Brevibacterium linens* BL2. J. Dairy Sci. 84: 1773–1782.

Upadhyay, V.K., PL.H. McSweeney, A.A.A. Magboul and P.F. Fox. 2004. Proteolysis in Cheese during Ripening. pp. 391–433. *In:* P.F. Fox, P.L.H. McSweeney, T.M. Cogan and T.P. Guinee [eds.]. Vol 1, 4th Ed. Elsevier Academic Press, London, UK.

[U.S. Food and Drug Administration. 2012. Generally Recognized as Safe (GRAS). http://www.fda.gov/Food/FoodIngredientsPackaging/GenerallyRecognizedasSafeGRAS/default.htm.

Valdes-Stauber, N. and S. Scherer. 1996. Nucleotide sequence and taxonomical distribution of the bacteriocin gene lin cloned from *Brevibacterium linens* M18. Appl. Environ. Microbiol. 62: 1283–1286.

Van de Guchte, M., S. Penaud, C. Grimaldi, V. Barbe, K. Bryson, P. Nicolas, C. Robert, S. Oztas, S. Mangenot, A. Couloux, V. Loux, R. Dervyn, R. Bossy, A. Bolotin, J.-M. Batto, T. Walunas, J.-F. Gibrat, P. Bessieres, J. Weissenbach, S.D. Ehrlich and E. Maguin. 2006. The complete genome sequence of *Lactobacillus bulgaricus* reveals extensive and ongoing reductive evolution. Microbiology 103: 9274–9279.

Varmanen, P., K. Savijoki, S. Eval, A. Palva and S. Tynkkynen. 2000. X-Prolyl dipeptidyl aminopeptidase gene (pepX) is part of the glnRA operon in *Lactobacillus rhamnosus*, J. Bacteriol. 182: 146–154.

Ventura, M. and R. Zink. 2002. Specific identification and molecular typing analysis of *Lactobacillus johnsonii* by using PCR-based methods and pulsed-field gel electrophoresis. FEMS Microbiol. Lett. 217: 141–154.

Vesanto, E., K. Savijoki, T. Rantanen, J.L. Steele and A. Palva. 1995. An X-prolyl-dipeptidyl aminopeptidase (pepX) gene from *Lactobacillus helveticus*. Microbiology 141: 3067–3075.

Wegmann, U., M. O'Connell-Motherway, A. Zomer, G. Buist, C. Shearman, C. Canchaya, M. Ventura, A. Goesmann, M.J. Gasson, O.P. Kuipers, D. van Sinderen and J. Kok. 2007. Complete genome sequence of the prototype lactic acid bacterium *Lactococcus lactis* subsp. *cremoris* MG1363. J. Bacteriol. 189: 3256–3270.

Zendo, T., J. Nakayama, K. Fujita and K. Sonomoto. 2008. Bacteriocin detection by liquid chromatography / mass spectrometry for rapid identification. Journal of Applied Microbiology 104: 499–507.

Zhong, W., K. Millsap, H. Bialkowska-Hobrzanska and G. Reid. 1998. Differentiation of *Lactobacillus* species by molecular typing. Appl. Environ. Microbiol. 64: 2418–2423.

CHAPTER 5

# Application of Probiotics in the Dairy Industry: The Long Way from Traditional to Novel Functional Foods

*Adele Costabile*[1,*] and *Simone Maccaferri*[2]

## INTRODUCTION

Dairy products have always been very important functional foods, and they currently represent an important sector in the "health and wellbeing" industry.

Dairy foods are generally divided in three categories (Saxelin 2008):

- Basic products (milk, fermented milks, cheeses, ice creams, etc.).
- Value-added products, in which the milk composition has been modified for enhancing some of the basic properties ascribable to the food. Examples of value-added products are low-lactose or lactose-free dairy products, sodium-reduced or calcium-enriched milk products, among others.
- Functional dairy products with proven health benefit. These kinds of products are based on the enrichment of the dairy food with a functional component. Most commonly, functional dairy foods are

[1] Research Fellow, Department of Food and Nutritional Sciences, The University of Reading, PO Box 226, Whiteknights, Reading RG6 6AP.

[2] Department of Pharmacy and Biotechnology, University of Bologna Via Belmeloro 6, 40126 Bologna, Italy.

* Corresponding author: a.costabile@reading.ac.uk

enriched with probiotics, but prebiotics-enriched dairy products also fall in this category. According to the most commonly acknowledged definition, probiotic foods are food products which contain one, or a combination of probiotic ingredients in a adequate matrix and in sufficient concentration, so that after their ingestion, the postulated effect is obtained and is beyond the usual nutrient suppliers (de Vrese et al. 2001).

In this chapter, we will explore the different applications of functional dairy products based on the utilization of probiotics, with discussion on the advancement of the research in this field, the main typologies of dairy probiotic foods and their related technological characteristics, health claims, and requirements to be met for commercialisation .

Notably, the dairy sector is the largest functional food market, accounting for nearly 33% of the broad market (Granato et al. 2010). To date, different functional dairy products containing probiotics have been developed, evaluated and commercialised these include: yogurts and fermented milks, cheeses, yog-ice creams, cheese-based dips, probiotic fermented lactic beverages, and dairy desserts (Di Criscio et al. 2010).

## Dairy Products

### *Yogurt and Fermented Milks*

Yogurt and fermented milks are considered to be the main carrier for the delivery of probiotics in the dairy industry. In recent years, since many consumers' associate yogurt—especially probiotics yogurt, with good health, per capita consumption of these dairy products has increased drastically (Hekmat and Reid 2006).

Traditionally, yogurts are prepared through a fermentation procedure of milk by specific pure cultures of lactic acid bacteria, such as *Streptococcus thermophilus* and *Lactobacillus bulgaricus*. However, even if the simple and standardized procedure at the basis of the yogurt production is widely established and commonly utilized, many factors influence the probiotics viability and its efficacy in promoting human health. Indeed, probiotic cell survival during the product's shelf life is a striking factor influenced by culture conditions, strain selection, level of inoculation, medium composition, the interactions among starter and probiotics species, final acidity, availability of nutrients and sugars, and storage and logistics conditions along the distribution chain (Talwalkar and Kailaspathy 2004, Dave and Shah 1997, Donkor et al. 2006, Vinderola et al. 2002, Ranadheera et al. 2010, Plessas et al. 2012).

For instance, while studying the survival of a *Bifidobacterium* strain in a yogurt matrix, it has been demonstrated that the concentration of milk fat is negatively correlated with the viability of probiotic cultures, thus resulting in higher fat concentration leading to sensible inhibitory effects (Vinderola et al. 2000). Similarly, a correlation has been shown between post storage pH in yogurts, presence of specific fruit pulp and the viability of probiotics. Mixed berry and passion fruits led to lower levels of viable probiotics lactobacilli in the tested yogurts with respect to plain-yogurt, whilst mango and strawberry had an enhancing activity. Therefore, taking into account the fruit mixtures as a paradigm of a single, simple exogenous factor influencing the probiotics viability in the food matrix, the importance of carefully studying the single environmental contributions which might reduce the viability of probiotic cultures, is easily understandable (Kailasapathy et al. 2008).

### *Probiotic Cheeses*

Cheeses are food products characterized by an intrinsic versatility in relation to the incorporation of probiotic bacteria, and their suitability for the delivery of probiotics in all age groups, from children to the elderly. Indeed, the consumption of cheese has increased in many countries during the past decade (da Cruz et al. 2009) and the design of novel probiotics foods should take this kind of product into account.

In fact, cheeses are considered much more advantageous for delivering viable probiotics rather than products of fermentation, as yogurt or fermented milk beverages. In particular, cheeses are commonly characterized by an higher pH, a higher buffering capacity, solid consistency and an higher fat content, thus resulting in greater protection to the probiotic cells during storage and passage through the upper gastrointestinal tract (Plessas et al. 2012). The higher pH is reflected in a more favourable environment for the viability of probiotic bacterial cells, as well as the fact that the higher fat content exerts a protective function against the peptic enzymes.

Similar to other dairy products, the probiotic bacteria included in cheese products mainly belong to the *Lactobacillus* and *Bifidobacterium* genera. Furthermore, some probiotics strains of *Enterococcus* and *Propionibacterium* have been characterized for their utilization in cheese manufacturing (Stanton et al. 1998).

To date, several studies are aimed at developing different types of probiotic cheeses, using cheeses such as Cheddar, Crescenza cheese, cottage cheese and fresh cheese (as reported in Table 1, modified from Plessas et al. 2012).

**Table 1.** Probiotics in several cheese types.

| **Cheese Type** | **Included Probiotics** |
|---|---|
| Cheddar | *L. salivarius, L. paracasei, L. acidophilus, L. casei, B. lactis, B. longum, B. infantis* |
| Gouda | *L. acidophilus, Bifidobacterium* spp. |
| Fresh cheese | *L. acidophilus, L. casei, B. longum, B. bifidum* |
| Feta cheese | *L. casei* |
| Crescenza cheese | *B. bifidum, B. longum* |

## *Probiotic Ice-creams*

Ice cream is a broad category including several related products, such as plain ice-cream, reduced- or low-fat ice-cream, puddings, variegated ice cream, mousse, frozen yogurt and sorbet (Cruz et al. 2009). These products have demonstrated a great potential as vehicles for probiotic cells, both for the socio-economics reasons underpinning the easy distribution of ice creams along the entire life span and for scientific and technological reasons. In fact, the ice-cream matrix, which is composed by milk proteins, fat and lactose, among others ingredients, represents a good probiotic micro-environment. Furthermore, ice cream is considered a suitable and supportive environment for acting as a probiotic carrier because of its lower storage temperature, and the strict maintenance of the cold chain and a minor risk of temperature abuse during frozen storage (Cruz et al. 2009). All of these factors lead to a higher viability in the high concentration of probiotics at the time of consumption.

However, since frozen storage temperature significantly affects different characteristics of lactic acid bacteria, as acid development and proteinase activity, a robust analysis aimed at assessing the efficacy of the supplemented probiotics in the small intestine must be undertaken. Indeed, it has been recently demonstrated that same probiotic strains incorporated in frozen food products exert better viability and activity during shelf-life, in comparison to non frozen foods (Heenan et al. 2004). At the same time, the damage to probiotic cells caused by freezing and thawing, as well as mechanical stresses that accompany manufacturing must be taken into account on a case by case basis, in order to avoid lower viability and negative functional effects on the consumer (Ranadheera et al. 2010).

While the incorporation of probiotics in cheeses has a longer history, the development of ice cream under the concept of functional food is a relatively new idea. To date, a limited number of studies have been conducted, and most of the research has been performed in order to bypass the technological hurdles for the incorporation of probiotic bacteria into the ice-cream, in order to lead to satisfactory products, both in terms of palatability and sensorial characteristics, as well as in term of functionality.

Different *Lactobacillus* spp. (*L. johnsonii* La1, *L. rhamnosus* GC) have been evaluated for their ability to easily survive as a probiotic supplement, without influencing the physical properties of the ice-creams (Alamprese et al. 2002, Alamprese et al. 2005). The utilization of well-evaluated probiotic strains, as well as the employment of the most appropriate technological methodologies, has been demonstrated to lead to the production of probiotic ice-creams with a good sensory acceptance (Vardar and Oksuz 2007).

## Micro-biotechnology Aids the Development of Better Probiotic Dairy Products

A dairy probiotic product must include a suitable level of probiotic cells in order to exert a health-promoting activity (usually $10^6$–$10^7$ CFU/g). At the same time, the included probiotics must not produce an unfavourable flavour, in order to gain a wide sensory acceptance by consumers. For example, probiotic bifidobacteria are known to be extensive acetate and lactate producers, which are potentially beneficial short-chain fatty acids but, however they are characterized by a very pungent off-flavour, which might require the concomitant utilization of flavouring agents, as well as additional synthetic molecules or natural products.

The evolution of knowledge in the field of nano- and micro-biotechnology in the food sciences are tremendously impacting the possibilities to promote a greater viability and to aid in avoiding the utilization of flavouring agents. One of the most effective strategies is represented by the addition of microencapsulated cells of probiotic cultures to dairy food products. Nowadays, different food products containing microencapsulated probiotics are marketed in the EU and US market. Mainly, they are dairy products, such as ice cream, fermented milk or chocolate snacks (Burgain et al. 2011).

Microencapsulation is the envelopment of small solid particles, including bacterial cells, in a coating. In the recent years, the incorporation of natural ingredients, polyphenols, volatile additives, and bacteria in a small capsule, in order to create a preserved and stable environment, protected from any potential exogenous factor influencing the health effects of the encapsulated agent.

Whilst allowing the protection of these bacteria during their transit in the upper gastrointestinal tract, to entrap probiotic microorganisms in a microcapsule is further useful in order to prevent interfacial inactivation, stimulation of production and excretion of secondary metabolites and their continuous utilization. Microparticles used in the dairy industry are water-insoluble, in order to maintain their structural integrity and are thus produced in order to allow progressive liberation of the probiotic cells during their presence in the large intestine (Ding and Shah 2007, Nazzaro et al. 2012). The most commonly used polymers used in the food industry to produce probiotic microcapsules—characterized by being GRAS, biocompatible and deriving by natural products—are chitosan, alginate, carrageenan, whey protein, pectin, poly-L-lysine and starch (Nazzaro et al. 2012).

## Evaluation of the Probiotics to be Included in Dairy Products: Safety and Functional Assessments

The exertion of widely acknowledged probiotics characteristics is pivotal for promoting the health characteristics ascribable to the plethora of dairy products to date which have been commercialised for restoring a number of conditions described previously in this chapter.

In order to be beneficial to human health, probiotics must fulfil several criteria. The joint Food and Agriculture Organisation of the United Nations/ World Health Organisation Expert Consultation established guidelines for the Evaluation of Health and Nutritional Properties of Probiotics in Food (Food and Agriculture Organisation of the United Nations/World Health Organisation 2001), with the aim of identifying and establishing the minimum requirements needed for the definition of probiotics (Fig. 1).

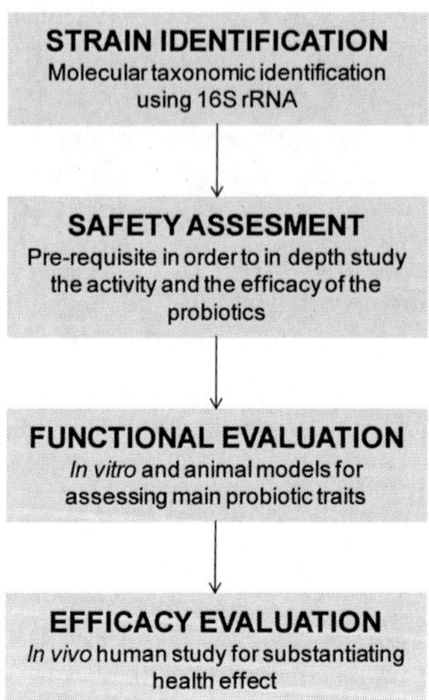


**Figure 1.** Scheme of the guidelines for the Evaluation of Health and Nutritional Properties of Probiotics in Food (Food and Agriculture Organisation of the United Nations/World Health Organisation 2001) (After: Collado et al. 2009).

### Strain Identification

A probiotic products correct identification must be fully qualified in relation to the incorporated bacterial species. Molecular taxonomy tools, based on the 16S rRNA sequencing, must be used in order to ensure the most reliable and univocal identification of the probiotic strain, which will be futher characterized for its safety and functional activities. Indeed, strain identification is of crucial importance in order to link a probiotic strain to a specific health effect (Gueimonde and Salminen 2006, Collado et al. 2009). Whilst the importance of correct taxonomic identification is an established concept, described and taken into account by European QPS (Qualified Presumption of Safety) strategy, the source of the probiotic strain has lately been debated. Indeed, even if one of the main recommendations is that probiotic strains have to be autochthonous of the ecosystem where they will be a part once ingested (Kõll et al. 2010), the probiotic potential of bacteria from other ecosystems would not be a priori to be excluded. For this reason, the number of studies on fermented foods (e.g., sourdoughs, cheeses, yoghurt and pickled vegetables) as alternative sources of novel probiotic candidates is increasing (Cammarota et al. 2009, Vitali et al. 2012).

### Safety Assessment

This is considered an essential phase in the selection and evaluation of probiotics for their utilization in the dairy industry, since probiotics strains must be safe for human consumption. Commonly utilized lactobacilli and bifidobacteria have a long history of safe use, and they are categorized as "generally recognized as safe" (GRAS). However, for every bacterial strain in use, some evaluation procedures must be undertaken. In particular, the probiotic strains considered as candidates for use need a series of *in vitro* analysis to verify the absence of beta-hemolytic activity and other harmful enzymatic activities, such as beta-glucosidase, N-acetyl-beta-glucosaminidase and beta-glucuronidase. Finally, since the importance of the upsurge of antibiotic resistance bacteria, from a medical perspective, it is mandatory to include the susceptibility to the main antibiotics as a requisite for the safety of a probiotics strain before it is commercialised (Gueimonde and Salminen 2006).

### Functional Activity

#### Tolerance to gastrointestinal conditions

Maintaining proper viability along the entire gastrointestinal tract is important to ensure the optimal functionality of probiotics. After ingestion, probiotic bacteria must be able to overcome the acidic environment of the stomach and not be altered by the bile secretion in the duodenum. For this reason, *in vitro* digestion methods have been employed to evaluate the ability of probiotic strains to survive the passage through the gastrointestinal tract. These experimental approaches are mainly based on screening for acid pH tolerance and bile effects (Gueimonde and Salminen 2006).

### *Adhesion*

The adhesion to the intestinal mucosa is often regarded as a prerequisite for the colonisation of the gut lumen, and is an important probiotic trait related to the ability of the strain to cross-talk with the immune system exerting a beneficial immune modulation. Different *in vitro* cell models have been developed for evaluating the adhesion of probiotic strains, mainly based on the utilization of HT-29 and Caco2 cells. Adhesion is one of the most strain-dependent physiological characteristics, since the adhesion levels of the probiotic strains show a great variability amongst genus and species (Candela et al. 2008, Collado et al. 2005, Maccaferri et al. 2012).

### *Antimicrobial substances*

The capability of probiotic bacteria to produce antimicrobial metabolites is of immense interest for enhancing the ecological characteristics of the environment in which the probiotic is meant to exert its activity. Two groups of antimicrobial or bacteriostatic substances have been described so far: i) low molecular mass compounds, as organic acids, which are characterized by a broad spectrum of action; ii) antimicrobial proteins, as bacteriocins, which have an higher molecular mass and which have a relatively higher specificity against groups of microorganisms (Collado et al. 2009, Chen and Hoover 2003). The utilization of probiotics- or LAB-produced bacteriocins is of particular interest for the food and dairy industry. In fact, bacteriocins may be considered natural preservatives used to antagonize the growth of undesired microorganisms in foods to enhance food safety and to extend the product shelf life (Chen and Hoover 2003, Schillinger and Holzapfel 1996).

### *Immune modulation*

Probiotics can interact with mucosa-associated lymphoid tissues and bind to epithelial surface receptors, inducing humoral and cellular immune responses. The establishment and maintenance of a well-balanced ratio between pro- and anti-inflammatory cytokines are crucial for human health. Therefore, study of the dynamic cytokine modulation elicited by a microorganism is a hot topic in the selection of novel probiotic strains. A wide strain-specific variation in the immune responses stimulated by probiotics has been described, and several *in vitro* cell models have been developed to evaluate their immunomodulatory effects (Delcenserie et al. 2008). Even if these cellular models lack the complexity of the human immune system, they aid in clarifying the mechanisms involved in different means of bacterial sensing by human colonocytes and immunocompetent cells (Boirivant and Strober 2007). Several *in vitro* and *in vivo* studies demonstrated the two main effects of probiotics on host immunity: (i) strengthening of the immunological barrier by stimulating the development and maintaining the state of alert of the innate and adaptive immune system, and (ii) decreasing of the immune responsiveness to unbalanced inflammatory conditions. Both of these health-promoting activities are accomplished through an effective modulation of the balance of pro- and anti-inflammatory cytokine production (Vanderpool et al. 2008). Many probiotic species have been demonstrated to share a relatively common immune pattern, such as a reduction in Th2 cytokines (i.e., IL-4, IL-5, IL-6, IL-10, and IL-13) or a shift toward Th1-mediated immunity (i.e., IL-2, TNF-$\alpha$, and IFN-$\gamma$ production).

## The Health-Promoting Probiotic Activities: Facts, Trends and Scientific Substantiation

Increasing evidence is supporting the importance of the role of diet and nutritional status among the most important modifiable determinants of human health, through a plethora of presumptive mechanisms among which intestinal microbiota-mediated processes are thought to be essential.

The concept of functional foods, which provide additional health benefits and may reduce the risk of disease and/or promote general well-being, is steadily growing (Granato et al. 2010). Among the several functional foods marketed worldwide, probiotics represent the most widely established and studied. Therefore, it is not surprising that the dairy sector, which is strongly linked to probiotics, holds the largest share in the functional food market, accounting for nearly one third of the market. It has been reported that the consumer market for probiotics foods is > 1.4 billion Euros in Western Europe (Saxelin 2008).

The majority of probiotic products currently marketed contain species of *Lactobacillus* and *Bifidobacterium*, which are the main bacterial genera characterized as probiotics (Wassenaar and Klein 2008), accordingly with the definition of Food and Agriculture Organisation of the United Nations/ World Health Organisation . Probiotics are defined as "live microorganisms which when administered in adequate amounts confer a health benefit on the host" (Food and Agriculture Organisation of the United Nations/World Health Organisation 2001). To be considered as probiotics, microorganisms should fulfill the following criteria: i) being non-pathogenic and non-toxic; ii) being able to survive through the GIT; iii) being stable during the intended product shelf life and contain an adequate number of viable cells to confer health benefit to the host.

Besides the most established bifidobacteria and lactobacilli, other lactic acid bacteria (LAB) with probiotic properties are *Enterococcus faecalis, E. faecium, Sporolactobacillus inulinus*. Similarly, non-LAB probiotics or nonlactic microorganisms with putative probiotic traits have been described in literature, as *Propionibacterium freudenrichii, Saccharomyces cerevisiae, S. boulardii, Kluyveromyces marxianus, Lactococcus lactis* and *Lecuonostoc mesenteroides*. Table 2 reports the most commonly studied probiotic strains in humans *in vivo* studies for the characterization of their health-promoting activities (Boyle and Tang 2006, Senok et al. 2005, Shah 2007).

Conversely, *L. delbrueckii* spp. *bulgaricus* and *S. thermophilus* are found in a number of preparations with presumptive health-promoting activities, such as traditional yogurt, but since these microorganisms are not expected to survive and grow in the host's intestinal tract, they are not commonly classified as probiotics (Senok et al. 2005).

To date, probiotics microorganisms are mainly incorporated into dairy products, such as cheese, yoghurt, ice cream and dairy desserts, in order to be included as part of the normal Western diet. Generally, high concentrations of viable microorganisms (about $10^7$–$10^9$ CFU/g) are included into alimentary probiotics formulation and are required in order to provoke the desired health-promoting effect (Ranadheera et al. 2010).

Probiotics have been demonstrated to exert health promoting effects through several proposed mechanisms (Fig. 2), which rely on microbe-gut epithelium, microbe-immune system and microbe-microbe interactions.

**Table 2.** Most commonly used probiotic species.

| ***Lactobacillus* spp.** | ***Bifidobacterium* spp.** | **Other bacterial genera** | **Probiotic yeasts** |
|---|---|---|---|
| *L. acidophilus* | *B. bifidum* | *E. coli* Nissle | *S. cerevisiae* |
| *L. casei* | *B. breve* | *E. francium* | *S. boulardii* |
| *L. crispatus* | *B. infantis* | *E. faecium* | *K. marxianus* |
| *L. fermentum* | *B. lactis* | *E. faecalis* | |
| *L. gasseri* | *B. longum* | *P. freudenrichii* | |
| *L. johnsonii* | *B. adolescentis* | *L. lactis* | |
| *L. paracasei* | *B. essensis* | *L. mesenteroides* | |
| *L. plantarum* | | *P. acidilactici* | |
| *L. reuteri* | | | |
| *L. rhamnosus* | | | |
| *L. helveticus* | | | |

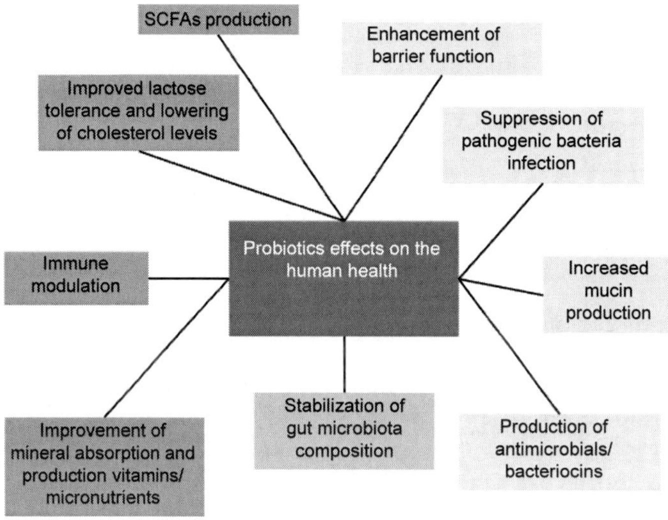


**Figure 2.** Effects of probiotics on the human health. Microbe-microbe interactions are represented in light blue, microbe-host interactions are represented in orange; interactions leading to effects directed either to host and microbes are represented in pink.

*Color image of this figure appears in the color plate section at the end of the book.*

These mechanisms include: i) SCFAs production and enhancement of the barrier function of the intestinal epithelium; ii) suppression of growth and binding of pathogenic bacteria; iii) increased mucin production; iv) induction of antimicrobial and heat-shock protein production; v) alteration of the immune activity of the host through modulation of host signaling pathways; vi) improvement in absorption of minerals and production of vitamins/micronutrients; vii) reduction of cholesterol and improvement of lactose tolerance (Aragon et al. 2010, De Vrese et al. 2001, Ventura et al. 2009, Kumar et al. 2011, Thomas and Versalovic 2010). Furthermore, probiotics can alter colonic fermentation and stabilize the symbiotic microbiota (Spiller 2008), improving the dynamic interplay between the resident bacterial community and the host.

### *Application of Probiotics to Promote Human Health*

Many studies indicate probiotics as promising in the treatment of irritable bowel syndrome, allergies and maintenance of remission in inflammatory bowel diseases (Floch et al. 2011, Preidis and Versalovic 2009).

### *Probiotics in Irritable Bowel Syndrome*

Irritable bowel syndrome (IBS) is a very common functional gastrointestinal disorder defined by the coexistence of abdominal discomfort or pain associated with alterations in bowel habits. Nowadays, the estimated prevalence of IBS in industrialised countries is of 10–15% and, despite its prevalence and impact on quality of life, few therapies have been found to be effective for treating IBS. Different studies indicated that the aetiology of IBS is most likely multifactorial, due to abnormalities in intestinal motility, visceral hypersensitivity, altered brain-gut interaction, food intolerance, unbalanced gut microbiota composition, and persistence of chronic low-grade inflammatory conditions (Brenner et al. 2009).

Due to the effects in modulating the immune function, motility, secretion and gut sensation, probiotics have been suggested to have the potential to exert a beneficial role in managing IBS symptoms (Camilleri 2008). Furthermore, the utilization of novel "omics" approaches, as metabolomics, suggested that IBS patients could be characterized by a potential dysregulation in energy homoestasis and liver function, resulting in unbalanced levels of serum glucose and tyrosine, which may be improved through probiotics supplementation (Hong et al. 2011).

Recently, Clarke et al. 2012 performed an extensive review of relevant literature describing the clinical trials for the assessment of the efficacy of probiotics based containing LAB in the management of IBS symptoms; 42 clinical trials have to date been reported in literature. In general, clinical trials involving LAB have reported improvement in abdominal pain, discomfort, abdominal bloating and distension as their main endpoints. Furthermore, beside these more robust endpoints, benefits over placebo have been also indicated in several clinical trials using global quality of life indices.

The most convincing evidence supporting the beneficial role of bifidobacteria in IBS management came from trials involving *B. bifidum* 3562, which have been demonstrated as effective in reducing abdominal pain/discomfort and bloating, as well as an improved composite severity score both in two trials (O'Mahony et al. 2005, Whorwell et al. 2006). *B. bifidum* 3562 has recently been demonstrated to promote immunoregulatory responses (Koniecza et al. 2012) and to induce an increase in plasma levels of tryptophan, a precursor of serotonin which is known to play a role as neurotransmitter in the brain-gut axis (Desbonnet et al. 2008). Another *Bifidobacterium* strain, *B. bifidum* MIMb75 have been demonstrated to significantly reduce pain/discomfort, bloating and global IBS symptoms (Guglielmetti et al. 2011).

While few studies investigated bifidobacteria, a larger number of clinical trials aimed at evaluating the efficacy of *Lactobacillus* strains in IBS. To date, *L. acidophilus* SDC 2012, 2013 (Sinn et al. 2008), *L. paracasei*

B2106 (Andriulli et al. 2008) and *L. plantarum* 299V (Niedzieilin et al. 2001) have been demonstrated to reduce abdominal pain/discomfort and *L. plantarum* 299V was further demonstrated to improve the overall IBS symptoms (Niedzieilin et al. 2001). A broad number of studies investigated the mechanistic insight at the basis of the probiotic activity of *Lactobacillus* strains. Notably, it has been demonstrated that *L. acidophilus* can produce visceral analgesic effects by up-regulating the expression of opioid and cannabinoid receptors in colonic epithelial cell line and *in vivo* in murine models (Rousseaux et al. 2007). *Lactobacillus paracasei* was shown to attenuate gut muscle hypercontractility in animal models of post-infectious IBS (Verdu et al. 2008). Furthermore, certain *Lactobacillus* species and strains affected epithelial integrity of colonic epithelial cell monolayers as measured by trans-epithelial electrical resistance (Parassol et al. 2005).

In addition to monostrain probiotic formulations, multispecies probiotics have been evaluated for their efficacy in ameliorating from IBS symptoms. Indeed, it has been recently indicated that probiotic mixtures appear to show greater efficacy than single strains, including strains that are components of the mixtures themselves (Chapman et al. 2011). However, it is still unclear whether this difference is due to the synergistic interactions between strains.

One of the best characterized multispecies probiotics is VSL#3, a mixture containing strains of three lyophilised species, *Bifidobacterium*, *Lactobacillus* and *Streptococcus*, which was found to be effective at alleviating abdominal pain and reducing bloating in adults and children suffering from IBS, as well as in improving their quality of life (Guandalini et al. 2010, Kim et al. 2003, 2005, Michail and Kenche 2011). Kajander et al. (2005) demonstrated that a probiotic mixture containing *L. rhamnosus* GG, *L. rhamnosus* LC705, *B. breve* 99 and *P. freudenreichii* ssp. *shermani* JS was effective in improving a composite IBS symptom score, which included abdominal pain, distension and flatulence.

Dairy probiotics other than bacterial strains have also been suggested to be useful in IBS management, even if they are not yet widely in use. *S. boulardii*, a species of yeast which has been described as a biotherapeutic agent able to decrease the expression of inflammation-associated cytokine IL-8, IL-6, IL-1b, TNF-α and IFN-γ (Zanello et al. 2009), improved the quality of life of IBS patients better than a placebo but was not a superior substitute for individual IBS (Choi et al. 2011). Similarly, *K. marxianus* B0399, which has very recently been demonstrated to be a probiotic strain with immune-modulatory activity and able to impact on the composition and functional activity of the human gut microbiota (Maccaferri et al. 2012), in combination with *B. animalis* subsp. *lactis* Bb12 was demonstrated to be effective in reducing abdominal pain and bowel movements abnormalities in IBS patients (Lisotti et al. 2011).

### *Probiotics in Inflammatory Bowel Disease*

Inflammatory bowel disease (IBD) is a complex and heterogeneous group of pathological conditions that involves an interaction between genetic, immunologic and environmental factors. An altered composition of the gut microbiota, defective clearance of bacteria and enhanced mucosal uptake, resulting in increased immune stimulation are characterizing of IBD (Packey and Sartor 2009, Ewaschuk et al. 2006). From a clinical perspective, IBD includes ulcerative colitis (UC) and Crohn's disease (CD). Whilst corticosteroids, mesalazine and antibiotics are the most commonly used therapeutic approaches to treat IBD, different probiotic trials are substantiating the potential beneficial role of probiotics in UC, CD and pouchitis. Notably, a broad array of studies have been performed in order to indicate the role of probiotics in alleviating particular symptoms of each of the three pathological conditions (Haller 2010, Ewaschuk et al. 2006).

Probiotic microorganisms used in clinical trials aimed at inducing or maintaining remission of CD included *S. boulardii* (Guslandi et al. 2000), *L. johnsonii* LA1 (Marteau et al. 2006, Van Gossum et al. 2007) and *L. rhamnosus* GG (Prantera et al. 2002). Guslandi et al. (2000) demonstrated that *S. boulardii* was more effective than mesalazine in maintaining clinical remission of CD. Successively, Vilela et al. (2008) demonstrated that *S. boulardii* was effective acting through the improvement of the CD-like abnormal intestinal barrier function. McCarthy et al. (2001) demonstrated that oral administration of *L. salivarius* UCC118 was decreasing disease activity in a cohort of patients suffering from mild to moderately active CD. Conversely, clinical trials studying the efficacy of *L. rhamnosus* GG and *L. johnsonii* LA1 did not find differences among probiotics treatment and placebo in maintaining clinical or endoscopic remission of CD.

Unlike the paucity of data and human studies in CD, a larger number of trials reported the effects of probiotics treatments in UC. In particular, two large studies demonstrated that *E. coli* Nissle 1917 was equivalent to mesalazine in maintaining remission in patients with UC (Rembacken et al. 1999, Kruis et al. 2004). The potential clinical relevance of *E. coli* Nissle 1917 was first demonstrated in several animal models in which probiotics resulted in a significant reduction in the secretion of pro-inflammatory cytokines and other markers of intestinal inflammation as well as improved histological findings (Kamada et al. 2005). Furthermore, *E. coli* Nissle 1917 was demonstrated to strengthen the mucosal barrier through the induction of epithelial B-defensin 2, a human antimicrobial peptide, and zonula occudens-2, a prominent member of the tight junctions, and to induce a potent anti-inflammatory response (Schlee et al. 2007, Zyrek et al. 2007, Helwig et al. 2006).

The use of LAB as unique probiotic strains did not show relevant results in the management of UC. Zocco et al. (2006) did not show any efficacy of probiotic *L. rhamnosus* GG in maintaining UC remission, whereas Fujimori et al. (2009) demonstrated that the utilization of *B. longum* was effective in improving general IBS symptoms only when administered within a synbiotic formulation.

In addition to a single strain probiotics formulation, multispecies probiotics were investigated and tested in clinical trials. Sood et al. (2009) demonstrated that VSL#3 probiotic mixture was effective in inducing UC remission in adults suffering from mild to moderate UC, whereas Tursi et al. (2010) demonstrated a significant clinical response in the group of UC patients subjected to probiotics therapy. Recently, Miele et al. (2009) further demonstrated that VSL#3 was useful in the maintenance of UC remission. Notably, the clinically active probiotic combination VSL#3 has been demonstrated to exert a potent induction of IL-10 by intestinal and blood dendritic cells and inhibited generation of pro-inflammatory Th1 cells (Hart et al. 2004).

### *Probiotics in Allergies and Atopic Dermatitis*

In the last decades, a rapid rise in the prevalence of allergic and autoimmune disorders has been observed (Bach 2002). In particular, atopic dermatitis (AD) is the most common chronic inflammatory skin disease in infancy. Allergic diseases are generally associated with an imbalance in the $T_H1/T_H2$ cytokine response and with stimulation of IgE and IgA synthesis (Winkler et al. 2007). Since different probiotic strains have been demonstrated to induce a reduction in $T_H2$ cytokines (i.e., IL-4, IL-5, IL-6, IL-10, IL-13) or a shift towards $T_H1$-mediated immunity (i.e., IL-2, TNF-$\alpha$, IFN-$\gamma$ production) (Vanderpool et al. 2008, Thomas and Versalovic 2010), a potential role for probiotics in the management of allergies and AD could be suggested.

Most studies assessing probiotic effects in the treatment of allergic diseases have focused on AD with or without associated food allergies in infants and children. Studies aimed at investigating the role of probiotics in adult-type AD are less frequent. Trials evaluating probiotic efficacy have investigated on either the treatment or prevention of AD and, to date, only a few rigorous randomized controlled studies have been performed.

The majority of studies have evaluated probiotic formulations containing *Lactobacillus* species, alone or in combination with *Bifidobacterium* species. The primary outcome of these trials was the change in the score for the evaluation of AD severity (SCORAD).

First pioneering studies from Majamaa and Isolauri (1997) and Isolauri et al. (2000), performed in small numbers of children who were administered with *L. rhamnosus* GG, demonstrated that probiotic treatment resulted in a rapid significant improvement in SCORAD with respect to placebos. In contrast to these initial studies, more recent and larger trials have failed to confirm beneficial effects of probiotics in the treatment of AD.

Recently, three meta-analyses summarized and evaluated, with some controversial conclusion, the clinical efficacy of probiotics in AD management (Boyle et al. 2008, Lee et al. 2008, Michail et al. 2008). Lee et al. (2008) and concluded that current evidence is more convincing for the efficacy of probiotics in the prevention rather than in the treatment of paediatric AD. Boyle et al. (2008) indicated that probiotics do not appear to be effective for the treatment of AD and that no sufficient evidences are reported to support their use for this condition. Vice versa, Michail et al. (2011) demonstrated that, even if the utilization of probiotics significantly reduced the SCORAD severity index score, it was associated with modest clinical effects and rapidly discontinued after the cessation of probiotic treatment.

Similar controversial findings were provided more recently by other clinical trials. A longitudinal study by Gore et al. (2012) performed administering *L. paracasei* CNCM I-2116 and *B. lactis* CNCM I-3446 to infants with AD aged 3–6 months demonstrated no beneficial effects of probiotics in the treatment of AD. Furthermore, probiotics did not affect the progression of AD from age 1 to 3 years.

Conversely, Drago et al. (2011) demonstrated that *L. salivarius* LS01 provoked a significant improvement in SCORAD index in a double-blind placebo-controlled study on a cohort of 38 adult patients with AD. Similar results were found by Moroi et al. (2011), who administered *L. paracasei* K71 to 34 adult-type AD subjects, demonstrating a significant decrease of skin severity scores after treatment. Whilst not taking into account clinical parameters, Roessler et al. (2012) demonstrated that a probiotic mix containing *L. paracasei* Lpc-37, *L. acidophilus* 74-2 and *B. animalis* subsp. *Lactis* DGCC420 lowers the genotoxic potential of faecal water in adult AD patients.

Taken together, research on the efficacy of probiotics in the management of AD, especially concerning prevention, has to be sustained in order to translate promising mechanistic insights related to their immune modulation and anti-genotoxic activity into meaningful health claims.

Besides the utilization of probiotics to manage AD in infants, children and adults, increasing evidence is supporting the administration of probiotics already during pregnancy and within the first months of life, in order to reduce the risk of developing AD. Doege et al. (2012) performed a meta-analysis on the randomised, double-blind, placebo-controlled trials aimed at assessing the impact of probiotics intake during pregnancy on the development of AD in children. Doege et al. 2012 concluded that probiotics significantly reduce the risk of development of AD, in particular when

*Lactobacillus* strains were administered. Conversely, no significant effects of mixed probiotics formulation including lactobacilli and bifidobacteria were demonstrated.

## References Cited

Alamprese, C., R. Foschino, M. Rossi, C. Pompei and L. Savani. 2002. Survival of Lactobacillus johnsonii La1 and influence of its addition in retail-manufactured ice cream produced with different sugar and fat concentrations. Int. Dairy J. 12: 201–208.

Alamprese, C., R. Foschino, M. Rossi, C. Pompei and S. Corti. 2005. Effects of Lactobacillus rhamnosus GG addition in ice cream. Int. J. Dairy Technol. 58: 200–206.

Andriulli, A., M. Neri, C. Loguercio, N. Terreni, A. Merla, M.P. Cardarella, A. Federico, F. Chilovi, G.L. Milandri, M. De Bona, S. Cavenati, S. Gullini, R. Abbiati, N. Garbagna, R. Cerutti and E. Grossi. 2008. Clinical trial on the efficacy of a new symbiotic formulation, Flortec, in patients with irritable bowel syndrome: a multicenter, randomized study. J. Clin. Gastroenterol. 42 suppl. 3: S218–223.

Aragon, G., D.B. Graham, M. Borum and D.B. Doman. 2010. Probiotic therapy for Irritable Bowel Syndrome. Gastroenterol. Hepatol. 6: 39–44.

Bach, J.F. 2002. The effect of infections on susceptibility to autoimmune and allergic diseases. New Engl. J. Med. 347: 911–920.

Boirivant, M. and W. Strober. 2007. The mechanism of action of probiotics. Curr. Opin. Gastroenterol. 23(6): 679–692.

Boyle, R. J. and M.L.K. Tang. 2006. The role of probiotics in the management of allergic disease. Clin. Exp. Allergy. 36: 568–576.

Boyle, R.J., F.J. Bath-Hextall, J. Leonardi-Bee, D.F. Murrell and M.L.K. Tang. 2008. Probiotics for treating eczema. Cochrane Database of Systematic Reviews (4): CD006135.

Brenner, D.M., M.J. Moeller, W.D. Chey and P.S. Schoenfeld. 2009. The utility of probiotics in the treatment of irritable bowel syndrome: a systematic review. Am. J. Gastroenterol. 104: 1033–1049.

Burgain, J., C. Gaiani, M. Linder and J. Scher. 2011. Encapsulation of probiotic living cells: from laboratory scale to industrial applications. J. Food Eng. 104: 467–483.

Cammarota, M., M. De Rosa, A. Stellavato, M. Lamberti, I. Marzaioli and M. Giuliano. 2009. *In vitro* evaluation of Lactobacillus plantarum DSMZ 12028 as a probiotic: emphasis on innate immunity. Int. J. Food Microbiol. 135: 90–98.

Camilleri, M. 2008. Probiotics and irritable bowel syndrome: rationale, mechanisms, and efficacy. J. Clin. Gastroenterol. 42 Suppl. 3: S123–125.

Candela, M., F. Perna, P. Carnevali, B. Vitali, R. Ciati, R. Gionchetti, F. Rizzello, M. Campieri and P. Brigidi. 2008. Interaction of probiotic Lactobacillus and Bifidobacterium strains with human intestinal epithelial cells: adhesion properties, competition against enteropathogens and modulation of IL-8 production. Int. J. Food Microbiol. 125: 286–292.

Chapman, C.M.C., G.R. Gibson and I. Rowland. 2011. Health benefits of probiotics: are mixtures more effective than single strains? Eur. J. Nutr. 50: 1–17.

Chen, H. and D.G. Hoover. 2003. Bacteriocins and their food applications. Comp. Rev. Food Sci. Food Safety 2: 82–100.

Choi, C.H., S.Y. Jo, H.J. Park, S.K. Chang, J.-S. Byeon and S.-J. Myung. 2011. A randomized, double-blind, placebo-controlled multicenter trial of Saccharomyces boulardii in irritable bowel syndrome: effect on quality of life. J. Clin. Gastroenterol. 45: 679–683.

Clarke, G., J.F. Cryan, T.G. Dinan and E.M. Quigley. 2012. Review article: probiotics for the treatment of irritable bowel syndrome—focus on lactic acid bacteria. Aliment. Pharmacol. Ther. 35(4): 403–413.

Collado, M.C., M. Gueimonde, M. Hernandez, Y. Sanz and S. Salminen. 2005. Adhesion of selected Bifidobacterium strains to human intestinal mucus and the role of adhesion in enteropathogen exclusion. J. Food Prot. 68: 2672–2678.

Collado, M.C., E. Isolauri, S. Salminen and Y. Sanz. 2009. The impact of probiotic on gut health. Curr. Drug Metab. 10: 68–78.

Cruz, A.G., A.E.C. Antunes, A.L.O.P. Sousa, J.A.F. Faria and S.M.I. Saad. 2009. Ice-cream as a probiotic food carrier. Food Res. Internat. 49: 1233–1239.

da Cruz, A.G., E.H.M. Walter, R.S. Cadena, J.A.F. Faria, H.M.A. Bolini and A.M. Frattini Fileti. 2009. Monitoring the authenticity of low-fat yogurts by an artificial neural network. J. Dairy Sci. 92: 4797–4804.

Dave, R.I. and N.P. Shah. 1997. Viability of yoghurt and probiotic bacteria in yoghurts made from commercial starter cultures. Int. Dairy J. 7: 31–41.

Delcenserie V., D. Martel, M. Lamoureux, J. Amiot, Y. Boutin and D. Roy. 2008. Immunomodulatory effects of probiotics in the intestinal tract. Curr. Issues Mol. Biol. 10: 37–54.

de Vrese, M., A. Stegelmann, B. Richter, S. Fenselau, C. Laue and J. Schrezenmeir. 2001. Probiotics—Compensation for lactase insufficiency. Am. J. Clin. Nutr. 73: 421S–429S.

Desbonnet, L., L. Garrett, G. Clarke, J. Bienenstock and T.G. Dinan. 2008. The probiotic Bifidobacteria infantis: An assessment of potential antidepressant properties in the rat. J. Psychiatric Res. 43: 164–174.

Di Criscio, T., A. Fratianni, R. Mignogna, L. Cinquanta, R. Coppola, E. Sorrentino and G. Panfili. 2010. Production of functional probiotic, prebiotic, and synbiotic ice creams. J. Dairy Sci. 93: 4555–4564.

Ding, W. K. and N.P. Shah. 2007. Acid, bile, and heat tolerance of free and microencapsulated probiotic bacteria. J. Food Sci. 72: 446–450.

Doege, K., D. Grajecki, B.-C. Zyriax, E. Detinkina, C. Zu Eulenburg and K.J. Buhling. 2012. Impact of maternal supplementation with probiotics during pregnancy on atopic eczema in childhood-a meta-analysis. Br. J. Nutr. 107: 1–6.

Donkor, O.N., A. Henriksson, T. Vasiljevic and N.P. Shah. 2006. Effect of acidification on the activity of probiotics in yoghurt during cold storage. Int. Dairy J. 16: 1181–1189.

Drago, L., E. Iemoli, V. Rodighiero, L. Nicola, E. De Vecchi and S. Piconi. 2011. Effects of Lactobacillus salivarius LS01 (DSM 22775) treatment on adult atopic dermatitis: a randomized placebo-controlled study. Int. J. Immunopathol. Pharmacol. 24: 1037–1148.

Ewaschuk, J.B., Q.Z. Tejpar, I. Soo, K. Madsen and R.N. Fedorak. 2006. The role of antibiotic and probiotic therapies in current and future management of inflammatory bowel disease. Curr. Gastroenterol. Rep. 8: 486–498.

FAO/WHO: Report on Joint FAO/WHO Expert Consultation on Evaluation of Health and Nutritional Properties of Probiotics in Food Including Powder Milk with Live Lactic Acid Bacteria. 2001, ftp://ftp.fao.org/es/esn/food/probio_report_en.pdf.

Floch, M.H., W.A. Walker, K. Madsen, M.E. Sanders, G.T. Macfarlane, H.J. Flint, L. Dieleman, Y. Ringel, S. Guandalini, C.P. Kelly and L.J. Brandt. 2011. Recommendations for probiotic use-2011 update. J. Clin. Gasteonterol. 45: S168–171.

Food and Agriculture Organisation of the United Nations/World Health Organisation. Report on Joint FAO/WHO Expert Consultation on Evaluation of Health and Nutritional Properties of Probiotics in Food Including Powder Milk with Live Lactic Acid Bacteria. 2001. ftp://ftp.fao.org/es/esn/food/probio_report_en.pdf

Fujimori, S., K. Gudis, K. Mitsui, T. Seo, M. Yonezawa, S. Tanaka, A. Tatsuguchi and C. Sakamoto. 2009. A randomized controlled trial on the efficacy of synbiotic versus probiotic or prebiotic treatment to improve the quality of life in patients with ulcerative colitis. Nutr. 25: 520–525.

Gore, C., A. Custovic, G.W. Tannock, K. Munro, G. Kerry, K. Johnson, C. Peterson, J. Morris, C. Chaloner, C.S. Murray and A. Woodcock. 2012. Treatment and secondary prevention effects of the probiotics Lactobacillus paracasei or Bifidobacterium lactis on early infant eczema: randomized controlled trial with follow-up until age 3 years. Clin. Exp. Immunol. 42: 112–122.

Granato, D., G.F. Branco, A.G. Cruz, J. de A.F. Faria and N.P. Shah. 2010. Probiotic Dairy Products as Functional Foods. Comp. Rev. Food Sci. Food Safety 9: 455–470.

Guandalini, S., G. Magazzù, A. Chiaro, V. La Balestra, G. Di Nardo, S. Gopalan, A. Sibal, C. Romano, R.B. Canani, P. Lionetti and M. Setty. 2010. VSL#3 improves symptoms in children with irritable bowel syndrome: a multicenter, randomized, placebo-controlled, double-blind, crossover study. J. Pediatr. Gastroenterol. Nutr. 51: 24–30.

Gueimonde, M. and S. Salminen. 2006. New methods for selecting and evaluating probiotics. Dig. Liver Dis. 38 Suppl 2: S242–247.

Guglielmetti, S., D. Mora, M. Gschwender and K. Popp. 2011. Randomised clinical trial: Bifidobacterium bifidum MIMBb75 significantly alleviates irritable bowel syndrome and improves quality of life—a double-blind, placebo-controlled study. Aliment. Pharmacol. Ther. 33: 1123–1132.

Guslandi, M., G. Mezzi, M. Sorghi and P.A. Testoni. 2000. Saccharomyces boulardii in maintenance treatment of Crohn's disease. Dig. Dis. Sci. 45: 1462–1464.

Haller, D. 2010. Nutrigenomics and IBD: the intestinal microbiota at the cross-road between inflammation and metabolism. J. Clin. Gastroenterol. 44 Suppl. 1: S6–9.

Hart, A.L., K. Lammers, P. Brigidi, B. Vitali, F. Rizzello, P. Gionchetti, M. Campieri, M.A. Kamm, S.C. Knight and A.J. Stagg. 2004. Modulation of human dendritic cell phenotype and function by probiotic bacteria. Gut. 53: 1602–1609.

Heenan, C.N., M.C. Adams, R.W. Hosken and G.H. Fleet. 2004. Survival and sensory acceptability of probiotic microorganisms in a nonfermented frozen vegetarian dessert. Food Sci. Technol. 37: 461–466.

Hekmat, S. and G. Reid. 2006. Sensory properties of probiotic yogurt is comparable to standard yogurt. Nutr. Res. 26: 163–166.

Helwig, U., K.M. Lammers, F. Rizzello, P. Brigidi, V. Rohleder, E. Caramelli, P. Gionchetti, J. Schrezenmeir, U.R. Foelsch, S. Schreiber and M. Campieri. 2006. Lactobacilli, bifidobacteria and E. coli nissle induce pro- and anti-inflammatory cytokines in peripheral blood mononuclear cells. World J. Gastroenterol. 12: 5978–5986.

Hong Y.-S., K.S. Hong, M.-H. Park, Y.-T., Ahn, J.-H. Lee, C.-S. Huh, J. Lee, I.-K. Kim, G.-S. Hwang and J.S. Kim. 2011. Metabonomic understanding of probiotic effects in humans with irritable bowel syndrome. J. Clin. Gastroenterol. 45: 415–25.

Isolauri, E. 2000. The use of probiotics in paediatrics. Hosp. Med. 61: 6–7.

Kailasapathy, K., I. Harmstorf and M. Phillips. 2008. Survival of Lactobacillus acidophilus and Bifidobacterium animalis ssp. lactis in stirred fruit yogurts. Food Sci. Technol. 41: 1317–1322.

Kajander, K., K. Hatakka, T. Poussa, M. Färkkilä and R. Korpela. 2005. A probiotic mixture alleviates symptoms in irritable bowel syndrome patients: a controlled 6-month intervention. Aliment. Pharmacol. Ther. 22: 387–394.

Kamada, N., N. Inoue, T. Hisamatsu, S. Okamoto, K. Matsuoka, T. Sato, H. Chinen, K.S. Hong, T. Yamada, Y. Suzuki, T. Suzuki, N. Watanabe, K. Tsuchimoto and T. Hibi. 2005. Non-pathogenic Escherichia coli strain Nissle1917 prevents murine acute and chronic colitis. Inflamm. Bow. Dis. 11: 455–463.

Kim, H.J., M. Camilleri, S. McKinzie, M.B. Lempke, D.D. Burton, G.M. Thomforde and A.R. Zinsmeister. 2003. A randomized controlled trial of a probiotic, VSL#3, on gut transit and symptoms in diarrhoea-predominant irritable bowel syndrome. Aliment. Pharmacol. Ther. 17: 895–904.

Kim, H.J., M.I. Vazquez Roque, M. Camilleri, D. Stephens, D.D. Burton, K. Baxter and G. homforde. 2005. A randomized controlled trial of a probiotic combination VSL#3 and placebo in irritable bowel syndrome with bloating. Neurogastroenterol. Motil. 17(5): 687–696.

Koll, O., S. von Wallpach and M. Kreuzer. 2010. Multi-Method research on consumer—brand associations: Comparing free associations, storytelling, and collages. Psychol. Market. 27(6): 584–602.

Konieczna, P., D. Groeger, M. Ziegler, R. Frei, R. Ferstl, F. Shanahan and E.M.M. Quigley. 2012. Bifidobacterium infantis 35624 administration induces Foxp3 T regulatory cells in human peripheral blood: potential role for myeloid and plasmacytoid dendritic cells. Gut. 61(3): 354–366.

Kruis, W., P. Fric, J. Pokrotnieks, M. Lukás, B. Fixa, M. Kascák and M.A. Kamm. 2004. Maintaining remission of ulcerative colitis with the probiotic Escherichia coli Nissle 1917 is as effective as with standard mesalazine. Gut. 53(11): 1617–1623.

Kumar, R., S. Grover and V.K. Batish. 2011. Hypocholesterolaemic effect of dietary inclusion of two putative probiotic bile salt hydrolase-producing Lactobacillus plantarum strains in Sprague-Dawley rats. Br. J. Nutr. 105(4): 561–573.

Lee, J., D. Seto and L. Bielory. 2008. Meta-analysis of clinical trials of probiotics for prevention and treatment of pediatric atopic dermatitis. J. Allergy Clin. Immunol. 121(1).

Lisotti, A., G.L. Cornia, A.M. Morselli-Labate, A. Sartini, L. Turco and V. Grasso. 2011. Effects of a fermented milk containing Kluyveromyces marxianus B0399 and Bifidobacterium lactis Bb12 in patients with irritable bowel syndrome. Minerva Gastroenterol. Dietol. 57: 1–12.

Maccaferri, S., A. Klinder, P. Brigidi, P. Cavina and A. Costabile. 2012. Potential Probiotic Kluyveromyces marxianus B0399 Modulates the immune response in caco-2 cells and peripheral blood mononuclear cells and impacts the human gut microbiota in an *in vitro* colonic model system. Appl. Environment. Microbiol. 78: 956–964.

Majamaa, H. and E. Isolauri. 1997. Probiotics: a novel approach in the management of food allergy. J. Allergy Clin. Immunol. 99(2): 179–185.

Marteau, P., M. Lémann, P. Seksik, D. Laharie, J.F. Colombel, Y. Bouhnik and G. Cadiot. 2006. Ineffectiveness of Lactobacillus johnsonii LA1 for prophylaxis of postoperative recurrence in Crohn's disease: a randomised, double blind, placebo controlled GETAID trial. Gut. 5(6): 842–847.

McCarthy, J., L. O'Mahony, C. Dunne, P. Kelly, M. Feeney, B. Kiely, G. O'Sullivan, J.K. Collins and F. Shanahan. 2001. An open trial of a novel probiotic as an alternative to steroids in mild/moderately active Crohn's disease. Gut. 49: A2447.

Michail, S.K., A. Stolfi, T. Johnson and G.M. Onady. 2008. Efficacy of probiotics in the treatment of pediatric atopic dermatitis: a meta-analysis of randomized controlled trials. Ann. Allergy Asthma Immunol. 101(5): 508–516.

Michail, S. and H. Kenche. 2011. Gut microbiota is not modified by Randomized, Double-blind, Placebo-controlled Trial of VSL#3 in Diarrhea-predominant Irritable Bowel Syndrome. Probiotics Antimicrobial Proteins. (1): 1–7.

Miele, E., F. Pascarella, E. Giannetti, L. Quaglietta, R.N. Baldassano and A. Staiano. 2009. Effect of a probiotic preparation (VSL#3) on induction and maintenance of remission in children with ulcerative colitis. Am. J. Gastroenterol. 104(2): 437–443.

Moroi, M., S. Uchi, K. Nakamura, S. Sato, N. Shimizu, M. Fujii and T. Kumagai. 2011. Beneficial effect of a diet containing heat-killed Lactobacillus paracasei K71 on adult type atopic dermatitis. J. Dermatol. 38(2): 131–139.

Nazzaro, F., F. Fratianni, P. Orlando and R. Coppola. 2012. Biochemical traits, survival and biological properties of the probiotic Lactobacillus plantarum grown in the presence of prebiotic inulin and pectin as energy source. Pharmaceuticals 5(5): 481–492.

Niedzielin, K., H. Kordecki and B. Birkenfeld. 2001. A controlled, double-blind, randomized study on the efficacy of Lactobacillus plantarum 299V in patients with irritable bowel syndrome. Eur. J. Gastroenterol. Hepatol. 13(10): 1143–1147.

O'Mahony, L., J. McCarthy, P. Kelly, G. Hurley, F. Luo, K. Chen and G.C. O'Sullivan. 2005. Lactobacillus and bifidobacterium in irritable bowel syndrome: symptom responses and relationship to cytokine profiles. Gastroenterology 128(3): 541–551.

Packey, C.D. and R.B. Sartor. 2009. Commensal bacteria, traditional and opportunistic pathogens, dysbiosis and bacterial killing in inflammatory bowel diseases. Curr. Opin. Infect. Dis. 22(3): 292–301.

Parassol, N., M. Freitas, K. Thoreux, G. Dalmasso, R. Bourdet-Sicard and P. Rampal. 2005. Lactobacillus casei DN-114 001 inhibits the increase in paracellular permeability of enteropathogenic Escherichia coli-infected T84 cells. Res. Microbiol. 156(2): 256–262.

Plessas, S., L. Bosnea, A. Alexopoulos and E. Bezirtzoglou. 2012. Potential effects of probiotics in cheese and yogurt production: A review. Eng. Lif. Sci. 12(4): 433–440.

Prantera, C., M.L. Scribano, G. Falasco, A. Andreoli and C. Luzi. 2002. Ineffectiveness of probiotics in preventing recurrence after curative resection for Crohn's disease: a randomised controlled trial with Lactobacillus GG. Gut. 51(3): 405–409.

Preidis, G.A. and J. Versalovic. 2009. Targeting the human microbiome with antibiotics, probiotics, and prebiotics: Gastroenterology Enters the Metagenomics Era. Gastroenterology 136: 2015–2031.

Ranadheera, R.D.C.S., S.K. Baines and M.C. Adams. 2010. Importance of food in probiotic efficacy. Food Res. Internat. 43(1): 1–7.

Rembacken, B.J., A.M. Snelling, P.M. Hawkey, D.M. Chalmers and A.T. Axon. 1999. Non-pathogenic Escherichia coli versus mesalazine for the treatment of ulcerative colitis: a randomised trial. Lancet. 354(9179): 635–639.

Roessler, A., S.D. Forssten, M. Glei, A.C. Ouwehand and G. Jahreis. 2012. The effect of probiotics on faecal microbiota and genotoxic activity of faecal water in patients with atopic dermatitis: a randomized, placebo-controlled study. Clin. Nutr. 31(1): 22–29.

Rousseaux, C., X. Thuru, A. Gelot, N. Barnich, C. Neut, L. Dubuquoy, C. Dubuquoy et al. 2007. Lactobacillus acidophilus modulates intestinal pain and induces opioid and cannabinoid receptors. Nat. Med. 13(1): 35–37.

Saxelin, M. 2008. Probiotic formulations and applications, the current probiotics market, and changes in the marketplace: a European perspective. Clin. Infect. Dis. 46(Suppl. 2): S76–79.

Schillinger, U. and W.H. Holzapfel. 1996. Guidelines for manuscripts on bacteriocins of lactic acid bacteria. Int. J. Food Microbiol. 33: 3–4.

Schlee, M., J. Wehkamp, A. Altenhoefer, T.A. Oelschlaeger, E.F. Stange and K. Fellermann. 2007. Induction of human beta-defensin 2 by the probiotic Escherichia coli Nissle 1917 is mediated through flagellin. Infection and Immunity 75(5): 2399–2407.

Senok, A.C., A.Y. Ismaeel and G.A. Botta. 2005. Probiotics: facts and myths. Clin. Microbiol. Infect. 11(12): 958–966.

Shah, S. 2007. Dietary factors in the modulation of inflammatory bowel disease activity. Med. Gen. Med. 9(1): 60.

Sinn, D.H., J.H. Song, H.J. Kim, J.H. Lee, H.J. Son, D.K. Chang and Y.-H. Kim. 2008. Therapeutic effect of Lactobacillus acidophilus-SDC 2012, 2013 in patients with irritable bowel syndrome. Dig. Dis. Sci. 53(10): 2714–2718.

Sood, A., V. Midha, G.K. Makharia, V. Ahuja, D. Singal, P. Goswami and R.K. Tandon. 2009. The probiotic preparation, VSL#3 induces remission in patients with mild-to-moderately active ulcerative colitis. Clin. Gastroenterol. Hepatol. 7(11): 1202–1209.

Spiller, R. 2008. Review article: probiotics and prebiotics in irritable bowel syndrome. Aliment. Pharmacol. Ther. 28(4): 385–396.

Stanton C., G. Gardiner, P.B. Lynch, J.K. Collins, G. Fitzgerald and R.P. Ross. 1998. Probiotic Cheese. Int. Dairy J. 8(5): 6.

Talwalkar, A. and K. Kailasapathy. 2004. The role of oxygen in the viability of probiotic bacteria with reference to L. acidophilus and Bifidobacterium spp. Curr. Iss. Intest. Microbiol. 5(1): 1–8.

Thomas, C.M. and J. Versalovic. 2010. Probiotics-host communication: Modulation of signaling pathways in the intestine. Gut Microbes 1(3): 148–163.

Tursi, A., G. Brandimarte, A. Papa, A. Giglio, W. Elisei, G.M. Giorgetti and G. Forti. 2010. Treatment of relapsing mild-to-moderate ulcerative colitis with the probiotic VSL#3 as adjunctive to a standard pharmaceutical treatment: a double-blind, randomized, placebo-controlled study. Am. J. Gastroenterol. 105(10).

Van Gossum, A., O. Dewit, E. Louis, G. de Hertogh, F. Baert, F. Fontaine and M. DeVos. 2007. Multicenter randomized-controlled clinical trial of probiotics (Lactobacillus johnsonii, LA1) on early endoscopic recurrence of Crohn's disease after ileo-caecal resection. Inflamm. Bowel. Dis. 13(2): 135–142.

Vanderpool, C., F. Yan and D.B. Polk. 2008. Mechanisms of probiotic action: Implications for therapeutic applications in inflammatory bowel diseases. Inflamm. Bowel. Dis. 14(11): 1585–1596.

Vardar, N.B. and ve Ö. Öksüz. 2007. Artisan strawberry ice-cream made with supplementation of Lactococci or Lactobacillus acidophilus. Italian J. Food Sci. 19(4): 403–411.

Ventura, M., S. O'Flaherty, M.J. Claesson, F. Turroni, T.R. Klaenhammer, D. van Sinderen and P.W. O'Toole. 2009. Genome-scale analyses of health-promoting bacteria: probiogenomics. Nat Rev Microbiol. 7(1): 61–71.

Verdu, E.F., P. Bercik, X.X. Huang, J. Lu, N. Al-Mutawaly, H. Sakai and T.A. Tompkins. 2008. The role of luminal factors in the recovery of gastric function and behavioral changes after chronic Helicobacter pylori infection. Am. J. Physiol. Gastrointest. Liver. Physiol. 295(4): G664–70.

Vilela, G.E., M. De Lourdes De Abreu Ferrari, H. Oswaldo Da Gama Torres, A. Guerra Pinto, A. Carolina Carneiro Aguirre, F. Paiva Martins, E. Marcos Andrade Goulart et al. 2008. Influence of Saccharomyces boulardii on the intestinal permeability of patients with Crohn's disease in remission. Scand. J. Gastroenterol. 43(7): 842–848.

Vitali, B., G. Minervini, C.G. Rizzello, E. Spisni, S. Maccaferri, P. Brigidi, M. Gobbetti and R. Di Cagno. 2012. Novel probiotic candidates for humans isolated from raw fruits and vegetables. Food Microbiol. 31(1): 116–125.

Vinderola, C.G., P. Mocchiutti and J.A. Reinheimer. 2002. Interactions among lactic acid starter and probiotic bacteria used for fermented dairy products. J. Dairy Sci. 85(4): 721–729.

Vinderola C.G., N. Bailo and J.A. Reinheimer. 2000. Survival of probiotic microflora in Argentinian yoghurts during refrigerated storage. Food Res. Int. 33(2): 6.

Wassenaar, T.M. and G. Klein. 2008. Safety aspects and implications of regulation of probiotic bacteria in food and food supplements. J. Food Protec. 71(8): 1734–1741.

Whorwell, P.J., L. Altringer, J. Morel, Y. Bond, D. Charbonneau, L. O'Mahony and B. Kiely. 2006. Efficacy of an encapsulated probiotic Bifidobacterium infantis 35624 in women with irritable bowel syndrome. Am. J. Gastroenterol. 101(7): 1581–1590.

Winkler, P., D. Ghadimi, J. Schrezenmeir and J.-P. Kraehenbuhl. 2007. Molecular and cellular basis of microflora-host interactions. J. Nutr. 137(3 Suppl. 2): 756S–772S.

Zanello, G., F. Meurens, M. Berri and H. Salmon. 2009. Saccharomyces boulardii effects on gastrointestinal diseases. Curr. Iss. Microbiol. 11(1): 47–58.

Zocco, M.A., L.Z. dal Verme, F. Cremonini, A.C. Piscaglia, E.C. Nista, M. Candelli and M. Novi. 2006. Efficacy of Lactobacillus GG in maintaining remission of ulcerative colitis. Aliment. Pharmacol. Ther. 23(11): 1567–1574.

Zyrek, A.A., C. Cichon, S. Helms, C. Enders, U. Sonnenborn and M.A. Schmidt. 2007. Molecular mechanisms underlying the probiotic effects of Escherichia coli Nissle 1917 involve ZO-2 and PKCzeta redistribution resulting in tight junction and epithelial barrier repair. Cell. Microbiol. 9(3): 804–816.

CHAPTER 6

# Application of Molecular Methods for Microbial Identification in Dairy Products

*Paul A. Lawson*[1,*] and *Dimitris Tsaltas*[2]

## INTRODUCTION

Although traditional approaches such as cultivation, physiological and chemo taxonomic methods are the cornerstone of the isolation and characterization of individual organisms and complex communities; molecular methods have made, and continue to make, incredible contributions to the study of microbial diversity. A major advantage of molecular methods is the ability to process large numbers of samples simultaneously and have been termed high-throughput methods. Principle investigators and students alike, therefore, favor such methods due to the huge amount of data that can be generated in a relatively short period of time in a cost-effective manner. Indeed many would argue that molecular methods have surpassed the more traditional methods, but this viewpoint is unwise. No one method can answer all questions and even the most powerful approaches such as genome analysis must be complemented by physiological investigations to provide a comprehensive polyphasic approach (Rainey 2011). For example, a gene may well be present in the genome but is it expressed at all, and if

[1] Department of Microbiology and Plant Biology, University of Oklahoma, OK 73019, USA. Email: paul.lawson@ou.edu

[2] Department of Agricultural Sciences, Biotechnology and Food Science, Cyprus University of Technology, P.O. Box 50329, 3603 Limassol, CYPRUS. Email: dimitris.tsaltas@cut.ac.cy

* Corresponding author

so, when? These questions are particularly important when considering the microbial community as a whole. Therefore, it is essential that high-throughput molecular methods be used in tandem with more traditional methods for a comprehensive investigation of microorganisms present and their potential roles in food spoilage as food pathogens, food additives, etc. With respect to food-borne organisms and the associated pathogens, in addition to cultivation and enzyme-linked immunosorbent assay (ELSA) three main approaches using molecular tools may be employed. The first are methods based on the polymerase chain reaction (PCR) (Hayden 2004) with *Escherichia coli* (Tsai et al. 1993, Naravaneni and Jamil 2005) *Salmonella* (Rahn et al. 1992) *Shigella* (Frankel et al. 1990) *Yersinia* (Ibrahim et al. 1992) *Vibrio cholera* (Shangkuan et al. 1995), *Vibrio parahaemolyticus* (Tada et al. 1992) *Vibrio vulnificus* (Brauns et al. 1991), *Listeria monocytogenes* (Simon et al. 1996), and *Staphylococcus aureus* (Wilson et al. 1991). A further refinement was the introduction of Real-Time PCR that is now the most commonly used technology for quantification of specific DNA fragments (Wittwer and Kusukawa 2004). The amount of product synthesized during the PCR is measured in real time by detection of the fluorescent signal produced as a result of specific amplification. The PCR methods are rapid and sensitive, but care should be taken with appropriate controls as false-positive and false-negative results can lead to misleading conclusions.

The second and most recent group is the microarray-based techniques, which is an extension of checkerboard hybridization methods. These methods allow for the simultaneous identification of the increasing number of food-borne pathogens worldwide in a single reaction (Sergeev et al. 2004). The basic idea is that many selected probes are attached spot wise in an array format to a solid surface, and each spot contains numerous copies of a probe. The array is subsequently hybridized with DNA isolated from the sample of interest labeled with fluorescence. During the hybridization phase, the labeled fragments will bind to the spotted probes based on DNA complementarity. As a high-throughput method, microarray-based techniques have some advantages, such as informative, highly repeatable, and potential to combine detection, identification, and effect quantification of unlimited number of food-borne pathogens in a single experiment.

The third approach is the direct sequencing of genes that are then used in a phylogenetic analysis. The most extensively used for this purpose is the 16S rRNA gene (Ludwig et al. 2011). This technology can be used for the rapid identification of pure culture isolates or for a community wide approach where all the 16S genes present are sequenced from a single DNA sample (Quigley et al. 2012b). Innovations to automated DNA sequencing and Next Generation Sequencing are occurring at an almost unbelievable rate (Loman et al. 2012) and in addition to other hardware improvements, the development of computer software to handle huge amounts of data generated by genomic and proteomic approaches facilitated the expansion of these technologies away from the specialist research laboratories to many clinical and diagnostic facilities. In particular the Basic Local Alignment Search Tool (BLAST) (Altschul et al. 1990), provide a rapid mean of sequence identification, coupled with sequence alignment programs such as CLUSTAL (Thompson et al. 1994) and MEGA (Tamura et al. 2007). The Chunlab (http://www.chunlab.com/), and EzTaxone [http://eztaxon-e.ezbiocloud.net/(Kim et al. 2012)] also have an excellent suit of user-friendly software packages that are now widely used. Subsequent manipulation of data and the construction of phylogenetic trees have contributed to the proliferation of genomic based methods. Although some components such as the hardware for the aforementioned methods can be beyond the financial capabilities of many laboratories, an increasing phenomenon is the "out-sourcing" of these methods to university or commercially based facilities. Indeed, the past decade has seen a large number of biotechnology companies offering high-throughput and rapid processing of bacterial and fungal isolates or extracted DNA submitted and within days a result is returned normally in electronic format. However, although this allows rapid and cost efficient identifications with a reduced need for specialist training, care is still required for the correct and accurate analysis of data received. Without a doubt, out-sourcing of this type has facilitated the proliferation of these powerful molecular methods into many laboratories dealing with food microbiology, often confirming the results of strains identified by more classical methods including miniaturized biochemical/enzymatic kits that are extremely useful in the diagnostic laboratory.

Another very strong driving force for the development and deployment of molecular methods to characterize the microbial flora of foods is the denomination of Protected Designation of Origin (PDO) that is well documented through links among the areas of origin, the procedures and the final products. Sensorial and texture characteristics of dairy products, are not only attributed to the microbial population of lactic acid bacteria and enterococci (Randazzo et al. 2009). Therefore, characterizing the microbial population of fermented products contributes understanding the biochemical "evolution" of these products and provides information on their identity (for PDO purposes) and technological development. Recently O'Sullivan and coworkers (2013) have reviewed nucleic acid based methods investigating the microbial related cheese quality defect problems, showing another very useful contribution by these methods. In addition, they have an extensive reference to next generation sequencing, giving good introductory information for its use in food microbiology. Neviani and coworkers (2013) have also reviewed the case of Grana Padano and Parmigiano Reggiano cheeses research, via molecular methods done over the last decade. In the same context Rodrigues and coworkers (2012) reviewed the analytical strategies for characterization and validation of functional dairy foods. The probiotic content leading to the transformation of milk to a variety of healthy products (yogurt, cheese, dairy drinks, etc.) via glycolysis, proteolysis and lipolysis can very well be characterized using molecular methods in parallel with analytical chemistry and classic microbiology ones.

It is beyond the scope of this chapter to describe in detail all molecular methods but it will outline some of the most useful molecular approaches applied to dairy microbiology. A basic knowledge of microbiology and molecular biology will be assumed, for each method discussed, a short introduction outlining the principle will be given and extensive use of references will provide the reader with additional resource material. Experience shows that in addition to access to these methods it is extremely beneficial to visit colleagues where such methods are running routinely in order for a smooth transition to their introduction into the laboratory. The text will focus on the methods that are the most accessible to majority of laboratories with a modest budget. Table 1 shows the methods and principles behind those that will be presented in this chapter.

**Table 1.** Description and principles of molecular-based methods.

| **Method** | **Principle** |
|---|---|
| 16S rDNA gene sequencing | Sequence determination of the 16S rDNA gene that contains both conserved regions to allow sequence alignment and variable regions that allow diagnostic identification and provide evolutionary relationships. |
| Random Amplified Polymorphic DNA (RAPD) | Employs short arbitrary primers and low stringency hybridisation to randomly amplify DNA fragments which are separated on agarose to give a fingerprint pattern. |
| Denaturing or Temporal Temperature Gradient Gel Electrophoresis (DGGE or TTGE) | Small PCR amplicons (distinguished by differences in their DNA sequences) are separated from a low to high gradient. DGGE uses a chemical gradient (urea or formamide) while TTGE has a temperature gradient and a constant concentration of denaturing agents. |
| Real-Time PCR (qPCR) | Uses species-specific primers to target a gene/organism. A fluorescent probe or dye is used to monitor the amplification of the target DNA in real time enabling quantification of a target organism. |
| Terminal Restriction Fragment Length Polymorphisms (tRFLP) | Fluorescent, end labeled, PCR products are digested with restriction endonucleases and separated by electrophoresis. The end-labelled terminal restriction fragments are compared with DNA size standards. Different groups have a difference in the number and location of restriction sites giving rise to different fragment lengths. |
| Intergenic Transcribed Spacer Analysis (ITS) | Analyses the bacterial ITS region located between the 16S and 23S ribosomal genes. Allows the differentiation between strains of the same species or closely related species. |

When choosing a technique, one must first address the question of what information we wish to derive from the sample being investigated. For instance, do we wish to study the general microbial diversity of an ecosystem (Duthoit et al. 2003, Callon et al. 2007, Bonetta et al. 2008) or identify specific microorganisms present (Delbes and Montel 2005) or both (Martin-Platero et al. 2009)? Our endeavor is to answer these questions in good faith, accurately and directly.

## Genomic Methods

### *Isolation of DNA*

The recovery of good quality DNA is important to the outcome of most DNA-based molecular approaches but is especially important in culture-independent methods. It is crucial that DNA extracted from a mixed community is truly representative of all the organisms present and is of sufficient quality and concentration. In addition to insufficient or preferential cell lysis, the presence of compounds such as fats, carbohydrates, proteins and salts that may inhibit downstream manipulations should be eliminated (Wilson 1997). However, many laboratories now use a range of commercially available DNA extraction kits that have mainly overcome many earlier problems. Added advantages are the elimination of harmful chemicals such as phenol and little required prior knowledge of complex molecular methods. However, the classical phenol/chloroform is useful as it introduces the novice to the basic principles of the extraction process and it has the benefit of the observation of DNA during the extraction process unlike kit systems.

In the case of dairy products it is a common problem acquiring non degraded and inhibitor free DNA as described, tested and reviewed by Pirondini and coworkers (Pirondini et al. 2010). Similarly Quigley with coworkers (2012a) have compared five commercial kits and two in-house, concluding that best results were obtained by the two commercial (solid-phase/column extraction—Milk Bacterial Isolation Kit from Norgen Biotek and Power Food Microbial DNA Isolation Kit from MoBio Laboratories) and an in-house method based on liquid-liquid extraction with purification steps using phenol-chloroform and ethanol. The authors favour the use of the Power Food Microbial DNA Isolation Kit from MoBio Laboratories because it is faster and less laborious being in a ready-to-use format. Conclusively, we can suggest that before the use of commercial kits for DNA extraction from raw milk, but not cheeses, one should consider the multifactorial environment of the fermented product and tests should be made in each case before the final decision is made.

**Table 2.** Commonly used DNA extraction methods.

| Method | Extraction type | Principle |
|---|---|---|
| In-House Lytic Method (Quigley et al. 2012a) | Liquid–liquid extraction | Cell lyses using, chaotropic agents, enzymes and mechanical force. Purification relies on using phenol-chloroform and ethanol purification. |
| QIAamp®DNA stool mini kit (Qiagen Ltd.) | Solid-phase/column extraction | Cell lyses using chaotrophic agents, detergents, proteinase K and heating, uses an exclusive adsorption resin to remove impurities. DNA purification uses a silica-gel membrane. |
| Chemagic Food Basic kit (Chemagen BiopolymerTechnologie) | Mobile solid-phase/ magnetic bead extraction | Cell lyses using chaotrophic agents and RNase A. Magnetic beads as solid-phase for binding target DNA. |
| Wizard®Magnetic DNA Isolation kit (Promega Inc.) | Mobile solid-phase/ magnetic bead extraction | Cell lyses using chaotrophic agents and RNase A. DNA bound and purified using magnetic beads as solid-phase support. |
| FastPrep®Kit (MP Biomedicals) | Solid-phase/column extraction | Cell lyses using chaotropic buffers. DNA is purified by a silica-based GeneClean® procedure. |
| PowerFood™ Microbial DNA Isolation kit (MoBio Laboratories Inc.) | Solid-phase/column extraction | Cell lyses based on chaotrophic agents, mechanical lyses and inhibitor removal technology. DNA binding is based on silica membrane spin column. |
| Guanidine Thiocyanate method (Duthoit et al. 2003) | Liquid–liquid extraction | Cell lyses using detergents, chaotrophic agents, mechanical lysis plus heat. DNA extraction using phenol-chloroform and ethanol purification. |

### *DNA extraction—Protocol 1*

1. Resuspend cells (1–2 loopfuls or approximately 1 ml of actively growing culture) in 500 µl TES buffer (0.05 M Tris, 0.05 M NaCl, 0.005 M EDTA, pH 8.0). Add 5 µl lysozyme (10 mg/ml), incubate at 37°C for 15–30 min.
   (N.B. Some cells are easier to break open and therefore the lysozyme step may be omitted. However if the cells prove to be resistant the lysosyme concentration and incubation time can be increased. If problems persist an alternative enzyme such as lysostatin or mutolysin may be used).

2. Add 8 µl each of Proteinase K and RNase (10 mg/ml), mix. Incubate for 1h at 65°C. (The increased temperature inhibits non-specific nucleases, which would otherwise degrade the DNA).
3. Add 120 µl 10% SDS and return immediately to the 65°C water-bath for a further 10 min. When the cells lyse the solution may become clear.
4. Remove from water-bath and leave to cool. Add an equal volume of phenol/chloroform mix until emulsion forms. Centrifuge on low-speed setting 1500–6500 rpm for 10 min.
5. Three layers should form, a lower solvent layer, an upper aqueous layer containing the DNA and a layer of protein/cell debris separating the two other layers. Carefully take off the top layer into a clean microcentrifuge tube using a wide bore blue tip (i.e., cut with scissors and smooth with bunsen flame). If the solution is too concentrated the solution may be very cloudy. Add TES until the solution clears and repeat the phenol/chloroform step until no more protein material remains.
6. To the DNA solution add 2.5 volumes of 100% ethanol (–20°C). Mix gently, DNA strands should start to precipitate. Centrifuge on high-speed for 5 min. Air-dry for 20 min (or speed vacuum for 5 min) and resuspend in 50–500 µl (depending on the yield) sterile water or TE buffer (10 mM Tris, 1 mM EDTA, pH 8.0). Keep the solution concentrated and fairly viscous (Even if no DNA is seen, continue with the protocol as very small amounts of DNA can be amplified by PCR).
7. Run 5 µl of the DNA solution on a 0.8% agarose gel with 1 µg lambda standard to get an approximate concentration. Usually very high quality DNA is seen as a tight band almost as good as the Lambda standard. Make a dilution 10 ng/µl for PCR.

*DNA extraction—Protocol 2 (for PCR)*

Although the first step for many molecular methods is the extraction of high quality DNA that requires the lysis of bacterial cells prior to downstream manipulations, for amplification of genes by PCR, a simple boiling method is often satisfactory.

1. Colonies are suspended on 100 µl of sterile water and incubated at 95–100°C for 10 min.
2. Centrifuge at 12000 rpm for 5 min to remove cell debris, transfer supernatant to a clean tube making sure none of the cell pellet is disturbed.

   (If cells prove to be more recalcitrant to lysis, a simple lysis buffer (0.5 M NaOH, 0.05 M sodium citrate) may be used. After 5 min incubation at room temperature, samples are pelleted and the supernatant transferred to a clean tube and treated as above).

Many research groups also apply the so called "Colony PCR" method where very small amount of cells are picked up via a pipette tip and immediately transferred to the PCR reaction mix. Although it is a very successful technique for the most common bacteria, special care and practice is required in order to acquire very small amount of cells.

*DNA extraction—Protocol 3* (Quigley et al. 2012a)

1. DNA is isolated by resuspending pellet obtained from 1 ml milk or 1 ml homogenized cheese in 500 µl of breaking buffer for enzymatic lysis (20 mmol $l^{-1}$ Tris HCl (pH8), 2 mmol $l^{-1}$ EDTA, 2% Triton X100, 50 µg $ml^{-1}$ lysozyme, 100 U mutanolysin) and incubated at 37°C for 1 h.
2. Protein digestion is then performed by adding 250 µg $ml^{-1}$ proteinase K and incubating at 55°C for 1 h.
3. The suspension is transferred to a 2 ml microcentrifuge tube containing 0.3 g zirconium beads and shaken for 90 sec in a bead beater, twice and centrifuged at 12,000 g x 10 min.
4. The supernatant is transferred to a clean tube and combined with equal volume of phenol:chloroform:isoamylalcohol (25:24:1), mixed gently and centrifuged at 12,000 g x 2 min.
5. The top aqueous phase is transferred to a clean tube and one-tenth the volume of 3 mol $l^{-1}$ sodium acetate and 2 volumes of 100% ice cold ethanol are added. The suspension is mixed gently and stored at –20°C overnight.
6. The sample is centrifuged at 14,000 g x 10 min, the supernatant removed and the pellet is washed with 70% ice cold ethanol followed by centrifugation at 12,000 g x 5 min and the pellet dried.
7. The pellet is re-suspended in 100 µl TE buffer.

### *Polymerase Chain Reaction (PCR)*

PCR is a method for synthesizing multiple copies of (amplifying) a specific piece of DNA. DNA polymerase copies strands of DNA (Mullis and Faloona 1987). Four basic components are required:

- A DNA template containing the target sequence that is to be amplified. For pathogen detection this sequence must be highly specific to the organism concerned, often a single gene, such as a virulence gene or the 16S rRNA gene.
- Primers—a pair of short single-stranded DNA sections, which are exactly complementary for specific parts of the target sequence.

- A heat-stable DNA-polymerase enzyme, usually *Taq* polymerase from a thermophilic bacterium, which catalyzes the reaction.
- Free nucleotides that are used as the building blocks for multiple copies of the DNA template.

A number of companies now offer "Mastermix" kits where all the necessary components are supplied and only the DNA is required.

The first stage in the PCR process is to raise the temperature to about 90–95°C. This causes the double stranded DNA to denature, or melt, into single strands. The temperature is then reduced to about 50–65°C to allow the two primers to bind, or anneal, at specific points on the single-stranded DNA of the target sequence. Finally, the temperature is raised to 70–74°C and the DNA-polymerase enzyme catalyzes the duplication of the target sequence, starting at the annealed primers on each single strand, in a process known as extension. This, results in two double-stranded DNA fragments that are identical copies of the original target sequence. The temperature cycling process is then repeated a number of times, typically 25–35, creating a theoretical doubling of the number of copies of the target sequence at each cycle. This gives an exponential increase in target DNA concentration and produces sufficient DNA for reliable detection from a single target sequence in a few hours.

Multiplex PCR is also an approach of many molecular microbiology publications and is of interest to molecular food microbiology as well (del Rio et al. 2007, Senan et al. 2008, Cremonesi et al. 2011, Pal et al. 2012, Bottari et al. 2013). Similarly, in a study analyzing the bacterial population, structure and dynamics of treated and untreated cold stored milk, Rasolofo and coworkers (2010) followed via real time PCR *Staphylococcus aureus, Aerococcus viridans, Acinetobacter cslcoaceticus, Streptococcus uberis, Corynebacterium variabile* and *Pseudomonas fluorescens*. In order to better approach this issue, a careful optimization using appropriate control organisms should be undertaken to resolve potential problems such as unspecific amplifications and/or primer competition.

An extensive list of primers for a variety of PCR based analysis methods is presented together with other analytical advances in food microbiology in the review by Juste and coworkers (2008).

### *Real Time PCR*

Rapid, sensitive and accurate methods for microbial identification have been expanded with the introduction of real time PCR or also called quantitative real time PCR. As a procedure, real time PCR follows the same principle of PCR with the difference of detecting the end product of the reaction in real time using fluorescent dyes (Mackay 2004). These dyes are either nonspecific DNA intercalating or sequence specific DNA probes labeled with fluorescent reporter molecules. Very recently Boyer and Combrisson (2013) have reviewed the opportunities of quantitative PCR in dairy microbiology. Using DNA intercalating dyes live and dead cells can also discriminated (Moreno et al. 2006, Nocker and Camper 2006, 2009, Nocker et al. 2006, 2007a, 2007b, 2010, Kramer et al. 2009, Rodrigues et al. 2012).

### *Real Time PCR with dsDNA binding dyes*

Fluorescent DNA binding dyes bind to all double stranded DNA thus DNA increased product during PCR leads to increased fluorescence intensity measured at each cycle. The most commonly used dye is SYBR Green and is sold either alone or incorporated in master mixes by a wide range of companies.

Reactions are prepared as usual with the addition of the dye and runs are performed on special PCR instruments carrying UV or LED lamps for excitation of the dye and sensitive detectors/cameras for the measurement of emitted light.

Some major pitfalls the reader should have in mind and seriously consider are the following:

- **Poor Primer Design.** The use of primer design software is strongly recommended. Most primer design software includes adjustable parameters for optimal primer design. These parameters consider primer melting temperature (Tm), complementarity, and secondary structure as well as amplicon size. The primer melting temperature (Tm) of each PCR primer should be between 58–60°C and the Tm of both primers should be within 1°C. Regions of low-complexity sequence can be problematic in designing a unique primer. The best option would be to select an alternative region and if that is not possible, choosing longer primer sequences with higher Tm or optimization of the thermal cycling protocol may be necessary to help reduce nonspecific binding. Designing primers that generate very long amplicons may lead to poor amplification efficiency. Ideally, amplicon length should be 50 to 150 bases for optimal PCR efficiency. In cases in which longer amplicons are necessary, optimization of the thermal cycling protocol and reaction components may be necessary.
- **Poor Quality DNA.** Degraded or impure DNA can limit the efficiency of the sensitive PCR reaction and reduce yield. It not uncommon to have a positive PCR reaction and negative real time PCR reaction using the same set of primers. Residues of cell (proteins, polysaccharides, etc.) and phenol and/or salts from DNA isolation method used are common PCR inhibitory substances.

- **Incorrect Concentration of Primers.** Primers should be reconstituted into working stock concentrations accurately. It is important to take into account the volumes that will be routinely pipetted and for this reason in-house or commercial master mixes should be considered when setting up real-time PCR assays. A common range of working stock concentrations for primers is 10–100 μM.
- **Baseline, Threshold, Efficiency and Standard Curves.** To obtain accurate threshold cycle (Ct) values, the baseline needs to be set two cycles earlier than the Ct value for the most abundant sample. For real-time PCR data to be meaningful, the threshold should be set when the product is in exponential phase. Typically this is set at least 10 standard deviations from the baseline. The efficiency (Eff) of the reaction can be calculated by the following equation:

  $\text{Efficiency} = 10^{(-1/\text{slope})} - 1$

  The efficiency of the PCR should be 90–110% and could be affected by a number of variables. These factors can include length of the amplicon, secondary structure, primer design, etc. Since real time PCR can provide quantitative meaningful results, standard curves should be prepared. The standard curve should extend above and below the expected abundance of target.

### *Real Time PCR with fluorescent reporter molecules*

Fluorescent reporter probes will detect only DNA containing the probe sequence. This significantly increases specificity because detection and quantification does not take place in non-specific DNA amplification. In addition fluorescent probes labeled with different color flurophores can be used in multiplex assays detecting several genes in one reaction. The reactions are as those commonly use in normal PCR. Typical probes used in many laboratories are TaqMan probes which are hydrolysis probes. TaqMan probes consist of a fluorophore covalently attached to the 5′–end of the oligonucleotide probe and a quencher at the 3′–end. Several different fluorophores (FAM, TET, HEX, Cy3, Cy5, JOE, VIC, ROX, and Texas Red) and quenchers (TAMRA, MGB) are available. The quencher molecule quenches the fluorescence emitted by the fluorophore when excited by the cycler's light source. As long as the fluorophore and the quencher are in proximity, quenching inhibits any fluorescence signals. During the reaction, primers and probe anneal to DNA target and once polymerase reaches the probe its exonuclease degrades the probe releasing the fluorescent reporter from the quenching molecule also attached to the probe and as a result fluorescence is detected.

Similarly to Real Time PCR with dsDNA binding dyes, there are additional major pitfalls that the reader should consider for Real Time PCR with fluorescent reporter molecules:

- Poor probe design. TaqMan® probe Tm should be ~10°C higher than the primer Tm.
- Concentration of probes is incorrect. A common range of working stock concentrations for probes is 2–10 µM.
- Ordering a probe labeled with a dye not calibrated or supported on the real-time PCR instrument being used.
- Combination of flurophores and quenchers. A well-chosen combination of flurophores and quenchers is very important for maximum fluorescence, minimum background, maximum signal to noise ratio and maximum sensitivity (Marras 2006). Good advice for compatibility of instruments and the combination of fluorophores/quenchers is usually provided on instrument and molecular probes manufactures websites.

### *Dead or Live Discrimination*

Viability assessment has been well explored using propidium iodide, propidium monoazide (PMA) staining (Nocker et al. 2007a, 2009, Kramer et al. 2009, Boyer and Combrisson 2013, O'Sullivan et al. 2013). Accumulating data support better results from propidium monoazide due to smaller probability to enter intact cell membranes as well as new applications of these type of dyes (Nocker et al. 2009, 2010). Elizaquivel et al. (2014) have recently reviewed the developments in the use of viability dyes and quantitative PCR in the food microbiology field. The article concludes that novel approaches, assessing metabolic activity towards preferential detection of viable cells could complement the use of viability dyes.

## ***Methods with Electrophoretic Output***

### *Restriction enzyme based*

*Restriction Enzyme Analysis—Pulsed Field Gel Electrophoresis (REA-PFGE).* Restriction enzyme analysis via pulsed field gel electrophoresis (REA-PFGE) is performed by rare cutting endonucleases (*Sma*I, *Apa*I, *Not*I, and *Sal*I). The approximate number of fragments ranges from 10 to 30 for the corresponding genome sizes. Pulsed field gel electrophoresis uses a periodically reoriented electric field that helps large DNA molecules to move in agarose gels (usually 1%). Generally, analysis using one restriction enzyme provides good and reliable differentiation as demonstrated with

*Lactobacillus* in olive fermentations using *Apa*I (Argyri et al. 2014, Blana et al. 2014). However, Vancanneyt and coworkers (2006) suggested the use of 2–3 restriction enzymes for strains of *Lactobacillus*.

Although low in cost (except initial equipment investment), REA PFGE is time consuming and therefore tends to be used as a supplementary technique for the purpose of confirming or improving results (Coppola et al. 2008). More information on PFGE can be found on PulseNet International www.pulsenetinternational.org. Finally, statistical analysis is required post electrophoresis for clustering purposes. Software like GelCompar (Applied Maths, Sint-Martens-Latem, Belgium, http://www.applied-maths.com/gelcompar-ii) is required for clustering analysis. Sample preparation and handling is one of the most important parameters the reader should consider. This is because high molecular weight DNA is easily cleaved through shearing and imparts very high solution viscosity. For these reasons, DNA samples for PFGE are generally prepared by embedding in gel medium as "gel plugs". Cellular source material is suspended in low gelling agarose and the gel suspension is poured into molds. All subsequent manipulations (cell lysis, DNA purification and restriction digestion) are performed by diffusing reagents into the resultant gel plugs. The processed gel plugs are then carefully loaded into wells of an agarose gel used for PFGE.

*Restriction Fragment Length Polymorphism (RFLP).* During Restriction Fragment Length Polymorphism analysis, the DNA sample is digested by restriction enzymes and the resulting restriction fragments are separated according to their sizes by gel electrophoresis. RFLP analysis was the first DNA profiling technique with widespread application due to low cost. Modifications of RFLP are PCR-RFLP, ARDRA-PCR (Amplified Ribosomal DNA Restriction Analysis) and ISR-RFLP-PCR (Intergenic Spacer Region Restriction Fragment Length Polymorphism PCR). The initial step of DNA amplification via PCR and the following digestion is common in all cases. In ARDRA and ISR-RFLP-PCR we use, in particular, ribosomal DNA regions as have been used by Aquilanti and coworkers (2006) and others [(Moschetti et al. 1998), computerized databases with LAB fingerprints from (Blaiotta et al. 2002, Chan et al. 2003, Fortina et al. 2003, Mora et al. 2003, Moreira et al. 2005)].

It is pertinent to note the possibility of over or under estimation of the microbial community in the dairy environment since every organism shows different cell lysis resistance, genome size and GC content and as a result leading to differential amplification (Sanchez et al. 2006a, 2006b).

Similar to this method but with added features is Terminal Restriction Fragment Length Polymorphism (T-RFLP) that is based on variation of the 16S rRNA gene. This technique has been used for both characterizations of microbial populations and structure and their dynamics (Rademaker et al. 2005, 2006, Sanchez et al. 2006a). Analysis is based on the restriction endonuclease digestion of fluorescently end labeled PCR products using a genetic analyzer and appropriate software (Applied Biosystems 2005).

### *PCR Based Methods*

#### *Randomly Amplified Polymorphic DNA (RAPD)*

Randomly amplified polymorphic DNA technique is PCR based where arbitrary primers (8–12 nucleotides) with low stringency hybridization are randomly amplifying DNA fragments separated via polyacrylamide gel electrophoresis. RAPD is an inexpensive and powerful typing method for many bacterial species including microbial populations from the dairy environment (Corroler et al. 1998, Baruzzi et al. 2000, Suzzi et al. 2000, Albenzio et al. 2001, Bouton et al. 2002, Mannu and Paba 2002, Andrighetto et al. 2004, Psoni et al. 2006, Sanchez et al. 2006a, Aquilanti et al. 2007, Ercolini et al. 2009).

Reproducibility of RAPDs is slightly problematic but as long as there is adequate experience, optimization and standardization, the technique is very useful and cost effective. In particular all PCR related parameters (primer to template ratio, DNA polymerase and $MgCl_2$ concentration, and thermal cycles) are required to be optimized.

Other limitation of RAPDs are that there is no possibility to distinguish from single copy to multiple copies of the targeted locus, while co-dominant RAPD markers observed as different-sized DNA segments amplified from the same locus, are detected rarely. Mismatches between the primer and the template may result in the total absence of PCR product as well as in a decreased amount of product.

Alternative methods to proceed after RAPDs is the development of locus-specific, co-dominant markers, isolation of bands, cloning and sequencing. From the acquired sequence new, longer and specific primers are designed and we proceed with the so called Sequenced Characterized Amplified Region Marker (SCAR) (NCBI 2014).

Extensive lists of primers used for RAPDs in dairy microorganisms can be found in Coppola et al. (2008). In these lists the reader can also find primers for yeast characterization (see also recent publication by Fadda et al. 2010). Finally, Moschetti and coworkers (1998) demonstrated how statistical analysis of the results can allow grouping of *Streptococcus thermophilus* strains. Most recent application of RAPDs in conjunction with 16S rDNA pyrosequencing was described by Cruciata et al. (2014). From this work it is apparent than when new species or strains appear, an array of techniques (at least two) are categorically required in order to increase confidence or results.

### *Amplified Fragment Length Polymorphism (AFLP)*

Amplified Fragment Length Polymorphism uses the combination of restriction enzymes and PCR. Restriction enzymes are used to digest genomic DNA, followed by ligation of adaptors to the sticky ends of the restriction fragments. A subset of the restriction fragments is then selected to be amplified by using primers complementary to the adaptor sequence, the restriction site sequence and a few nucleotides inside the restriction site fragments. The amplified fragments are separated and visualized on denaturing polyacrylamide gels or via automated capillary sequencing instruments.

The digestion is performed with two different restriction endonucleases, one with an average cutting frequency and a second with higher cutting frequency (*Eco*RI – *Mse*I, *Taq*I). Following digestion, double-stranded nucleotide adapters are usually ligated to the DNA fragments serving as primer binding sites for PCR amplification. The use of PCR primers complementary to the adapter and the restriction site sequence yields strain-specific amplification patterns (Amor et al. 2007, Goering 2013).

AFLP has mostly been employed in clinical studies, but its successful application for strain typing of the *Lactobacillus acidophilus* group, *L. johnsonii* isolates and *L. plantarum* group, has been reported (Gancheva et al. 1999, Torriani et al. 2001, Ventura and Zink 2002). Improvements in AFLP include the use of multiple enzyme adapters and fluorescent labeled primers thus increasing throughput.

Analysis software such as GeneScan (Applied Biosystems, Foster City, CA, USA) may be used for compiling data, analysis and presentation. Worth mentioning is that the resulting data are not scored as length polymorphisms, but instead as presence-absence polymorphisms. The AFLP method is time consuming and laborious, requiring either good quality large polyacrylamide gels or expensive equipment such as sequencing instruments and for these reasons is not suggested unless experience is already established.

### *Denaturing Gradient Gel Electrophoresis (DGGE)*

Denaturing gradient gel electrophoresis is a technique using a chemical gradient to denature the sample as it moves across an acrylamide gel. Denaturing agents (usually urea and/or formamide) are capable of inducing DNA to melt at various stages. As a result of this melting, the DNA can be analyzed for single components, even those as small as 200–700 base pairs. In practice, in DGGE, the DNA is subjected to increasingly denaturing conditions causing most fragments to melt in a step-wise process. In this way we may discriminate differences in DNA sequences or mutations of various genes. By placing two samples side-by-side on the gel and allowing them to denature together, we can easily see even the smallest differences in two samples or fragments of DNA. Main disadvantages to this technique are the small reproducibility due to difficulties preparing the acrylamide gels and the difficulties handling and preparing polyacrylamide. These problems are partially addressed by TGGE/TTGE (Temperature Gradient Gel Electrophoresis or Temporal Temperature Gel Electrophoresis), which uses temperature, rather than chemical, gradient to denature the sample. In this case the denaturing agent is added at a constant concentration.

DGGE has been used by an extensive number of dairy microbiologists worldwide Randazzo et al. 2002, 2009, Chen et al. 2008, Dolci et al. 2008a, 2010, Gala et al. 2008, Nikolic et al. 2008, Rantsiou et al. 2008b, Van Hoorde et al. 2008, Alegria et al. 2009), and the same happened for TGGE (Henri-Dubernet et al. 2004, 2008, Abriouel et al. 2008).

Resolution problems shown by all electrophoretic methods could be resolved by the addition of a GC-clamp to one of the primers increasing resolution when using DGGE (Sheffield et al. 1989, Cocolin et al. 2001, Chen et al. 2008). Also, fragments with identical migration are strongly suggested to be analyzed by sequencing since closely related species might co-migrate (Ogier et al. 2004, Parayre et al. 2007, Giannino et al. 2009, Masoud et al. 2011). Confirmatory results may be acquired via DNA sequencing of the same or other appropriate PCR products (Chen et al. 2008, Masoud et al. 2011).

In order to reveal metabolically active microbiota of dairy products scientists have used RNA processed through reverse transcription and analyzed also by PCR-DGGE (Randazzo et al. 2002, Florez and Mayo 2006, Rantsiou et al. 2008a, Masoud et al. 2011). In a recent attempt by Porcellato et al. (2012a, 2012b) combining DGGE with high resolution melt analysis of DGGE bands, the authors compared reference strain bands with unknown bands, concluding in high accuracy results of identification.

### *Single Strand Conformation Polymorphism-PCR (SSCP-PCR)*

Single stranded conformation polymorphism method uses an acrylamide gel or a capillary electrophoresis for the separation of denatured PCR products. Microbial communities have been analyzed by amplifying the V4 region of 18S rRNA for yeasts and the V2 and V3 regions of 16S rRNA for bacteria. Overall SSCP follows DGGE/TTGE methods in applications and could be characterized as a good profiling method for microbial populations but is not the most appropriate for identification of microbial species due to co-migration of amplicons of different species (Saubusse et al. 2007, Coppola et al. 2008, Verdier-Metz et al. 2009). RNA based SSCP profiles have also been used for characterizing sensorial properties of dairy products (Giraffa and Neviani 2001, Duthoit et al. 2005).

### *Multi Locus VNTR Analysis (MLVA)*

Multiple Locus VNTR Analysis is a method for genetic analysis that takes advantage of the polymorphism of tandemly repeated DNA sequences. VNTR stands for Variable Number of Tandem Repeats. This method is well known in forensic science since it is the basis of DNA fingerprinting in humans. When applied to bacteria, it contributes to forensic microbiology through which the source of a particular strain might eventually be traced back. For this reason it has been extensively used for pathogenic bacteria (Stefano et al. 2008, Chen et al. 2011, Radtke et al. 2012, Seale et al. 2012, Tilburg et al. 2012). In MLVA a number of well-selected and characterized (in terms of mutation rate and diversity) loci are amplified by polymerase chain reaction (PCR) so that the size of each locus can be measured. From this size, the number of repeat units at each locus can be deduced. Repeat unit sizes and repeat sequences can vary when multiple loci are examined in a number of different isolates of an individual microbial species. It has been documented on many occasions that the number of repeat units per locus is a strain-defining parameter. Consequently, there is isolate specificity in the number of repeats per locus, when different strains of a given bacterial species are compared. The resulting information is a code which can be easily compared to reference databases (http://www.mlva.eu/). MLVA had limited use in dairy microbiology up to now (Diancourt et al. 2007, Matamoros et al. 2011) but has found extensive use in pathogenic and food spoilage microorganisms (Duffy 2009). Well-designed multiplex PCR primers producing MLVA banding patterns in lactic acid bacteria or other dairy microflora microorganisms will provide ample new data. A cornerstone for the adoption of this method will be the development of open access reference online databases with MLVA patterns of such organisms.

Variable Tandem Repeats can be found using online tools; Tandem Repeats Finder [http://tandem.bu.edu/trf/trf.html (Benson 1999)] or the Microorganism Tandem Repeat Database [http://minisatellites.u-psud.fr/GPMS/(Denoeud and Vergnaud 2004)].

## ***DNA Sequence Based Methods***

### *Ribosomal DNA sequencing for identification and classification of prokaryotes and fungi*

Without doubt the single most important advance both in the identification of individual species/strains and in the characterization of communities within complex ecosystems is the application of 16S rRNA sequencing. Since the work of Woese and Fox (1977) most community surveys are focusing on RNA genes and intergenic spacers. For the microbial world 16S, 23S and 5S rRNA genes (bacteria) and 18S, 28S and 5.8S rRNA genes (fungi) are sequenced and catalogued in online databases. Ribosomal RNA genes show extremely high sequence homogeneity within species and this is because of concerted evolution thus repeated rRNA genes are treated as one locus. As a result rRNA genes are used in phylogeny studies and species identification (Ludwig et al. 2011, Rainey 2011). However, the 16S rRNA gene has a number of weaknesses in that recent speciation events cannot be recognized resulting in a lack of resolution between closely related species. However, this can be overcome by utilizing alternative chronometers or a number of house-keeping genes (De Vos 2011).

Bacterial 16S rRNA genes comprise nine hypervariable regions (V1–V9) exhibiting significant sequence diversity among species (Baker et al. 2003). V3 region has mostly been used, although different sets of primers provide different areas to be analyzed within it. In fungi the rRNA genes are showing reduced taxonomic resolution while the internal transcriber spacers (ITS) provide the analyst with a higher discriminatory power. The ITS region is located between the 18S (also called Small Subunit—SSU) and the 28S rRNA genes. 25–28S regions are also called Large Subunit—LSU. The 5.8S rRNA gene splits the ITS region into two parts; ITS1 and ITS2. In an excellent review of the culture independent methods for microbial ID in cheeses Jany and Barbier (2008) recommend using other targets in addition to ITS when analyzing cheeses microbial communities because the dairy environment hosts many difficult to discriminate genuses. Schoch and coworkers (2012) concluded that among the regions of the ribosomal cistron, the internal transcribed spacer (ITS) region has the highest probability of successful identification for the broadest range of fungi, with the most clearly defined barcode gap between inter- and intraspecific variation. The LSU had superior species resolution in some taxonomic groups, such as the early diverging lineages and the ascomycete yeasts, but was otherwise slightly inferior to the ITS. The SSU has poor species-level resolution in fungi. The authors are clearly proposing that ITS should be adopted as the primary fungal barcode marker. Other genes for fungal IDing are *CO1, RPB1, EF-1α, BenA* and *GPD* (Berbee et al. 1999, Einax and Voigt 2003, James et al. 2006, Seifert et al. 2007, Seifert 2009).

The progression from RNA cataloging whereby short oligonucleotide sequences were laboriously generated to produce the first phylogenetic classification frameworks to the use of almost full length 16S rRNA gene sequences via the use of reverse transcriptase sequencing of rRNA is well documented (Collins et al. 1991, Weisburg et al. 1991, Williams et al. 1991). Further advances came with the incorporation of the direct sequencing of

PCR-generated DNA amplicons, automated DNA sequencing and ever-improving computer software have made the use of 16S gene sequencing routine for many laboratories. Indeed, there appears to be no end to technical improvements with next generation sequencing methods being reported almost before the previous advance reaches the majority of laboratories. Technically, PCR reactions are performed using universal primer sets amplifying V1–V2 or V1–V3 hypervariable segments of the 16S rRNA gene (Hunt et al. 2011, Ercolini et al. 2012). Primers used to amplify the 16S rRNA genes and internal sequences used to derive the entire sequence are given in Table 3. Although the primers are designed towards conserved regions within the 16S molecule, variation between taxa (especially at the 5′ end of the molecule) is sometimes observed and a number of alternative primers are provided. For rapid screening often only a single primer (R536) which covers a number of diagnostic variable regions is required for a rapid identification. Sequences that may be candidates for novel taxa may be subjected to a full phylogenetic analysis. Work presented by Kumar et al. (2011) using pyrosequencing, concluded that averaging V1–V3 and

**Table 3.** Shows commonly used 16S rRNA primers for amplification and complete 16S gene sequencing. The numbering refers to the positions of the 16S rRNA of *Escherichia coli* (Brosius et al. 1978).

| Primer | Sequence | Target | *E. coli* position |
|---|---|---|---|
| GM3F[a] | AGAGTTTGATCCTGGC | *Bacteria* | 8–24 |
| F27[b] | AGAGTTTGATCCTGGCTCAG | *Bacteria* | 8–27 |
| F399[b] | ACTGCTGCCTCCCGTAGGAG | *Bacteria* | 361–342 |
| R536[b] | CAGCAGCCGCGGTAATAC | *Bacteria* | 518–536 |
| F786[b] | GATTAGATACCCTGGTAG | *Bacteria* | 786–803 |
| R947[b] | TTCGAATTAAACCACATGC | *Bacteria* | 965–947 |
| GM4R[a] | TACCTTGTTACGACTT | *Bacteria* | 1492–1507 |
| R1542[b] | AAGGAGGTGATCCAGCCGCA | *Bacteria* | 1542–1522 |
| A2Fa[c] | TTCCGGTTGATCCYGCCGGA | *Archaea* | 7–26 |
| A2Fb[d] | TTCCGGTTGATCCTGCCGGA | *Archaea* | 7–26 |
| A3Fa[e] | TCCGGTTGATCCYGCCGG | *Archaea* | 8–27 |
| A109F[f] | ACKGCTCAGTAACACGT | *Archaea* | 109–128 |
| Ab127R[g] | CCACGTGTTACTSAGC | *Archaea* | 112–127 |
| A348R[h] | CCCCGTAGGGCCYGG | *Archaea* | 335–349 |
| A934R[g] | GTGCTCCCCCGCCAATTCCT | *Archaea* | 915–934 |
| A1098F[c] | GGCAACGAGCGMGACCC | *Archaea* | 1098–1114 |
| A1115R[c] | GGGTCTCGCTCGTTG | *Archaea* | 1100–1114 |
| fD1[i] | AGAGTTTGATCCTGGCTCAG | *Bacteria* | 8–17 |
| rD1[i] | AAGGAGGTGATCCAGCC | *Bacteria* | 1540–1524 |

[a](Muyzer et al. 1995), [b](Hutson et al. 1993), [c](Reysenbach and Pace 1995), [d](Lopez-Garcia et al. 2001), [e](McInerney et al. 1995), [f](Whitehead and Cotta 1999), [g](Achenbach and Woese 1995), [h](Barns et al. 1994), [i](Weisburg et al. 1991, Ercolini et al. 2009)

V7–V9 regions provides similar results to Sanger sequencing while allowing significantly greater depth of coverage than the Sanger method.

The explosion of microbial identification due to modern genomic tools such as high-throughput capillary and next generation sequencing created an era of "omics" and "microbiomes". Today a huge number of publication, almost on a monthly basis, demonstrate the communities of microorganisms that share our body space, soil, plants, animals, inert surfaces and our food matrixes. The term "microbiome" was originally coined by Joshua Lederberg (Lederberg and Mccray 2001), who argued the importance of microorganisms inhabiting the human body in health and disease. It appears with the development of the tools and the metagenomic approach that the unknown microcosm simply knows no boundaries.

Commercial kits and services now make the use of this powerful tool accessible to most laboratories. However, one such protocol for a commonly used kit and apparatus is given below. It is advisable to use sterile, UV treated bench space and irradiate all tubes, tips, pipettes for 10 min in order to eliminate any possible contamination of extraneous DNA which will be easily amplified from the universal primers. If possible avoid bringing amplified PCR products into the PCR setup area.

### *Protocol—DNA Sequencing with BigDye Terminators Ver 3.1 [Recourse material: (Alcorn and Anderson 2004, Kolbert et al. 2011)]*

#### *Sequencing Reactions*

For each reaction add the following:

- Template DNA (see below) 2 to 5 µl
- Primer (1.6 µM) 2.0 µl
- Sequencing dilution buffer or TM (2.5X) 2.0 µl
- BigDye Mix 0.65 µl
- Water to total 10.0 µl

Mix well and spin briefly.

#### *PCR Program*

Use hot lid

Initial denaturation 96°C for 30 sec

45 cycles of:

96°C for 10 sec

55°C for 15 sec

60°C for 4 min

Reactions are light sensitive and will degrade. Wrap in foil or put in box to limit effects of light on samples.

*DNA template quantities:*

PCR product

500–1000 bp: 25–35 ng (typically 2–3 μl)
1000–2000 bp: 40–70 ng (typically 3–4 μl)
>2000 bp: ~70–200 ng (4–6 μl depending on concentration).

*Removal of Unincorporated Dye-Terminators*

Unincorporated dye can cause serious problems with peak detection and must be removed.

1. Briefly spin plates/tubes at 1000 rpm to bring down any condensation. Remove mat/tape/lid from sequencing reaction plate/tube.
2. Add 20 μl 95% ethanol/3M sodium acetate solution (19:1) to each 10 μl reaction. Note: the final concentration of ethanol should be 60%.
3. Seal with mat/tape (or cap tubes) and mix by vortexing.
4. Let sit for 20 min at room temperature to precipitate. Note: less than 15 min may result in lost signal.
5. Place plate in support rack and spin the centrifuge with plate rotor at 3600×g for 30 min (similarly for tubes).
6. Fold two paper towels to size of plate. Carefully remove covering and invert plate onto paper towel. Then insert towel-side-down in centrifuge. Spin at 900 rpm for 1 min.
7. Wash pellets by adding 100 μl of 70% ethanol.
8. Spin in support rack at 3600×g for 10 min, and repeat step 6.
9. Dry by leaving plate uncovered on bench-top for 25 min, or until no trace of ethanol remains. Pellets will not be visible. Cover for storage.

*Prepare Samples for Capillary Sequencer*

1. Add 20 μl Hi-Di™ (deionized Formamide) to each sample.
2. Cap tubes or cover plates. Vortex thoroughly, and centrifuge briefly if there are droplets on tube walls.
3. Heat for 3 min at 95°C in a thermal cycler, then chill samples in a cold block or on ice for 3 min.
4. Vortex.
5. Spin briefly in a centrifuge to bring down any condensate.
6. Run on ABI 3130xl with 36 cm capillary array using POP7 Polymer for 46 min at 8.5 kV and an injection time of 18 sec at 1.2 kV.

Whether using in-house or commercial services, data is normally retrieved in an electronic format and electropherograms can be viewed using a number of programs for each of the commonly used operating systems (4 Peaks by A. Griekspoor and Tom Groothuis, mekentosj.com for Macintosh and Chromas Litehttp://technelysium.com.au/for Windows). Although trimming of terminal sequence data and the combination of sequences to form a complete sequence may be automated; it is essential to check this data or manually edit such data. Analysis of sequences and accurate identification depends on the use of high quality data.

### *MicroSeq®*

The MicroSeq® Full Gene 16S rDNA and 500 16S rDNA Bacterial Identification Kit provide all the reagents necessary for determining the sequence of the 16S rDNA or a part of it. The resulting DNA sequence is analyzed and compared to a library of 16S rDNA bacterial gene sequences using MicroSeq® ID Analysis Software and the MicroSeq® ID 16S rDNA Full Gene Library. MicroSeq® ID Analysis Software enables you to analyze sequences obtained with any of the MicroSeq® 500 16S rDNA Bacterial Identification Sequencing Kit, MicroSeq® Full Gene 16S rDNA Bacterial Identification Sequencing Kit and the MicroSeq® D2 rDNA Fungal Sequencing Kit. The MicroSeq® ID 16S rDNA Full Gene Library (v1.0) includes over 1200 validated 16S rDNA sequences. Fungal identification, Applied Biosystems offers MicroSeq® D2 rDNA Fungal Identification Sequencing Kit which contains reagents for amplifying and sequencing the D2 expansion segment region of the nuclear large-subunit (LSU) ribosomal RNA gene. Variation within this region is sufficient to identify most organisms at the species level. More than 1070 validated nuclear large-subunit (LSU) ribosomal RNA gene sequences are included in the library.

## ***Data Analysis***

For rapid comparisons with reference sequences a number of DNA databases are used for identification purposes. BLAST [http://www.ncbi.nlm.nih.gov/BLAST), FASTA (http://www.ebi.ac.uk/Tools/sss/fasta/nucleotide.html, http://fasta.bioch.virginia.edu/fasta_www2/fasta_list2.shtml, http://www.genome.jp/tools/ fasta/(Pearson and Lipman 1988)] and the Ribosomal Database Projects [http://rdp.cme.msu.edu (Maidak et al. 1999, Cole et al. 2009)], Chunlab (http://www.chunlab.com/), EzTaxon-e[http://eztaxon-e.ezbiocloud.net/(Kim et al. 2012)] and RIDOM [http://www.ridom-rdna.de (Harmsen et al. 1994, 2002)]. Searches are returned from a search and alignment algorithm typically as a series of pair-wise alignments of decreasing similarities. Identical or similar sequences corresponding to known species can then be compared with phenotypic, biochemical and chemo taxonomic information available. Sequences with no close matches (3% sequence dissimilarity is often used as a cutoff point) may represent novel species and a full phylogenetic reconstruction may be performed.

Partial or full sequences derived from the use of multiple primers (Table 3) should be subjected to quality checking of the raw data. Sequences can then be aligned and phylogenetic reconstructions performed using a number of software programs such as MEGA, SeqTools or on-line tools such as those provided by the Ribosomal Database Project.

For many laboratories a result showing the nearest relatives and a percent sequence similarity may be sufficient. The organism from which the sequence was derived can then be compared with biochemical and phenotypic data. When the sequence is returned with a low percentage, this may indicate an organism may have been isolated that represents a novel taxon at either the genus or species level or, on occasion, higher taxonomic levels. A number of databases were established in the 1980s and 90s (EMBL, Genbank, RDP) but one must be aware that little quality checking was performed and erroneous data is still present. More recent databases such as the SILVA [http://www.arb-silva.de (Pruesse et al. 2007)] and Greengenes [http://greengenes.lbl.gov (DeSantis et al. 2006)] have proved to be very useful and are based more on taxonomic frameworks.

For additional reference material and more detailed explanation of methods and computer software discussed, the reader is directed to the following references (Lepp and Relman 2011, Ludwig et al. 2011). Such programs require investment of time to become fully acquainted with their capabilities, each have excellent help and tutorial sections; many individuals now prefer the on-line analysis tools where software pipelines allow the user to more easily navigate programs for phylogenetic analyses.

### *Phylogenetic analysis*

A phylogeny is an estimate of the evolutionary relationships between taxa or genes. This phylogeny is based on incomplete information with no direct information from the past and with real evolutionary processes not completely known. The most popular methods include distance methods, maximum parsimony, and maximum likelihood. Algorithms attempt to convert molecular data consisting of variables (A, T, C, and G) into continuous variable represented by a branch length. Therefore "phylogenetic inferences" are only a "best estimate" of evolutionary history and a number of methods have been developed, each with advantages and disadvantages. Phylogenetic relationships are typically presented as radial dendograms commonly referred to as "trees" (Fig. 1). In this format, sequences are linked to internal branching nodes via vertical and horizontal lines, the vertical lines only give structure to the tree whereas the horizontal lines represent the phylogenetic distance often denoted as a percent or number of nucleotides substituted per 100 bases. In order to provide some confidence to the robustness of tree topologies generated,

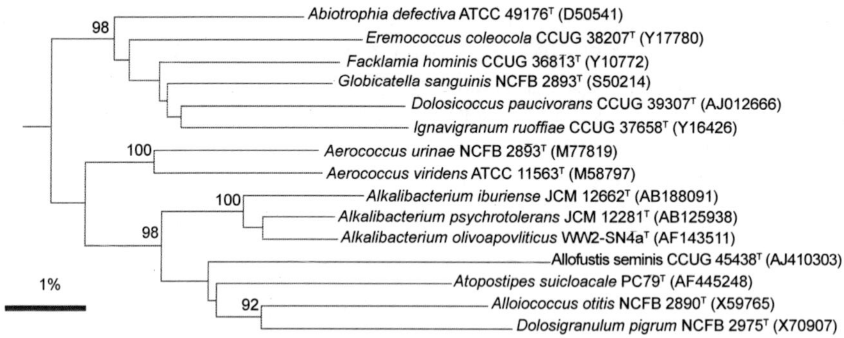


**Figure 1.** Phylogenetic associations constructed using the neighbor-joining method.

resampling procedures are used. This Bootstrapping, put simplistically, generates 100–1000 slightly different data sets presenting different orders of sequences added to the algorithm and a consensus tree is generated. The values obtained (often given at the branching nodes, Fig. 1) represent the percent of how often two sequences (or clusters of sequences) are obtained by that particular method. Given the shortcoming of all the models used it is encouraged that at least two methods be compared.

It is not appropriate to discuss the details of each method and the models of evolution and it can be overwhelming for a novice when faced with different alignment programs, phylogenetic algorithms and models; indeed this is why many now prefer the online-tools that take the uploaded sequence and with a few key strokes a phylogenetic tree is presented to the user! However, it is important for individuals to understand these programs and the underlining principles employed. The reader is encouraged to consult the following references (Swofford et al. 1996, Hall 2004, Lepp and Relman 2011, Ludwig et al. 2011).

### *Next generation sequencing in dairy microbiology*

Although sequencing can also be performed via classic Sanger method (Rasolofo et al. 2010) and is often preferred for very sensitive analysis or checking sequences determined using high through-put methods, Next Generation Sequencing (NGS) methods are rapidly gaining popularity due to their shear capacity to generate a huge amount of data in one operational run (Loman et al. 2012). Total isolated DNA is amplified using 16S rDNA primers and cloned on appropriate vector (pGEM-T-Easy, Promega, USA). Inserts can be sequenced either by previously used primers or vector universal primers in a bidirectional way. NGS techniques offer three major advantages in dairy microbiology research. The first is to provide a thorough examination of the biodiversity of a sample using universal primers and the second relies on the thoroughness of the technique translating into analysis of more sequences increasing, in this way, the capacity to observe less abundant bacterial phylotypes. The third advantage is that sequence data and analysis help elucidating the molecular basis of how microorganisms respond to the food substrate and microenvironment (Solieri et al. 2013). In recent years an increasing number of research groups incorporate NGS in their dairy ecosystem investigations (Alegria et al. 2012, Ercolini et al. 2012, Lusk et al. 2012, Masoud et al. 2012, Quigley et al. 2012b, Ercolini 2013).

The major difference between capillary based sequencing and NGS is that one to ninety six samples can be analyzed simultaneously in the first, while millions of sequences can be analyzed by the second. In addition NGS technology does not require the generation of vector based libraries so it is free from cloning associated biases. On the other hand, NGS technology is computationally demanding which translates to increased cost of instrumentation and requirement of bioinformatics specialists/competent staff (Ercolini 2013).

As mentioned earlier, the elucidation of microbiomes in a variety of live and inert systems where life exists, has been a subject of frequent discussions over the last three years (2011–2013) (Bokulich and Mills 2013, Castro-Carrera et al. 2014). From Soggiu et al. (2013) we receive the discussion about the milk and cheese microbiome for safety and quality of dairy products and their reasons have led to dozens of published work on the microbiomes of PDO dairy products; it is of unprecedented scale as is the ease of acquiring them.

The application of metagenomics through an NGS platform is also of much interest. The term metagenomics was introduced by Handelsman and coworkers (1998) while referring to the study of genetic material recovered directly from environmental samples. This broad field may also be referred to as environmental genomics, ecogenomics or community genomics. Instead of solely targeting the 16S rDNA genes that is normally undertaken, all genes present in a habitat are now being sequenced due to the huge capacity of NGS methods.

### *Single locus sequence typing*

Single locus sequence typing has been developed by the use of 16S rRNA gene sequence analysis and later via the use of more discriminatory genes such as elongation factor Tu gene (*tuf*) (Jian et al. 2001), DNA repair recombinase (*recA*) (Ventura and Zink 2003), chaperonin Hsp60 (*Cpn*60) (Blaiotta et al. 2008), RNA polymerase β subunit (*rpoB*) (Rantsiou et al. 2004), and β subunit of DNA gyrase (*gyrB*) (Itoh et al. 2006).

### *Multi locus sequence typing*

Although single locus sequence typing (De Vos 2011) hasn't been extensively used in dairy microbiology, multilocus sequence typing (MLST) has become a popular approach in dairy fermented products for the characterization of the microflora. The robustness of MLST derives from the advantage that sequence data are far less ambiguous and easier to record and interpret than band patterns produced from all other electrophoresis based techniques (Spratt 1999).

PCR amplification and sequencing of internal regions of multiple housekeeping genes (usually seven) assigns numeric allelic designations and the individual strains are characterized by a seven digit MLST sequence type. Using algorithm we may identify a parent as the one with the greatest number of single locus variants. Various tools exist in the internet for graphical representation of clonal complexes which group a minimum of 5–6 of the 7 allelic designations.

Databases of MLST information are these days easily accessible offering portability of the data and facilitating global usage of them (http://www.mlst.net/databases/default.asp, http://pubmlst.org/databases/,http://www.pasteur.fr/recherche/genopole/PF8/mlst/, http://cge.cbs.dtu.dk/services/MLST/). Increased cost and time associated with sequencing is of limited importance since the introduction of chip based sequencing methods. An MLST analysis for *Lactococcus lactis* subsp. *lactis* and *cremoris* genotypes gave deeply branched trees (Fernandez et al. 2011). The genes used were *atpA, pheS, rpoA, bcaT, pepN* and *pepX*. For PCR conditions the reader should refer to Rademaker et al. (2007) and data analysis can be performed using MEGA software (Tamura et al. 2007).

### *Whole genome sequence typing*

Since the introduction of next generation sequencing approaches we are witnessing a tremendous race of competition for the fastest and cheapest production of genomes (Loman et al. 2012). We are currently heading to the 1000 euros genome sequence and it becomes obvious that we can also compare whole genome sequences with the same cost and easiness. Whole genome sequencing is of course the ultimate molecular typing approach and is following similar evolutionary speeds as informatics and bioinformatics (Chun and Rainey 2014). Common conflicting reports such as those published for the 2011 German *E. coli* outbreak that claimed different origins of the strain (Mellmann et al. 2011, Rasko et al. 2011) are most likely attributed to lagging bioinformatics tools for the interpretation of these "new" type of data.

Whole genome sequence techniques offer two major advantages in dairy microbiology research. The first is to provide a thorough examination of the total sequence of an organism, identifying all interesting genetic variation that explains unique physiological attributes useful for the development of new dairy products or the characterization of old ones. In addition any questionable genetic material (antibiotic resistance genes, pathogen related genes etc.) may be avoided. Second, whole genome sequencing provides ample amount of data, useful for future phylogenetic studies in a cost and time effective manner (Prajapati et al. 2011, 2012, Papadimitriou et al. 2012).

## Fluorescent *In Situ* Hybridization—FISH

Fluorescent *in situ* hybridization is a single cell analysis determining spatial arrangement and semi quantitative information. Fixed cells are hybridized with a fluorescently labeled DNA probe and visualized by epifluorescent microscopy. The samples can be isolated bacteria or food samples appropriately sectioned for microscopical observation (Juste et al. 2008, Rantsiou et al. 2008b). Studies by Ercolini and coworkers (2003a), (2003b) and Coppola and coworkers (2008) gave very interesting insights into the spatial arrangement of LABs and their possible role in the development of microenvironments within the cheese matrix. FISH is not an appropriate method for mapping total diversity as there are practical limits to how many probes can be used simultaneously.

Of a similar approach and use is the combination of fluorescently labeled cells and flow cytometry where quantitative results at the species level can be combined by exploring the physiological state of cells in a fast, efficient and reduced labor-intensive manner (Bianchi et al. 2004, Lahtinen et al. 2006a, Lahtinen et al. 2006b, Maukonen et al. 2006). As mentioned previously, in dead/live methods we can employ permanent and non-permanent DNA stains (Propidium Iodide, Propidium Monoazide, TOTO-1) to assess viability of cells (Bunthof et al. 2001, Bunthof and Abee 2002, Lahtinen et al. 2006a).

## Non Sequence-based Whole Cell Typing

Mass spectrometry detection of microbial nucleic acids was first reported by Ecker et al. (2005), showing a method for rapid identification and strain-typing of respiratory pathogens for epidemic surveillance. The use of electrospray ionization mass spectrometry and base composition analysis of PCR amplification products gave identification and quantification of pathogenic bacteria present in the samples. Most recently Massire et al. (2013) used and evaluated PCR coupled with electrospray ionization mass spectrometry (PCR/ESI-MS) as a novel means for identification of fungal pathogens while Kern et al. (2014) were differentiating *Lactobacillus brevis* strains using Matrix-Assisted-Laser-Desorption-Ionization-Time-of-Flight Mass Spectrometry with respect to their beer spoilage potential.

Currently, due to the mass spectrometry revolution for microbial diagnostics and the low rates of inter-laboratory reproducibility of some of the DNA based techniques there has been a boost in the efforts developing methods that are simple, reliable, with high specificity and uniformity while analyzing multiple groups of microorganisms. At the same time reducing the cost of analysis is a prerequisite. Robust and mass scale mass spectrometry methods and technologies exist in our daily life. Mass spectrometry instruments are the "absolute analyzers" from which we expect, globally, most if not all the answers for life on earth and beyond.

Matrix assisted laser desorption/ionization time of flight mass spectrometry (MALDI-TOF-MS) has been introduced in chemo taxonomy since 1994 (Cain et al. 1994, Böhme et al. 2010, 2011, 2012a, 2012b, Tanigawa et al. 2010). Doan et al. (2012) have shown the potential of MALDI-TOF MS-MS to replace molecular techniques based on genomic fingerprinting while an excellent review is published by Sandrin and coworkers (2012). Also, as any other existing fingerprinting method, it is of equal importance to develop reference databases with microbial spectra. Bohme et al. (2012a) have launched a publicly accessed MALDI TOF MS library of 79 bacterial spectra.

## Conclusion and Further Challenges

Culture independent molecular approaches are increasingly used in most microbiology laboratories worldwide, whether dealing with clinical, environmental or dairy-based. Methods and instrumentation continue to be developed or refined. In tandem with reduction in costs, commercially based out-sourcing and the abundant use of commercial kits for relatively complex procedures have made these methods very accessible to most laboratories. For example at the time of writing, the biology community was expecting the miniature sequencer of the size of a large USB memory stick called MinION (Oxford Nanopore Technologies Ltd.) and the production of a whole genome sequence in real-time and at a cost of under 1,000 US dollars. Indeed it is pertinent to note that a number of methods described in this chapter may well become obsolete in a relatively short period of time!

Microbial ecology is moving to the study of microbial function via metatranscriptomic approach (Cardenas and Tiedje 2008) and structure-function studies will provide new knowledge in dairy ecosystems (Irlinger and Mounier 2009). However, as powerful as molecular methods are, it should be noted that if we lose sight of a polyphasic approach employing a range of methods including cultivation, physiological and chemotaxonomic approaches it will be at our peril and to the detriment of microbiology as a whole.

## References Cited

Abriouel, H., A. Martin-Platero, M. Maqueda, E. Valdivia and M. Martinez-Bueno. 2008. Biodiversity of the microbial community in a Spanish farmhouse cheese as revealed by culture-dependent and culture-independent methods. Int. J. Food Microbiol. 127(3): 200–208.

Achenbach, L. and C. Woese. 1995. 16S and 23S rRNA-like primers. Archaea: A Laboratory Manual. K.R.a.S. Sower, H.J. New York, Cold Spring Harbor Laboratory Press 521–523.

Albenzio, M., M.R. Corbo, S.U. Rehman, P.F. Fox, M. De Angelis, A. Corsetti, A. Sevi and M. Gobbetti. 2001. Microbiological and biochemical characteristics of Canestrato Pugliese cheese made from raw milk, pasteurized milk or by heating the curd in hot whey. Int. J. Food Microbiol. 67(1-2): 35–48.

Alcorn, T.M. and S.M. Anderson. 2004. Automated DNA Sequencing. Molecular Microbiology: Diagnostic Principles and Practice. T.F.C. Persing D.H., Versalovic J., Tang Y.-W., Unger E.R., Relman D.A., White T.J., ASM Press 153–159.

Alegria, A., P. Alvarez-Martin, N. Sacristan, E. Fernandez, S. Delgado and B. Mayo. 2009. Diversity and evolution of the microbial populations during manufacture and ripening of Casin, a traditional Spanish, starter-free cheese made from cow's milk. Int. J. Food Microbiol. 136(1): 44–51.

Alegria, A., P. Szczesny, B. Mayo, J. Bardowski and M. Kowalczyk. 2012. Biodiversity in Oscypek, a traditional Polish cheese, determined by culture-dependent and -independent approaches. Appl. Environ. Microbiol. 78(6): 1890–1898.

Altschul, S.F., W. Gish, W. Miller, E.W. Myers and D.J. Lipman. 1990. Basic local alignment search tool. Journal of Molecular Biology 215(3): 403–410.

Amor, K.B., E.E. Vaughan and W.M. De Vos. 2007. Advanced molecular tools for the identification of lactic acid bacteria. Journal of Nutrition 137(3): 741S–747S.

Andrighetto, C., G. Marcazzan and A. Lombardi. 2004. Use of RAPD-PCR and TTGE for the evaluation of biodiversity of whey cultures for Grana Padano cheese. Letters in Applied Microbiology 38(5): 400–405.

Applied Biosystems. 2005. Terminal Fragment Length Polymorphism (T-RFLP) Analysis on Applied Biosystems Capillary Electrophoresis Systems. Retrieved 12/2005, from http://www3.appliedbiosystems.com/cms/groups/mcb_marketing/documents/generaldocuments/cms_042272.pdf.

Aquilanti, L., L. Dell'Aquila, E. Zannini, A. Zocchetti and F. Clementi. 2006. Resident lactic acid bacteria in raw milk Canestrato Pugliese cheese. Letters in Applied Microbiology 43(2): 161–167.

Aquilanti, L., G. Silvestri, E. Zannini, A. Osimani, S. Santarelli and F. Clementi. 2007. Phenotypic, genotypic and technological characterization of predominant lactic acid bacteria in Pecorino cheese from central Italy. J. Appl. Microbiol. 103(4): 948–960.

Argyri, A.A., A.A. Nisiotou, A. Mallouchos, E.Z. Panagou and C.C. Tassou. 2014. Performance of two potential probiotic Lactobacillus strains from the olive microbiota as starters in the fermentation of heat shocked green olives. International Journal of Food Microbiology 171: 68–76.

Baker, G.C., J.J. Smith and D.A. Cowan. 2003. Review and re-analysis of domain-specific 16S primers. J. Microbiol. Methods 55(3): 541–555.

Barns, S.M., R.E. Fundyga, M.W. Jeffries and N.R. Pace. 1994. Remarkable archaeal diversity detected in a Yellowstone National Park hot spring environment. Proc. Natl. Acad. of Sci. 91(5): 1609–1613.

Baruzzi, F., M. Morea, A. Matarante and P.S. Cocconcelli. 2000. Changes in the Lactobacillus community during Ricotta forte cheese natural fermentation. J. Appl. Microbiol. 89(5): 807–814.

Benson, G. 1999. Tandem repeats finder: a program to analyze DNA sequences. Nucleic acids Research 27(2): 573–580.

Berbee, M., M. Pirseyedi and S. Hubbard. 1999. Cochliobolus phylogenetics and the origin of known, highly virulent pathogens, inferred from ITS and glyceraldehyde-3-phosphate dehydrogenase gene sequences. Mycologia 964–977.

Bianchi, M.A., D. Del Rio, N. Pellegrini, G. Sansebastiano, E. Neviani and F. Brighenti. 2004. A fluorescence-based method for the detection of adhesive properties of lactic acid bacteria to Caco-2 cells. Letters in Applied Microbiology 39(3): 301–305.

Blaiotta, G., V. Fusco, D. Ercolini, M. Aponte, O. Pepe and F. Villani. 2008. Lactobacillus strain diversity based on partial hsp60 gene sequences and design of PCR-restriction fragment length polymorphism assays for species identification and differentiation. Applied and Environmental Microbiology 74(1): 208–215.

Blaiotta, G., O. Pepe, G. Mauriello, F. Villani, R. Andolfi and G. Moschetti. 2002. 16S-23S rDNA intergenic spacer region polymorphism of Lactococcus garvieae, Lactococcus raffinolactis and Lactococcus lactis as revealed by PCR and nucleotide sequence analysis. Syst. Appl. Microbiol. 25(4): 520–527.

Blana, V.A., A. Grounta, C.C. Tassou, G.-J.E. Nychas and E.Z. Panagou. 2014. Inoculated fermentation of green olives with potential probiotic Lactobacillus pentosus and Lactobacillus plantarum starter cultures isolated from industrially fermented olives. Food Microbiology 38: 208–218.

Böhme, K., I.C. Fernández-No, J. Barros-Velázquez, J.M. Gallardo, B. Cañas and P. Calo-Mata. 2010. Comparative analysis of protein extraction methods for the identification of seafood-borne pathogenic and spoilage bacteria by MALDI-TOF mass spectrometry. Analytical Methods 2(12): 1941–1947.

Böhme, K., I.C. Fernández-No, J. Barros-Velázquez, J.M. Gallardo, B. Cañas and P. Calo-Mata. 2011. Rapid species identification of seafood spoilage and pathogenic Gram-positive bacteria by MALDI-TOF mass fingerprinting. Electrophoresis 32(21): 2951–2965.

Böhme, K., I.C. Fernández-No, J. Barros-Velázquez, J.M. Gallardo, B. Cañas and P. Calo-Mata. 2012a. SpectraBank: An open access tool for rapid microbial identification by MALDI-TOF MS fingerprinting. Electrophoresis 33(14): 2138–2142.

Böhme, K., S. Morandi, P. Cremonesi, I.C. Fernández No, J. Barros-Velázquez, B. Castiglioni, M. Brasca, B. Cañas and P. Calo-Mata. 2012b. Characterization of Staphylococcus aureus strains isolated from Italian dairy products by MALDI—TOF mass fingerprinting. Electrophoresis 33(15): 2355–2364.

Bokulich, N.A. and D.A. Mills 2013. House microbiome drives microbial landscapes of artisan cheesemaking plants. Appl. Environ. Microbiol.

Bonetta, S., S. Bonetta, E. Carraro, K. Rantsiou and L. Cocolin. 2008. Microbiological characterisation of Robiola di Roccaverano cheese using PCR-DGGE. Food Microbiol. 25(6): 786–792.

Bottari, B., C. Agrimonti, M. Gatti, E. Neviani and N. Marmiroli. 2013. Development of a multiplex real time PCR to detect thermophilic lactic acid bacteria in natural whey starters. Int. J. Food Microbiol. 160(3): 290–297.

Bouton, Y., P. Guyot, E. Beuvier, P. Tailliez and R. Grappin. 2002. Use of PCR-based methods and PFGE for typing and monitoring homofermentative lactobacilli during Comte cheese ripening. Int. J. Food Microbiol. 76(1-2): 27–38.

Boyer, M.l. and J.r.m. Combrisson. 2013. Analytical opportunities of quantitative polymerase chain reaction in dairy microbiology. International Dairy Journal 30(1): 45–52.

Brauns, L.A., M.C. Hudson and J.D. Oliver. 1991. Use of the polymerase chain reaction in detection of culturable and nonculturable Vibrio vulnificus cells. Appl. Environ. Microbiol. 57(9): 2651–2655.

Brosius, J., M.L. Palmer, P.J. Kennedy and H.F. Noller. 1978. Complete nucleotide sequence of a 16S ribosomal RNA gene from Escherichia coli. Proc. Natl. Acad. Sci. USA 75(10): 4801–4805.

Bunthof, C.J. and T. Abee. 2002. Development of a flow cytometric method to analyze subpopulations of bacteria in probiotic products and dairy starters. Appl. Environ. Microbiol. 68(6): 2934–2942.

Bunthof, C.J., K. Bloemen, P. Breeuwer, F.M. Rombouts and T. Abee. 2001. Flow cytometric assessment of viability of lactic acid bacteria. Appl. Environ. Microbiol. 67(5): 2326–2335.

Cain, T.C., D.M. Lubman, W.J. Weber and A. Vertes. 1994. Differentiation of bacteria using protein profiles from matrix-assisted laser desorption/ionization time-of-flight mass spectrometry. Rapid Communications in Mass Spectrometry 8(12): 1026–1030.

Callon, C., F. Duthoit, C. Delbes, M. Ferrand, Y. Le Frileux, R. De Cremoux and M.C. Montel. 2007. Stability of microbial communities in goat milk during a lactation year: molecular approaches. Syst. Appl. Microbiol. 30(7): 547–560.

Cardenas, E. and J.M. Tiedje. 2008. New tools for discovering and characterizing microbial diversity. Current Opinion in Biotechnology 19(6): 544–549.

Castro-Carrera, T., P. Toral, P. Frutos, N. McEwan, G. Hervás, L. Abecia, E. Pinloche, S. Girdwood and A. Belenguer. 2014. Rumen bacterial community evaluated by 454 pyrosequencing and terminal restriction fragment length polymorphism analyses in dairy sheep fed marine algae. Journal of Dairy Science.

Chan, R.K., C.R. Wortman, B.K. Smiley and C.A. Hendrick. 2003. Construction and use of a computerized DNA fingerprint database for lactic acid bacteria from silage. J. Microbiol. Methods 55(3): 565–574.

Chen, H.C., S.Y. Wang and M.J. Chen. 2008. Microbiological study of lactic acid bacteria in kefir grains by culture-dependent and culture-independent methods. Food Microbiol. 25(3): 492–501.

Chen, S., J. Li, S. Saleh-Lakha, V. Allen and J. Odumeru. 2011. Multiple-locus variable number of tandem repeat analysis (MLVA) of Listeria monocytogenes directly in food samples. International Journal of Food Microbiology 148(1): 8–14.

Chun, J. and F.A. Rainey. 2014. Integrating genomics into the taxonomy and systematics of the Bacteria and Archaea. International Journal of Systematic and Evolutionary Microbiology 64(Pt 2): 316–324.

Cocolin, L., M. Manzano, C. Cantoni and G. Comi. 2001. Denaturing gradient gel electrophoresis analysis of the 16S rRNA gene V1 region to monitor dynamic changes in the bacterial population during fermentation of Italian sausages. Appl. Environ. Microbiol. 67(11): 5113–5121.

Cole, J.R., Q. Wang, E. Cardenas, J. Fish, B. Chai, R.J. Farris, A. Kulam-Syed-Mohideen, D. McGarrell, T. Marsh and G.M. Garrity. 2009. The Ribosomal Database Project: improved alignments and new tools for rRNA analysis. Nucleic Acids Research 37(suppl. 1): D141–D145.

Collins, M., U. Rodrigues, C. Ash, M. Aguirre, J. Farrow, A. Martinez-Murcia, B. Phillips, A. Williams and S. Wallbanks. 1991. Phylogenetic analysis of the genus Lactobacillus and related lactic acid bacteria as determined by reverse transcriptase sequencing of 16S rRNA. FEMS Microbiology Letters 77(1): 5–12.

Coppola, S., G. Blaiotta and D. Ercolini. 2008. Dairy Products. Molecular Techniques in the Microbial Ecology of Feremnted Foods. L. Cocolin and d. Ercolini, Springer 31–90.

Corroler, D., I. Mangin, N. Desmasures and M. Gueguen. 1998. An ecological study of lactococci isolated from raw milk in the camembert cheese registered designation of origin area. Appl. Environ. Microbiol. 64(12): 4729–4735.

Cremonesi, P., L. Vanoni, S. Morandi, T. Silvetti, B. Castiglioni and M. Brasca. 2011. Development of a pentaplex PCR assay for the simultaneous detection of Streptococcus thermophilus, *Lactobacillus delbrueckii* subsp. *bulgaricus*, *L. delbrueckii* subsp. *lactis*, L. helveticus, L. fermentum in whey starter for Grana Padano cheese. Int. J. Food Microbiol. 146(2): 207–211.

Cruciata, M., C. Sannino, D. Ercolini, M. L. Scatassa, F. De Filippis, I. Mancuso, A. La Storia, G. Moschetti and L. Settanni. 2014. Animal rennets as sources of dairy lactic acid bacteria. Appl. Environ. Microbiol. AEM. 03837–03813.

De Vos, P. 2011. Multilocus Sequence Determination and Analysis. Methods in Microbiology 38: 385–407.

del Rio, B., A.G. Binetti, M.C. Martin, M. Fernandez, A.H. Magadan and M.A. Alvarez. 2007. Multiplex PCR for the detection and identification of dairy bacteriophages in milk. Food Microbiol. 24(1): 75–81.

Delbes, C. and M.C. Montel. 2005. Design and application of a Staphylococcus-specific single strand conformation polymorphism-PCR analysis to monitor Staphylococcus populations diversity and dynamics during production of raw milk cheese. Lett. Appl. Microbiol. 41(2): 169–174.

Denoeud, F. and G. Vergnaud. 2004. Identification of polymorphic tandem repeats by direct comparison of genome sequence from different bacterial strains: a web-based resource. BMC Bioinformatics 5: 4.

DeSantis, T.Z., P. Hugenholtz, N. Larsen, M. Rojas, E.L. Brodie, K. Keller, T. Huber, D. Dalevi, P. Hu and G.L. Andersen. 2006. Greengenes, a chimera-checked 16S rRNA gene database and workbench compatible with ARB. Appl. Environ. Microbiol. 72(7): 5069–5072.

Diancourt, L., V. Passet, C. Chervaux, P. Garault, T. Smokvina and S. Brisse. 2007. Multilocus sequence typing of Lactobacillus casei reveals a clonal population structure with low levels of homologous recombination. Appl. Environ. Microbiol. 73(20): 6601–6611.

Doan, N.T.L., K. Van Hoorde, M. Cnockaert, E. De Brandt, M. Aerts, B. Le Thanh and P. Vandamme. 2012. Validation of MALDI-TOF MS for rapid classification and identification of lactic acid bacteria, with a focus on isolates from traditional fermented foods in Northern Vietnam. Letters in Applied Microbiology 55(4): 265–273.

Dolci, P., V. Alessandria, K. Rantsiou, M. Bertolino and L. Cocolin. 2010. Microbial diversity, dynamics and activity throughout manufacturing and ripening of Castelmagno PDO cheese. Int. J. Food Microbiol. 143(1-2): 71–75.

Dolci, P., V. Alessandria, K. Rantsiou, L. Rolle, G. Zeppa and L. Cocolin. 2008a. Microbial dynamics of Castelmagno PDO, a traditional Italian cheese, with a focus on lactic acid bacteria ecology. Int. J. Food Microbiol. 122(3): 302–311.

Dolci, P., V. Alessandria, G. Zeppa, K. Rantsiou and L. Cocolin. 2008b. Microbiological characterization of artisanal Raschera PDO cheese: analysis of its indigenous lactic acid bacteria. Food Microbiol. 25(2): 392–399.

Duffy, G. 2009. Detecting and tracking emerging pathogenic and spoilage bacteria from farm to fork. Safety of Meat and Processed Meat. F. Toldrá, Springer New York 447–459.

Duthoit, F., C. Callon, L. Tessier and M.C. Montel. 2005. Relationships between sensorial characteristics and microbial dynamics in "Regis ered Designation of Origin" Salers cheese. Int. J. Food Microbiol. 103(3): 259–270.

Duthoit, F., J.J. Godon and M.C. Montel. 2003. Bacterial community dynamics during production of registered designation of origin Salers cheese as evaluated by 16S rRNA gene single-strand conformation polymorphism analysis. Appl. Environ. Microbiol. 69(7): 3840–3848.

Ecker, D.J., R. Sampath, L.B. Blyn, M.W. Eshoo, C. Ivy, J.A. Ecker, B. Libby, V. Samant, K.A. Sannes-Lowery and R.E. Melton. 2005. Rapid identification and strain-typing of respiratory pathogens for epidemic surveillance. Proc. Natl. Acad. Sci. USA 102(22): 8012–8017.

Einax, E. and K. Voigt. 2003. Oligonucleotide primers for the universal amplification of β-tubulin genes facilitate phylogenetic analyses in the regnum fungi. Organisms Diversity & Evolution 3(3): 185–194.

Elizaquível, P., R. Aznar and G. Sánchez. 2014. Recent developments in the use of viability dyes and quantitative PCR in the food microbiology field. Journal of Applied Microbiology 116(1): 1–13.

Ercolini, D. 2013. High-throughput sequencing and metagenomics: steps ahead in the culture-independent analysis of food microbial ecology. Appl. Environ. Microbiol.

Ercolini, D., F. De Filippis, A. La Storia and M. Iacono. 2012. "Remake" by high-throughput sequencing of the microbiota involved in the production of water buffalo mozzarella cheese. Appl. Environ. Microbiol. 78(22): 8142–8145.

Ercolini, D., P.J. Hill and C.E. Dodd. 2003a. Bacterial community structure and location in Stilton cheese. Appl. Environ. Microbiol. 69(6): 3540–3548.

Ercolini, D., P.J. Hill and C.E. Dodd. 2003b. Development of a fluorescence *in situ* hybridization method for cheese using a 16S rRNA probe. J. Microbiol Methods 52(2): 267–271.

Ercolini, D., F. Russo, I. Ferrocino and F. Villani. 2009. Molecular identification of mesophilic and psychrotrophic bacteria from raw cow's milk. Food Microbiol. 26(2): 228–231.

Fadda, M.E., S. Viale, M. Deplano, M.B. Pisano and S. Cosentino. 2010. Characterization of yeast population and molecular fingerprinting of Candida zeylanoides isolated from goat's milk collected in Sardinia. Int. J. Food Microbiol. 136(3): 376–380.

Fernandez, E., A. Alegria, S. Delgado, M.C. Martin and B. Mayo. 2011. Comparative phenotypic and molecular genetic profiling of wild *Lactococcus lactis* subsp. lactis strains of the *L. lactis* subsp. lactis and *L. lactis* subsp. cremoris genotypes, isolated from starter-free cheeses made of raw milk. Appl. Environ. Microbiol. 77(15): 5324–5335.

Florez, A.B. and B. Mayo. 2006. Microbial diversity and succession during the manufacture and ripening of traditional, Spanish, blue-veined Cabrales cheese, as determined by PCR-DGGE. Int. J. Food. Microbiol. 110(2): 165–171.

Fortina, M.G., G. Ricci, A. Acquati, G. Zeppa, A. Gandini and P.L. Manachini. 2003. Genetic characterization of some lactic acid bacteria occurring in an artisanal protected denomination origin (PDO) Italian cheese, the Toma piemontese. Food Microbiology 20(4): 397–404.

Frankel, G., L. Riley, J.A. Giron, J. Valmassoi, A. Friedmann, N. Strockbine, S. Falkow and G.K. Schoolnik. 1990. Detection of Shigella in feces using DNA amplification. The Journal of Infectious Diseases 161(6): 1252–1256.

Gala, E., S. Landi, L. Solieri, M. Nocetti, A. Pulvirenti and P. Giudici. 2008. Diversity of lactic acid bacteria population in ripened Parmigiano Reggiano cheese. Int. J. Food Microbiol. 125(3): 347–351.

Gancheva, A., B. Pot, K. Vanhonacker, B. Hoste and K. Kersters. 1999. A polyphasic approach towards the identification of strains belonging to Lactobacillus acidophilus and related species. Syst. Appl. Microbiol. 22(4): 573–585.

Giannino, M.L., M. Marzotto, F. Dellaglio and M. Feligini. 2009. Study of microbial diversity in raw milk and fresh curd used for Fontina cheese production by culture-independent methods. Int. J. Food Microbiol. 130(3): 188–195.

Giraffa, G. and E. Neviani. 2001. DNA-based, culture-independent strategies for evaluating microbial communities in food-associated ecosystems. Int. J. Food Microbiol. 67(1-2): 19–34.

Goering, R.V. 2013. Molecular Typing Techniques: State of the Art. Advanced Techniques in Diagnostic Microbiology. Y.-W. Tang and C.W. Stratton, Springer US 239–261.

Hall, B.G. 2004. Phylogenetic trees made easy: a how-to manual, Sinauer Associates Sunderland.

Handelsman, J., M.R. Rondon, S.F. Brady, J. Clardy and R.M. Goodman. 1998. Molecular biological access to the chemistry of unknown soil microbes: a new frontier for natural products. Chemistry & Biology 5(10): R245–R249.

Harmsen, D., J. Heesemann, T. Brabletz, T. Kirchner and H. Müller-Hermelink. 1994. Heterogeneity among Whipple's-disease-associated bacteria. The Lancet 343(8908): 1288.

Harmsen, D., J. Rothgänger, M. Frosch and J. Albert. 2002. RIDOM: ribosomal differentiation of medical micro-organisms database. Nucleic Acids Research 30(1): 416–417.

Hayden, R.T. 2004. *In Vitro* Nucleic Acid Amplification Techniques. Molecular Microbiology Diagnostic Principles and Practice. D.H. Persing, F.C. Tenover, J. Versalovic, Y.-W. Tang, E.R. Unger, D.A. Relman and T.J. White, ASM Press 43–69.

Henri-Dubernet, S., N. Desmasures and M. Guegen. 2004. Culture-dependent and culture-independent methods for molecular analysis of the diversity of lactobacilli in "Camembert de Normandie" cheese. Lait 84(1-2): 179–189.

Henri-Dubernet, S., N. Desmasures and M. Gueguen. 2008. Diversity and dynamics of lactobacilli populations during ripening of RDO Camembert cheese. Can. J. Microbiol. 54(3): 218–228.

Hunt, K.M., J.A. Foster, L.J. Forney, U.M. Schutte, D.L. Beck, Z. Abdo, L.K. Fox, J.E. Williams, M.K. McGuire and M.A. McGuire. 2011. Characterization of the diversity and temporal stability of bacterial communities in human milk. PLoS One 6(6): e21313.

Hutson, R.A., D.E. Thompson and M.D. Collins. 1993. Genetic interrelationships of saccharolytic Clostridium botulinum types B, E and F and related clostridia as revealed by small-subunit rRNA gene sequences. FEMS Microbiol. Lett. 108(1): 103–110.

Ibrahim, A., W. Liesack and E. Stackebrandt. 1992. Polymerase chain reaction-gene probe detection system specific for pathogenic strains of Yersinia enterocolitica. Journal of Clinical Microbiology 30(8): 1942–1947.

Irlinger, F. and J. Mounier. 2009. Microbial interactions in cheese: implications for cheese quality and safety. Current Opinion in Biotechnology 20(2): 142–148.

Itoh, Y., Y. Kawamura, H. Kasai, M.M. Shah, P.H. Nhung, M. Yamada, X. Sun, T. Koyana, M. Hayashi and K. Ohkusu. 2006. dnaJandgyrBgene sequence relationship among species and strains of genus Streptococcus. Systematic and Applied Mmicrobiology 29(5): 368–374.

James, T.Y., F. Kauff, C.L. Schoch, P.B. Matheny, V. Hofstetter, C.J. Cox, G. Celio, C. Gueidan, E. Fraker and J. Miadlikowska. 2006. Reconstructing the early evolution of Fungi using a six-gene phylogeny. Nature 443(7113): 818–822.

Jany, J.L. and G. Barbier. 2008. Culture-independent methods for identifying microbial communities in cheese. Food Microbiol. 25(7): 839–848.

Jian, W., L. Zhu and X. Dong. 2001. New approach to phylogenetic analysis of the genus Bifidobacterium based on partial HSP60 gene sequences. International Journal of Systematic and Evolutionary Microbiology 51(5): 1633–1638.

Juste, A., B.P. Thomma and B. Lievens. 2008. Recent advances in molecular techniques to study microbial communities in food-associated matrices and processes. Food Microbiol. 25(6): 745–761.

Kern, C.C., R.F. Vogel and J. Behr. 2014. Differentiation of Lactobacillus brevis strains using Matrix-Assisted-Laser-Desorption-Ionization-Time-of-Flight Mass Spectrometry with respect to their beer spoilage potential. Food Microbiology 40(0): 18–24.

Kim, O.-S., Y.-J. Cho, K. Lee, S.-H. Yoon, M. Kim, H. Na, S.-C. Park, Y. S. Jeon, J.-H. Lee, H. Yi, S. Won and J. Chun. 2012. Introducing EzTaxon-e: a prokaryotic 16S rRNA gene sequence database with phylotypes that represent uncultured species. 62: 716–721. Pt 3.

Kolbert, C.P., P.N. Rys, M. Hopkins, D.T. Lynch, J.J. Germer, C.E. O'Sullivan, A. Trampuz and R. Patel. 2011. 16S Ribosomal DNA Sequence Analysis for Identification of Bacteria in a Clinical Microbiology Laboratory. Molecular Microbiology: Diagnostic Principles and Practice. D.H. Persing, F.C. Tenover, J. Versalovic, Y.-W. Tang, E.R. Unger, D.A. Relman and T.J. White, ASM Press.

Kramer, M., N. Obermajer, B. Bogovic Matijasic, I. Rogelj and V. Kmetec. 2009. Quantification of live and dead probiotic bacteria in lyophilised product by real-time PCR and by flow cytometry. Appl. Microbiol. Biotechnol. 84(6): 1137–1147.

Kumar, P.S., M.R. Brooker, S.E. Dowd and T. Camerlengo. 2011. Target region selection is a critical determinant of community fingerprints generated by 16S pyrosequencing. PLoS One 6(6): e20956.

Lahtinen, S.J., M. Gueimonde, A.C. Ouwehand, J.P. Reinikainen and S.J. Salminen. 2006a. Comparison of four methods to enumerate probiotic bifidobacteria in a fermented food product. Food Microbiology 23(6): 571–577.

Lahtinen, S.J., A.C. Ouwehand, J.P. Reinikainen, J.M. Korpela, J. Sandholm and S.J. Salminen. 2006b. Intrinsic properties of so-called dormant probiotic bacteria, determined by flow cytometric viability assays. Appl. Environ. Microbiol. 72(7): 5132–5134.

Lederberg, J. and A. Mccray. 2001. The Scientist:\'Ome Sweet\'Omics—A Genealogical Treasury of Words. The Scientist 17(7).

Lepp, P.W. and D.A. Relman. 2011. Molecular Phylogenetic Analysis. Molecular Microbiology: Diagnostic Principles and Practice. D.H. Persing, F.C. Tenover, J. Versalovic, Y.W. Tang, E.R. Unger, D.A. Relman and T.J. White, ASM Press 161–180.

Loman, N.J., C. Constantinidou, J.Z. Chan, M. Halachev, M. Sergeant, C.W. Penn, E.R. Robinson and M.J. Pallen 2012. High-throughput bacterial genome sequencing: an embarrassment of choice, a world of opportunity. Nature Reviews Microbiology 10(9): 599–606.

Lopez-Garcia, P., F. Rodriguez-Valera, C. Pedros-Alio and D. Moreira. 2001. Unexpected diversity of small eukaryotes in deep-sea Antarctic plankton. Nature 409(6820): 603–607.

Ludwig, W., F.O. Glockner and P. Yilmaz. 2011. The use of rRNA gene sequence data in the classification and identification of prokaryotes. Methods in Microbiology 38: 349–384.

Lusk, T.S., A.R. Ottesen, J.R. White, M.W. Allard, E.W. Brown and J.A. Kase. 2012. Characterization of microflora in Latin-style cheeses by next-generation sequencing technology. BMC Microbiol. 12: 254.

Mackay, I.M. 2004. Real-time PCR in the microbiology laboratory. Clinical Microbiology and Infection 10(3): 190–212.

Maidak, B.L., J.R. Cole, C.T. Parker, G.M. Garrity, N. Larsen, B. Li, T.G. Lilburn, M.J. McCaughey, G.J. Olsen and R. Overbeek. 1999. A new version of the RDP (Ribosomal Database Project). Nucleic Acids Research 27(1): 171–173.

Mannu, L. and A. Paba. 2002. Genetic diversity of lactococci and enterococci isolated from home-made Pecorino Sardo ewes' milk cheese. J. Appl. Microbiol. 92(1): 55–62.

Marras, S.E. 2006. Selection of fluorophore and quencher pairs for fluorescent nucleic acid hybridization probes. Fluorescent Energy Transfer Nucleic Acid Probes. V. Didenko, Humana Press. 335: 3–16.

Martin-Platero, A.M., M. Maqueda, E. Valdivia, J. Purswani and M. Martinez-Bueno. 2009. Polyphasic study of microbial communities of two Spanish farmhouse goats' milk cheeses from Sierra de Aracena. Food Microbiol. 26(3): 294–304.

Masoud, W., M. Takamiya, F.K. Vogensen, S.r. Lillevang, W.A. Al-Soud, S.r.J. S√∏rensen and M. Jakobsen. 2011. Characterization of bacterial populations in Danish raw milk cheeses made with different starter cultures by denaturating gradient gel electrophoresis and pyrosequencing. International Dairy Journal 21(3): 142–148.

Masoud, W., F.K. Vogensen, S. Lillevang, W. Abu Al-Soud, S.J. Sorensen and M. Jakobsen. 2012. The fate of indigenous microbiota, starter cultures, Escherichia coli, Listeria innocua and Staphylococcus aureus in Danish raw milk and cheeses determined by pyrosequencing and quantitative real time (qRT)-PCR. Int. J. Food Microbiol. 153(1-2): 192–202.

Massire, C., D.R. Buelow, S.X. Zhang, R. Lovari, H.E. Matthews, D.M. Toleno, R.R. Ranken, T.A. Hall, D. Metzgar and R. Sampath. 2013. PCR followed by electrospray ionization mass spectrometry for broad-range identification of fungal pathogens. Journal of Clinical Microbiology 51(3): 959–966.

Matamoros, S., P. Savard and D. Roy. 2011. Genotyping of *Bifidobacterium longum* subsp. *longum* strains by multilocus variable number of tandem repeat analysis. J. Microbiol. Methods 87(3): 378–380.

Maukonen, J., H.-L. Alakomi, L. Nohynek, K. Hallamaa, S. Leppämäki, J. Mättö and M. Saarela. 2006. Suitability of the fluorescent techniques for the enumeration of probiotic bacteria in commercial non-dairy drinks and in pharmaceutical products. Food Research International 39(1): 22–32.

McInerney, J.O., M. Wilkinson, J.W. Patching, T.M. Embley and R. Powell. 1995. Recovery and phylogenetic analysis of novel archaeal rRNA sequences from a deep-sea deposit feeder. Appl. Environ. Microbiol. 61(4): 1646–1648.

Mellmann, A., D. Harmsen, C.A. Cummings, E.B. Zentz, S.R. Leopold, A. Rico, K. Prior, R. Szczepanowski, Y. Ji and W. Zhang. 2011. Prospective genomic characterization of the German enterohemorrhagic Escherichia coli O104: H4 outbreak by rapid next generation sequencing technology. PloS One 6(7): e22751.

Mora, D., G. Ricci, S. Guglielmetti, D. Daffonchio and M.G. Fortina. 2003. 16S-23S rRNA intergenic spacer region sequence variation in Streptococcus thermophilus and related dairy streptococci and development of a multiplex ITS-SSCP analysis for their identification. Microbiology (Reading, England) 149(Pt 3): 807–813.

Moreira, J.L.S., R.M. Mota, M.F. Horta, S.M.R. Teixeira, E. Neumann, J.R. Nicoli and Ã.C. Nunes. 2005. Identification to the species level of Lactobacillus isolated in probiotic prospecting studies of human, animal or food origin by 16S-23S rRNA restriction profiling. BMC Microbiology 5.

Moreno, Y., M.C. Collado, M.A. Ferrús, J.M. Cobo, E. Hernández and M. Hernández. 2006. Viability assessment of lactic acid bacteria in commercial dairy products stored at 4°C using LIVE/DEAD® BacLightTM staining and conventional plate counts. International Journal of Food Science & Technology 41(3): 275–280.

Moschetti, G., G. Blaiotta, M. Aponte, P. Catzeddu, F. Villani, P. Deiana and S. Coppola. 1998. Random amplified polymorphic DNA and amplified ribosomal DNA spacer polymorphism: powerful methods to differentiate Streptococcus thermophilus strains. J. Appl. Microbiol. 85(1): 25–36.

Mullis, K.B. and F.A. Faloona. 1987. Specific synthesis of DNA *in vitro* via a polymerase-catalyzed chain reaction. Methods in Enzymology 155: 335–350.

Muyzer, G., A. Teske, C.O. Wirsen and H.W. Jannasch. 1995. Phylogenetic relationships of Thiomicrospira species and their identification in deep-sea hydrothermal vent samples by denaturing gradient gel electrophoresis of 16S rDNA fragments. Archives of Microbiology 164(3): 165–172.

Naravaneni, R. and K. Jamil. 2005. Rapid detection of food-borne pathogens by using molecular techniques. Journal of Medical Microbiology 54(Pt 1): 51–54.

NCBI. 2014. Random Amplified Polymorphic DNA (RAPD). Retrieved 1/2/2014, 2014, from http://www.ncbi.nlm.nih.gov/projects/genome/probe/doc/techRAPD.shtml.

Neviani, E., B. Bottari, C. Lazzi and M. Gatti. 2013. New developments in the study of the microbiota of raw-milk, long-ripened cheeses by molecular methods: the case of Grana Padano and Parmigiano Reggiano. Front Microbiol. 4: 36.

Nikolic, M., A. Terzic-Vidojevic, B. Jovcic, J. Begovic, N. Golic and L. Topisirovic. 2008. Characterization of lactic acid bacteria isolated from Bukuljac, a homemade goat's milk cheese. Int. J. Food Microbiol. 122(1-2): 162–170.

Nocker, A. and A.K. Camper. 2006. Selective removal of DNA from dead cells of mixed bacterial communities by use of ethidium monoazide. Appl. Environ. Microbiol. 72(3): 1997–2004.

Nocker, A. and A.K. Camper. 2009. Novel approaches toward preferential detection of viable cells using nucleic acid amplification techniques. FEMS Microbiol. Lett. 291(2): 137–142.

Nocker, A., C.Y. Cheung and A.K. Camper 2006. Comparison of propidium monoazide with ethidium monoazide for differentiation of live vs. dead bacteria by selective removal of DNA from dead cells. J. Microbiol. Methods 67(2): 310–320.

Nocker, A., A. Mazza, L. Masson, A.K. Camper and R. Brousseau. 2009. Selective detection of live bacteria combining propidium monoazide sample treatment with microarray technology. J. Microbiol. Methods 76(3): 253–261.

Nocker, A., T. Richter-Heitmann, R. Montijn, F. Schuren and R. Kort. 2010. Discrimination between live and dead cellsin bacterial communities from environmental water samples analyzed by 454 pyrosequencing. Int. Microbiol. 13(2): 59–65.

Nocker, A., P. Sossa-Fernandez, M.D. Burr and A.K. Camper. 2007a. Use of propidium monoazide for live/dead distinction in microbial ecology. Appl. Environ. Microbiol. 73(16): 5111–5117.

Nocker, A., K.E. Sossa and A.K. Camper. 2007b. Molecular monitoring of disinfection efficacy using propidium monoazide in combination with quantitative PCR. J. Microbiol. Methods 70(2): 252–260.

O'Sullivan, D.J., L. Giblin, P.L. McSweeney, J.J. Sheehan and P.D. Cotter. 2013. Nucleic acid-based approaches to investigate microbial-related cheese quality defects. Front Microbiol. 4: 1.

Ogier, J.C., V. Lafarge, V. Girard, A. Rault, V. Maladen, A. Gruss, J.Y. Leveau and A. Delacroix-Buchet. 2004. Molecular fingerprinting of dairy microbial ecosystems by use of temporal temperature and denaturing gradient gel electrophoresis. Appl. Environ. Microbiol. 70(9): 5628–5643.

Pal, K., O. Szen, A. Kiss and Z. Naar. 2012. Comparison and evaluation of molecular methods used for identification and discrimination of lactic acid bacteria. J. Sci. Food Agric. 92(9): 1931–1936.

Papadimitriou, K., S. Ferreira, N.C. Papandreou, E. Mavrogonatou, P. Supply, B. Pot and E. Tsakalidou. 2012. Complete genome sequence of the dairy isolate Streptococcus macedonicus ACA-DC 198. J. Bacteriol. 194(7): 1838–1839.

Parayre, S., H. Falentin, M.N. Madec, K. Sivieri, A.S. Le Dizes, D. Sohier and S. Lortal. 2007. Easy DNA extraction method and optimisation of PCR-Temporal Temperature Gel Electrophoresis to identify the predominant high and low GC-content bacteria from dairy products. J. Microbiol. Methods 69(3): 431–441.

Pearson, W.R. and D.J. Lipman. 1988. Improved tools for biological sequence comparison. Proc. Natl. Acad. Sci. USA 85(8): 2444–2448.

Pirondini, A., U. Bonas, E. Maestri, G. Visioli, M. Marmiroli and N. Marmiroli. 2010. Yield and amplificability of different DNA extraction procedures for traceability in the dairy food chain. Food Control 21(5): 663–668.

Porcellato, D., H. Grønnevik, K. Rudi, J. Narvhus and S.B. Skeie. 2012a. Rapid lactic acid bacteria identification in dairy products by high-resolution melt analysis of DGGE bands. Letters in Applied Microbiology 54(4): 344–351.

Porcellato, D., H.M. Ostlie, K.H. Liland, K. Rudi, T. Isaksson and S.B. Skeie. 2012b. Strain-level characterization of nonstarter lactic acid bacteria in Norvegia cheese by high-resolution melt analysis. J. Dairy Sci. 95(9): 4804–4812.

Prajapati, J., C. Khedkar, J. Chitra, S. Suja, V. Mishra, V. Sreeja, R. Patel, V. Ahir, V. Bhatt and M. Sajnani. 2012. Whole-genome shotgun sequencing of Lactobacillus rhamnosus MTCC 5462, a strain with probiotic potential. J. Bacteriol. 194(5): 1264–1265.

Prajapati, J.B., C.D. Khedkar, J. Chitra, S. Suja, V. Mishra, V. Sreeja, R.K. Patel, V.B. Ahir, V.D. Bhatt, M.R. Sajnani, S.J. Jakhesara, P.G. Koringa and C.G. Joshi. 2011. Whole-genome shotgun sequencing of an Indian-origin Lactobacillus helveticus strain, MTCC 5463, with probiotic potential. J. Bacteriol. 193(16): 4282–4283.

Pruesse, E., C. Quast, K. Knittel, B.M. Fuchs, W. Ludwig, J. Peplies and F.O. Glockner. 2007. SILVA: a comprehensive online resource for quality checked and aligned ribosomal RNA sequence data compatible with ARB. Nucleic Acids Research 35(21): 7188–7196.

Psoni, L., C. Kotzamanides, C. Andrighetto, A. Lombardi, N. Tzanetakis and E. Litopoulou-Tzanetaki. 2006. Genotypic and phenotypic heterogeneity in Enterococcus isolates from Batzos, a raw goat milk cheese. Int. J. Food Microbiol. 109(1-2): 109–120.

Quigley, L., O. O'Sullivan, T.P. Beresford, R. Paul Ross, G.F. Fitzgerald and P.D. Cotter. 2012a. A comparison of methods used to extract bacterial DNA from raw milk and raw milk cheese. J. Appl. Microbiol. 113(1): 96–105.

Quigley, L., O. O'Sullivan, T.P. Beresford, R.P. Ross, G.F. Fitzgerald and P.D. Cotter. 2012b. High-throughput sequencing for detection of subpopulations of bacteria not previously associated with artisanal cheeses. Appl. Environ. Microbiol. 78(16): 5717–5723.

Rademaker, J.L., H. Herbet, M.J. Starrenburg, S.M. Naser, D. Gevers, W.J. Kelly, J. Hugenholtz, J. Swings and J.E. van Hylckama Vlieg. 2007. Diversity analysis of dairy and nondairy Lactococcus lactis isolates, using a novel multilocus sequence analysis scheme and (GTG) 5-PCR fingerprinting. Appl. Environ. Microbiol. 73(22): 7128–7137.

Rademaker, J.L. W., J.D. Hoolwerf, A.A. Wagendorp and M.C. Te Giffel. 2006. Assessment of microbial population dynamics during yoghurt and hard cheese fermentation and ripening by DNA population fingerprinting. International Dairy Journal 16(5): 457–466.

Rademaker, J.L.W., M. Peinhopf, L. Rijnen, W. Bockelmann and W.H. Noordman. 2005. The surface microflora dynamics of bacterial smear-ripened Tilsit cheese determined by T-RFLP DNA population fingerprint analysis. International Dairy Journal 15(6-9): 785–794.

Radtke, A., T. Bruheim, J.E. Afset and K. Bergh. 2012. Multiple-locus variant-repeat assay (MLVA) is a useful tool for molecular epidemiologic analysis of Streptococcus agalactiae strains causing bovine mastitis. Veterinary Microbiology 157(3-4): 398–404.

Rahn, K., S.A. De Grandis, R.C. Clarke, S.A. McEwen, J.E. Galan, C. Ginocchio, R. Curtiss, 3rd and C.L. Gyles. 1992. Amplification of an invA gene sequence of Salmonella typhimurium by polymerase chain reaction as a specific method of detection of Salmonella. Molecular and Cellular Probes 6(4): 271–279.

Rainey, F.A. 2011. How to describe new species of prokaryotes. Methods in Microbiology 38: 7–14.

Randazzo, C.L., C. Caggia and E. Neviani. 2009. Application of molecular approaches to study lactic acid bacteria in artisanal cheeses. J. Microbiol. Methods 78(1): 1–9.

Randazzo, C.L., S. Torriani, A.D. Akkermans, W.M. de Vos and E.E. Vaughan. 2002. Diversity, dynamics, and activity of bacterial communities during production of an artisanal Sicilian cheese as evaluated by 16S rRNA analysis. Appl. Environ. Microbiol. 68(4): 1882–1892.

Rantsiou, K., V. Alessandria, R. Urso, P. Dolci and L. Cocolin. 2008a. Detection, quantification and vitality of Listeria monocytogenes in food as determined by quantitative PCR. Int. J. Food Microbiol. 121(1): 99–105.

Rantsiou, K., G. Comi and L. Cocolin. 2004. TherpoBgene as a target for PCR-DGGE analysis to follow lactic acid bacterial population dynamics during food fermentations. Food Microbiology 21(4): 481–487.

Rantsiou, K., R. Urso, P. Dolci, G. Comi and L. Cocolin. 2008b. Microflora of Feta cheese from four Greek manufacturers. Int. J. Food. Microbiol. 126(1-2): 36–42.

Rasko, D.A., D.R. Webster, J.W. Sahl, A. Bashir, N. Boisen, F. Scheutz, E.E. Paxinos, R. Sebra, C.-S. Chin and D. Iliopoulos. 2011. Origins of the E. coli strain causing an outbreak of hemolytic–uremic syndrome in Germany. New England Journal of Medicine 365(8): 709–717.

Rasolofo, E.A., D. St-Gelais, G. LaPointe and D. Roy. 2010. Molecular analysis of bacterial population structure and dynamics during cold storage of untreated and treated milk. Int. J. Food Microbiol. 138(1-2): 108–118.

Reysenbach, A.L. and N.R. Pace. 1995. Reliable amplification of hyperthermophilic archaeal 16S rRNA genes by PCR. Thermophiles. F.T. Robb and A. Place. New York, Cold Spring Habor Press 101–106.

Rodrigues, D., T.A.P. Rocha-Santos, A.C. Freitas, A.C. Duarte and A.M.P. Gomes 2012. Analytical strategies for characterization and validation of functional dairy foods. TrAC Trends in Analytical Chemistry 41(0): 27–45.

Sanchez, I., S. Sesena, J.M. Poveda, L. Cabezas and L. Palop. 2006a. Genetic diversity, dynamics, and activity of Lactobacillus community involved in traditional processing of artisanal Manchego cheese. Int. J. Food Microbiol. 107(3): 265–273.

Sanchez, J.I., L. Rossetti, B. Martinez, A. Rodriguez and G. Giraffa. 2006b. Application of reverse transcriptase PCR-based T-RFLP to perform semi-quantitative analysis of metabolically active bacteria in dairy fermentations. J. Microbiol. Methods 65(2): 268–277.

Sandrin, T.R., J.E. Goldstein and S. Schumaker. 2012. MALDI TOF MS profiling of bacteria at the strain level: a review. Mass Spectrometry Reviews.

Saubusse, M., L. Millet, C. Delbes, C. Callon and M.C. Montel. 2007. Application of Single Strand Conformation Polymorphism—PCR method for distinguishing cheese bacterial communities that inhibit Listeria monocytogenes. Int. J. Food Microbiol. 116(1): 126–135.

Schoch, C.L., K.A. Seifert, S. Huhndorf, V. Robert, J.L. Spouge, C.A. Levesque, W. Chen and F.B. Consortium. 2012. Nuclear ribosomal internal transcribed spacer (ITS) region as a universal DNA barcode marker for Fungi. Proc. Natl. Acad. Sci. USA 109(16): 6241–6246.

Seale, R.B., R. Dhakal, K. Chauhan, H.M. Craven, H.C. Deeth, C.J. Pillidge, I.B. Powell and M.S. Turnera. 2012. Genotyping of present-day and historical geobacillus species isolates from milk powders by high-resolution melt analysis of multiple variable-number tandem-repeat loci. Appl. Environ. Microbiol. 78(19): 7090–7097.

Seifert, K.A. 2009. Progress towards DNA barcoding of fungi. Molecular Ecology Resources 9(s1): 83–89.

Seifert, K.A., R.A. Samson, J. Houbraken, C.A. Lévesque, J.-M. Moncalvo, G. Louis-Seize and P.D. Hebert. 2007. Prospects for fungus identification using CO1 DNA barcodes, with Penicillium as a test case. Proc. Natl. Acad. Sci. USA 104(10): 3901–3906.

Senan, S., S. Grover and V.K. Batish. 2008. Comparison of specificity of different primer pairs for the development of multiplex PCR assays for rapid identification of dairy Lacrobacilli. International Journal of Science & Technology 3(2): 123–137.

Sergeev, N., D. Volokhov, V. Chizhikov and A. Rasooly. 2004. Simultaneous analysis of multiple staphylococcal enterotoxin genes by an oligonucleotide microarray assay. Journal of Clinical Microbiology 42(5): 2134–2143.

Shangkuan, Y.H., Y.S. Show and T.M. Wang. 1995. Multiplex polymerase chain reaction to detect toxigenic Vibrio cholerae and to biotype Vibrio cholerae O1. The Journal of Applied Bacteriology 79(3): 264–273.

Sheffield, V.C., D.R. Cox, L.S. Lerman and R.M. Myers. 1989. Attachment of a 40-base-pair G + C-rich sequence (GC-clamp) to genomic DNA fragments by the polymerase chain reaction results in improved detection of single-base changes. Proc. Natl. Acad. Sci. USA 86(1): 232–236.

Simon, M.C., D.I. Gray and N. Cook. 1996. DNA extraction and PCR methods for the detection of Listeria monocytogenes in cold-smoked salmon. Appl. Environ. Microbiol. 62(3): 822–824.

Soggiu, A., E. Bendixen, M. Brasca, S. Morandi, C. Piras, L. Bonizzi and P. Roncada. 2013. Milk and cheese microbiome for safety and quality of dairy products. Farm animal proteomics 2013. A. Almeida, D. Eckersall, E. Bencurova, S. Dolinska, P. Mlynarcik, M. Vincova and M. Bhide, Wageningen Academic Publishers 262–265.

Solieri, L., T. Dakal and P. Giudici. 2013. Next-generation sequencing and its potential impact on food microbial genomics. Ann. Microbiol. 63(1): 21–37.

Spratt, B.G. 1999. Multilocus sequence typing: molecular typing of bacterial pathogens in an era of rapid DNA sequencing and the internet. Current Opinion in Microbiology 2(3): 312–316.

Stefano, M., B. Milena, L. Roberta and B. Lorenzo. 2008. Molecular typing of Staphylococcus aureus isolated from Italian dairy products on the basis of coagulase gene polymorphism, multiple-locus variable-number tandem-repeat and toxin genes. The Journal of Dairy Research 75(4): 444–449.

Suzzi, G., M. Caruso, F. Gardini, A. Lombardi, L. Vannini, M.E. Guerzoni, C. Andrighetto and M.T. Lanorte. 2000. A survey of the enterococci isolated from an artisanal Italian goat's cheese (semicotto caprino). J. Appl. Microbiol. 89(2): 267–274.

Swofford, D.L., G.J. Olsen, P.J. Waddell and D.M. Hillis. 1996. Chapter 11: Phylogenetic inference. Molecular Systematics 2.

Tada, J., T. Ohashi, N. Nishimura, Y. Shirasaki, H. Ozaki, S. Fukushima, J. Takano, M. Nishibuchi and Y. Takeda. 1992. Detection of the thermostable direct hemolysin gene (tdh) and the thermostable direct hemolysin-related hemolysin gene (trh) of Vibrio parahaemolyticus by polymerase chain reaction. Molecular and Cellular Probes 6(6): 477–487.

Tamura, K., J. Dudley, M. Nei and S. Kumar. 2007. MEGA4: Molecular Evolutionary Genetics Analysis (MEGA) software version 4.0. Molecular Biology and Evolution 24(8): 1596–1599.

Tanigawa, K., H. Kawabata and K. Watanabe. 2010. Identification and typing of Lactococcus lactis by matrix-assisted laser desorption ionization-time of flight mass spectrometry. Appl. Environ. Microbiol. 76(12): 4055–4062.

Thompson, J.D., D.G. Higgins and T.J. Gibson 1994. CLUSTAL W: improving the sensitivity of progressive multiple sequence alignment through sequence weighting, position-specific gap penalties and weight matrix choice. Nucleic Acids Research 22(22): 4673–4680.

Tilburg, J.J., H.J. Roest, M.H. Nabuurs-Franssen, A.M. Horrevorts and C.H. Klaassen. 2012. Genotyping reveals the presence of a predominant genotype of Coxiella burnetii in consumer milk products. Journal of Clinical Microbiology 50(6): 2156–2158.

Torriani, S., F. Clementi, M. Vancanneyt, B. Hoste, F. Dellaglio and K. Kersters. 2001. Differentiation of Lactobacillus plantarum, L. pentosus and L. paraplantarum species by RAPD-PCR and AFLP. Syst. Appl. Microbiol. 24(4): 554–560.

Tsai, Y.L., C.J. Palmer and L.R. Sangermano. 1993. Detection of Escherichia coli in sewage and sludge by polymerase chain reaction. Appl. Environ. Microbiol. 59(2): 353–357.

Van Hoorde, K., T. Verstraete, P. Vandamme and G. Huys. 2008. Diversity of lactic acid bacteria in two Flemish artisan raw milk Gouda-type cheeses. Food Microbiol. 25(7): 929–935.

Vancanneyt, M., G. Huys, K. Lefebvre, V. Vankerckhoven, H. Goossens and J. Swings. 2006. Intraspecific genotypic characterization of Lactobacillus rhamnosus strains intended for probiotic use and isolates of human origin. Appl. Environ. Microbiol. 72(8): 5376–5383.

Ventura, M. and R. Zink. 2002. Specific identification and molecular typing analysis of Lactobacillus johnsonii by using PCR-based methods and pulsed-field gel electrophoresis. FEMS Microbiology Letters 217(2): 141–154.

Ventura, M. and R. Zink. 2003. Comparative sequence analysis of the tuf and recA genes and restriction fragment length polymorphism of the internal transcribed spacer region sequences supply additional tools for discriminating Bifidobacterium lactis from Bifidobacterium animalis. Applied and Environmental Microbiology 69(12): 7517–7522.

Verdier-Metz, I., V. Michel, C. Delbes and M.C. Montel. 2009. Do milking practices influence the bacterial diversity of raw milk? Food Microbiol. 26(3): 305–310.

Weisburg, W.G., S.M. Barns, D.A. Pelletier and D.J. Lane. 1991. 16S ribosomal DNA amplification for phylogenetic study. J. Bacteriol. 173(2): 697–703.

Whitehead, T.R. and M.A. Cotta. 1999. Phylogenetic diversity of methanogenic archaea in swine waste storage pits. FEMS Microbiol. Lett. 179(2): 223–226.

Williams, A., U. Rodrigues and M. Collins. 1991. Intrageneric relationships of< i> Enterococci</ i> as determined by reverse transcriptase sequencing of small-subunit rRNA. Research in Microbiology 142(1): 67–74.

Wilson, I. G. 1997. Inhibition and facilitation of nucleic acid amplification. Appl Environ Microbiol 63(10): 3741–3751.

Wilson, I.G., J.E. Cooper and A. Gilmour. 1991. Detection of enterotoxigenic Staphylococcus aureus in dried skimmed milk: use of the polymerase chain reaction for amplification and detection of staphylococcal enterotoxin genes entB and entC1 and the thermonuclease gene nuc. Appl. Environ. Microbiol. 57(6): 1793–1798.

Wittwer, C.T. and N. Kusukawa. 2004. Real-Time PCR. Molecular Microbiology Diagnostic Principles and Practice. T.F.C. Persing D.H., Versalovic J., Tang Y.-W., Unger E.R., Relman D.A., White T.J., ASM Press 71–84.

Woese, C.R. and G.E. Fox. 1977. Phylogenetic structure of the prokaryotic domain: the primary kingdoms. Proc. Natl. Acad. Sci. USA 74(11): 5088–5090.

CHAPTER 7

# Application of Food Safety Management Systems (FSMS) in the Dairy Industry

*Thomas Bintsis*

## INTRODUCTION

It is a generally accepted fact that milk from a healthy udder is sterile. Therefore, pathogens that enter the milk originate either from the milking parlor (i.e., equipment, environment, personnel), or from means of transportation (i.e., equipment) to the factory for processing. This chapter deals with the application of Hazard Analysis and Critical Control Points (HACCP), in dairy products (including Risk Assessment) and it highlights the importance of on-farm HACCP.

### HACCP in Dairy Products—Principles

The implementation of food hygiene and HACCP system has been reported to be an effective and cost-effective approach to food safety regulation (Mortimore 2001, Unnevehr and Jensen 1999). Although the HACCP system per se does not make food safe, its proper implementation can make a difference. Despite the fact that the food hygiene basic text from Codex Alimentarius (Codex Alimentarius Commission 2003, 2009) serves as the basis for all food safety management systems, a number of countries have

PhD, Dairy Science, 25 Kappadokias St., 55134 Thessaloniki, Greece.
Email: tbintsis@gmail.com

developed national standards for the supply of safe food and individual companies and groupings in the food sector have developed their own standards or programmes for auditing their suppliers such as BRC, IFS, Dutch HACCP, etc. (National Advisory Committee on Microbiological Criteria for Foods 1997, Bernard 1998, Forsythe and Hayes 1998, BRC 2005, Arvanitoyannis and Traikou 2005, Frost 2006). The plethora of more than 20 different such schemes worldwide generates risks of uneven levels of food safety, confusion over requirements, and increased cost and complication for suppliers that find themselves obliged to conform to multiple programmes (Papademas and Bintsis 2010). To fill this gap, the International Organization for Standardization (ISO) published a new food safety management system, the ISO 22000:2005—Food safety management systems—Requirements for any organization in the food chain (International Organization for Standardization 2005). These standards provide a framework of internationally harmonized requirements for the global approach to food safety issues. However, ISO 22000:2005 doesn't contain the non-exhaustive list of Good Manufacturing Practices present in the Global Food Safety Initiative (GFSI) guidance document (GFSI 2007). Thus, the Publicly Available Specification—PAS 220:2008 developed by British Standards Institution (BSI 2008) and published in Oct 2008 in association with major branded dairies (Unilevel, Kraft, Nestle, Danone) and is a new complementary standard to ISO 22000:2005. It specifies the prerequisite programmes requirements in detail to assist in controlling food safety hazards. In fact, it provides harmonization of prerequisite programmes and industry best practice for food manufacturing. GFSI agreed that the combination of ISO 22000:2005 and PAS 220:2008 contained adequate content for approval, but that an industry-owned scheme governing the combination of these two standards must exist. Consequently, the Foundation for Food Safety Certification developed the FSSC 22000, an auditable standard which incorporates food safety elements already known from previous standards such as HACCP, ISO 22000:2005, BRC (2005) and IFS as well as from specifications such as PAS 220:2008 (Sansawat and Muliyil 2009) which has approved by the GFSI as a global benchmark in food safety management. In addition, ISO, through the Technical Committee ISO/TC 34 and based on PAS: 2008, has published ISO 22002-1:2009—Prerequisite programmes on food safety—Part 1: Food manufacturing (International Organization for Standardization 2009). Thus, ISO 22002-1:2009, which has completed with the publication of ISO 22002-2:2013—Prerequisite programmes on food safety—Part 2: Catering (International Organization for Standardization 2013) and ISO 22002-3:2011—Prerequisite programmes on food safety—Part 3: Farming (International Organization for Standardization 2011), is now used in conjunction with ISO 22000:2005 throughout the whole food chain to manage the food safety aspects.

According to the European Legislation, the HACCP system has been mandatorily applied in the food industry since 1993, and has been incorporated in the new food hygiene package since January 2006 with the Regulations 852/2004 and 853/2004 (EC 2004a, EC 2004b). However, an exception was made for producers of primary products (e.g., dairy farmers), even though they should follow the principles of food safety and hygiene codes and the best way is to use HACCP-based systems (Maunsell and Bolton 2004). Well, all food businesses must have a documented food safety management system appropriate to its size and nature and this must be based upon the principles of HACCP. Operators have to identify and regularly review the critical points in their processes and ensure controls are applied at these points. In addition, management personnel responsible for HACCP must receive HACCP training. The procedure needs to be reviewed and any necessary changes made, when any modification is made in the product, process, or any other step.

According to the Regulation 852/2004 (Article 5), the HACCP principles consist of the following (EC 2004a):

a) identifying any hazards that must be prevented, eliminated or reduced to acceptable levels;
b) identifying the critical control points at the step or steps at which control is essential to prevent or eliminate a hazard or to reduce it to acceptable levels;
c) establishing critical limits at critical control points which separate acceptability from unacceptability for the prevention, elimination or reduction of identified hazards;
d) establishing and implementing effective monitoring procedures at critical control points;
e) establishing corrective actions when monitoring indicates that a critical control point is not under control;
f) establishing procedures, which shall be carried out regularly, to verify that the measures outlined in (a) to (e) are working effectively; and
g) establishing documents and records commensurate with the nature and size of the food business to demonstrate the effective application of the measures outlined in (a) to (f).

### *On Farm HACCP*

Although the responsibility lies with the manufacturer for ensuring that the dairy foods manufactured are safe and suitable, there is a continuum of effective effort or controls needed by other parties, including milk producers, to assure the safety and suitability of milk products. It is important to recognize that distributors, competent authorities and consumers also have a role in ensuring the safety and suitability of milk and milk products. Due to the special importance of the primary production to the safety of the dairy products, major branded dairies have introduced their own on-farm HACCP programmes. For example, Arla Foods has established a quality program entitled Arlagεrden ('the Arla farm') to be used by their farmers. The program specifies Arla Foods' requirements not only for food safety and milk composition, but also for animal welfare and environmental protection (Junedahl et al. 2008). The Canadian dairy industry has begun implementing an on-farm food-safety program called Canadian Quality Milk (Young et al. 2010). These quality assurance programmes starting at dairy farm level deals with food safety, animal health and animal welfare issues to take account of the demands of consumers and retailers (Noordhuizen and Metz 2005).

The HACCP concept, focused on risk management and prevention, appears to be very promising to control on-farm processes. It can be easily linked to both operational management and food chain quality assurance and is suitable for certification (Noordhuizen 2003, Heeschen and Bluthgen 2004, Noordhuizen and Jorritsma 2006). In fact, introduction of HACCP on dairy farms means nothing more than 'structuring and formalising what the truly good farmer would be doing anyway' (Ryan et al. 1997). Until now, the introduction of HACCP principles in on-farm management has hardly been tested in practice, due to many objectively immeasurable processes in an on-farm situation (Noordhuizen 2003).

Livestock species are an important reservoir of *C. jejuni*, shiga-toxin producing *E. coli*, *L. monocytogenes*, *Salmonella* spp., and *Y. enterocolitica* (Jayarao and Henning 2001, Murinda et al. 2002, Jayarao et al. 2006) and other pathogenic bacteria that have been implicated in a number of food-borne outbreaks (Papademas and Bintsis 2010). These pathogens have been recovered with various frequencies from dairy-cattle faeces, bulk milk tanks and the dairy-farm environment (Troutt 1995, Jayarao and Henning 2001, Murinda et al. 2002, Wiedmann 2006, Van Kessel et al. 2004, Srinivasan et al. 2005, Karns et al. 2007, Vissers and Driehuis 2009). Molecular epidemiological studies of E. coli O157:H7 have demonstrated that subtypes of the organism can persist on cattle farms for years (Hancock et al. 2001, Aspán and Eriksson 2010). The presence of food-borne pathogens in milk is due to direct contact with contaminated sources in the dairy farm environment and to excretion from the udder of an infected animal (Oliver et al. 2005, Kousta et al. 2010). Fox et al. (2009) demonstrated the prevalence of *L. monocytogenes* in the dairy farm environment and the need for good hygiene practices to prevent its entry into the food chain and Hussein and Sakuma (2005) described pre- and postharvest control measures to ensure safety of dairy cattle products. D'Amico et al. (2008) reported that the incidence of food-borne pathogens of concern in raw milk utilized for farmstead cheese production was very low, whereas, Danielsson-Tham et al. (2004) stated that the conditions on a summer farm can hardly fulfill the requirements for hygienic and strictly controlled conditions necessary for safe processing of fresh cheese.

Outbreaks due to the consumption of unpasteurized milk (Peterson 2003, Centers for Disease Control and Prevention 2007, Lind et al. 2008, Heuvelink et al. 2009, Lejeune and Rajala-Schultz 2009, Oliver et al. 2009), inadequately pasteurized milk (Fahey et al. 1995) and cheeses made from unpasteurized milk (Honish et al. 2005, Center for Science in the Public Interest 2008, 2009) continue to occur. Campylobacteriosis and Salmonellosis was the most common zoonotic diseases in humans in the European Union during 2008, but incidences of both have fallen, whereas, the number of cases of Verotoxigenic Escherichia coli (VTEC) rose by almost 9% and the number of listeriosis cases in humans decreased by 11.1% (European Food Safety Authority 2010). In 2008, there were 5,332 food-borne outbreaks in the EU, sickening over 45,000 people and causing 32 deaths. Some 35 per cent of these were triggered by *Salmonella* spp., with viruses and bacterial toxins detailed as the next most common causes.

When milk is intended to be used for the manufacture of raw milk products, hygienic conditions used at the primary production are one of the most important public health control measures, as a high level of hygiene of the milk is essential in order to obtain milk with a sufficiently low initial microbial load in order to enable the manufacturing of raw milk products that are safe and suitable for human consumption (Codex Alimentarius Commission 2004). In such situations, additional control measures may be necessary. In addition, increased emphasis in certain aspects of the production of milk for raw milk products (animal health, animal feeding, and milk hygiene monitoring) are specified and are critical to the production of milk that is safe and suitable for the intended purpose. Interestingly, the FDA/Health Canada risk assessment found that the risk of listeriosis from soft-ripened cheeses made with raw milk is estimated to be 50 to 160 times higher than that from soft-ripened cheese made with pasteurized milk (Food and Drug Administration/Health Canada 2012). This finding is consistent with the fact that consuming raw milk and raw milk products generally poses a higher risk from pathogens than do pasteurized milk and its products.

*E. coli, S. aureus, Corynebacterium bovis, Klebsiella* spp., or *Pseudomonas* spp., *Streptococcus agalactiae, Streptococcus dysgalactiae* and *Streptococcus uberis* may cause, under certain circumstances, clinical mastitis (Hahn 1996, Barkema et al. 1998). Special care should be taken for the use of antibiotics, and Sawant et al. (2007) found that enteric bacteria such as *E. coli* from healthy lactating cattle can be an important reservoir for tetracycline and other antimicrobial resistance determinants.

Antimicrobial residues and antimicrobial-resistant bacteria from milk and milk products can also pose potential health risks to consumers (Katz and Brady 2000, Straley et al. 2006). Contamination of milk via the exterior of the cows' teats occurs when teats, and subsequently milk, are contaminated with dirt consisting of faeces, bedding material, soil, or a combination of these. Vissers et al. (2007a) applied quantitative microbial risk analysis of the microbial contamination of farm tank milk for the amount of dirt transmitted to milk via the exterior of teats using spores of mesophilic aerobic bacteria as a marker for transmitted dirt. Silage was the main source of butyric acid bacteria and clostridia spores in cheese milk (Vissers et al. 2007a, Julien et al. 2008). When silage fermentation conditions are not prone to rapid pH decrease and maintenance of uniformly anaerobic conditions, germination of the spores and subsequent vegetative cell multiplication can occur. Vissers et al. (2006) applied a modeling approach to identify an effective control strategy at the farm level and found that contamination level of silage with the butyric acid bacteria was found to be the most important factor for the control of contamination of farm tank milk. In addition, Sanaa et al. (1993) found that poor quality of silage (pH > 4), inadequate frequency of cleaning the exercise area, poor animal cleanliness, insufficient lighting of milking barns and parlors, and incorrect disinfection of towels between milkings were significantly associated with raw milk contamination by *L. monocytogenes*.

Cleaning and disinfection of the udder and the teats with appropriate agents before and after milking can minimize infections during milking (Burgess et al. 1994). Temperature abuse should be avoided at all stages in the farm, and temperature should not increase above 6°C during transportation to the dairy (Dijkers et al. 1995). In recent years, improved standards of housing and use of separate milking parlors have reduced the risk of raw milk contamination. Probable control points on the farm for many of these human pathogens will be: 1) housing and bedding, 2) water and waste management areas, 3) hospital pens, 4) calving pens, 5) treatment areas, 6) bulk tank milk, and 7) young stock and cull animals (Cullor 1997).

As part of any on-farm HACCP system, cost-effective, accurate and reproducible tests should be used to monitor certain control points (Reybroeck 1996). However, the monitoring is limited by inadequacies and costs of existing testing methodologies (Gardner 1997).

The dairy industry can adapt and implement Good Dairy Farming Practice (GDFP) to aid in managing animal health problems and to begin addressing pathogens of concern for food-borne and waterborne illness. A joint guidance on GDFP was published from the International Dairy Federation and the Food and Agriculture Organization of the United Nations (International Dairy Federation/Food and Agriculture Organisation 2004). The objective is that milk should be produced on-farm from healthy animals under generally accepted conditions. To achieve this, dairy farmers need to apply Good Agricultural Practice in the following areas: a) animal health, b) milking hygiene, c) animal feeding and water, d) animal welfare, and e) environment.

In addition, Globalgap has recently published regulations for integrated farm assurance, which contains all the control points and compliance criteria that must be followed by the producer and which are audited to verify compliance (Globalgap 2007). For dairy farms, the protocol includes specifications for feed, housing and facilities, dairy health, milking, milking facilities (i.e., milking equipment, milking parlour, milk collection equipment), hygiene, cleaning agents and other chemicals.

## Application of HACCP System in Dairy Products

Several studies have been published concerning the implementation of HACCP on dairy products such as pasteurized, ultra high temperature (UHT) and condensed milk (Dijkers et al. 1995, Ali and Fischer 2002, Sandrou and Arvanitoyannis 2000a), yogurt (Sandrou and Arvanitoyannis 2000a), a variety of cheeses (Sandrou and Arvanitoyannis 2000b, Mauropoulos and Arvanitoyannis 1999, Arvanitoyannis and Mavropoulos 2000, Arvanotoyannis et al. 2009), ice cream (Mortimore and Wallace 1998, Papademas and Bintsis 2002, Arvanotoyannis et al. 2009) as well as cream and butter (Sandrou and Arvanitoyannis 2000a, Ali and Fischer 2005).

### *Risk Analysis*

More than 100 countries have signed the "Sanitary and Phytosanitary (SPS) Agreement" of the World Trade Organization (WTO). This agreement states that "whilst a country has the sovereign right to decide on the degree of protection it wishes for its citizens, it must provide, if required, the scientific evidence on which this level of protection rests." It follows that if a country sets a microbiological criterion—or any other limit—for a particular health hazard in a particular food product, they must be able to explain, based on scientific data, consideration of risk and societal considerations, the rationale and justification for the criterion. The "Technical Barriers to Trade (TBT) Agreement," also requires that a country must not ask for a higher degree of safety for imported goods than it does for goods produced in its own country (International Commission on Microbiological Specifications for Food 2006).

Risk assessment for food safety is part of the risk analysis framework, provided by the Codex Alimentarius Commission (Codex Alimentarius Commission 1999), which also includes risk management and risk communication as interdependent concepts. Risk analysis is used to develop an estimate of the risks to human health and safety, to identify and implement appropriate measures to control the risks, and to communicate with stakeholders about the risks and measures applied. It can be used to support and improve the development of standards, as well as to address food safety issues that result from emerging hazards or breakdowns in food control systems. It provides food safety regulators with the information and evidence they need for effective decision-making, contributing to better food safety outcomes and improvements in public health (Food and Agriculture Organisation/World Health Organisation 2004).

Risk assessment has been set as a top priority issue on the basic food legislation document, EC Regulation 178/2002 (EC 2002), and is defined as a "scientifically based process consisting of four steps: hazard identification, hazard characterization, exposure assessment and risk characterization". Risk is a "function of the probability of an adverse health effect and the severity of that effect, consequential to a hazard" (EC 2002). Thus, risk assessment requires the collection of scientific data regarding the nature, frequency and impact on public health in Europe of food safety hazards. Indeed, the severity of a food-borne illness caused by a biological hazard must be combined with its occurrence in humans to accurately define risk ( Food and Agriculture Organisation/World Health Organisation 2004).

The core elements of a risk assessment process are 1) hazard identification (i.e., the identification of biological, chemical and physical agents capable of causing adverse health effects which may be present in a particular food or group of foods), 2) hazard characterization (i.e., the qualitative and/or quantitative evaluation of the nature of the adverse health effects associated with biological, chemical and physical agents which may be present in food—for chemical agents a dose–response assessment should be performed, for biological or physical agents a dose-response assessment should be performed if the data are obtainable), 3) exposure assessment (i.e., the qualitative and/or quantitative evaluation of the likely intake of biological, chemical and physical agents via food as well as exposures from other sources if relevant) and 4) risk characterization (i.e., the qualitative and/or quantitative estimation, including attendant uncertainties, of the probability of occurrence and severity of known or potential adverse health effects in a given population based on hazard identification, hazard characterization and exposure assessment) (ILSI 2012) as presented in Fig. 1. Risk characterization brings together all of the qualitative or quantitative information of the previous steps to provide a soundly based estimate of risk for a given population.

Fully quantitative approaches are used nowadays and risks are expressed as, for example, the number of cases of food-borne disease per number of people per year, dose–response relationships and exposure

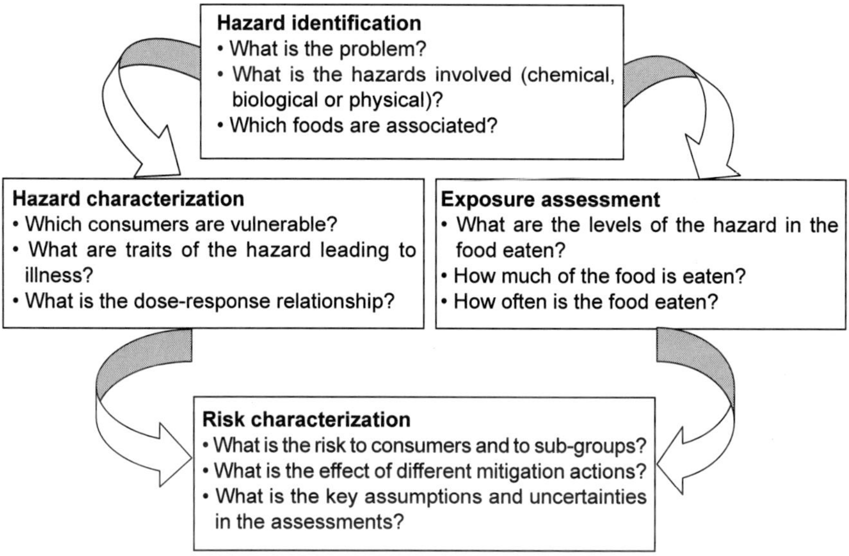


**Figure 1.** The basic steps in risk assessment (After: Codex Alimentarius Commission (1999)).

assessments. Guidelines for chemicals in foods will inevitably have to address the differences between safety evaluation and a genuine risk assessment approach. With respect to microbiological hazards, the unique problems associated with risk assessment of living organisms in food make it likely that the application of guidelines in the medium term will more commonly use qualitative approaches. As risk assessment increases and applied and internationally accepted guidelines become established, decision criteria for risk management arguably present the greatest challenge in establishing and maintaining quantitative SPS measures for food in international trade and in judging their equivalence (Hathaway 1997).

Mathematical/probabilistic modeling is employed to estimate the risk per serving of a specific food as described in FDA/Health Canada (2012). In addition, second order Monte-Carlo simulation is used (Frey 1992) and the variability in the risk (e.g., from serving to serving or from country to country) is estimated and the uncertainty can be evaluated.

The International Commission on Microbiological Specifications for Foods (ICMSF) has proposed a scheme for the management of microbial hazards for food that involves the concept of food safety objectives (FSOs), i.e., the maximum frequency and/or concentration of a hazard in a food at the time of consumption that provides or contributes to the appropriate level of protection (ALOP) (International Commission on Microbiological Specifications for Foods 2006). To ensure that an FSO is met, it is required to set performance objectives (POs), which correspond to the levels that must be met during the earlier steps in the food chain before consumption. The FSO gives flexibility to the food chain to use different operations and processing techniques that best suit their situation, as long as the maximum hazard level specified at consumption is not exceeded (e.g., the replacement of heat treatment with another equivalent technique, i.e., microfiltration). The position of these concepts appearing in the food chain can be seen in Fig. 2. Microbiological standards (Buchanan 1995) have been included in European Legislation (EC 2005).

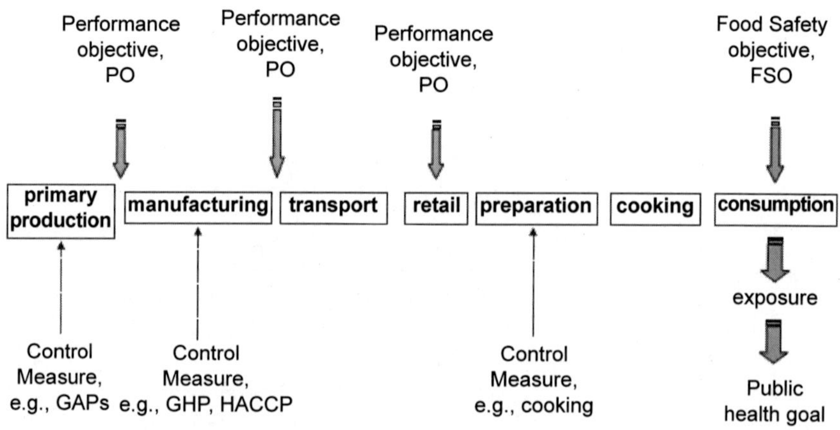


**Figure 2.** A presentation of a model food chain indicating the position of a Food Safety Objective and derived Performance Objectives (After: International Commission on Microbiological Specifications for Foods (2006)).

Risk communication between risk managers and stakeholders is simplified when the results of a qualitative assessment are available (Clough et al. 2006, Hauser et al. 2007). However, quantitative assessments are much more useful tools (Food and Drug Administration/Health Canada 2012, Peeler and Bunning 1994, Bemrah et al. 1999, Sanaa et al. 2004) for microbial risk assessments, and the Monte Carlo exposure assessment model for mycotoxins in dairy milk was developed by Coffie et al. (2009).

Brouillaud-Delattre et al. (1997) used predictive microbiology to study the influence of biological factors affecting the growth of Listeria monocytogenes in sterilized milk and raw dairy products and postulated that it was influenced greatly by bacterial interactions and physiological state of inoculum cells. Albert et al. (2005) described a Monte Carlo simulation that forecasts bacterial growth and exposure assessment for L. monocytogenes in milk. Predictive models are now essential part of risk assessments (McMeekin and Ross 2002, Notermans et al. 2002).

The development and use of a simple tool for food safety risk assessment has been described by Ross and Sumner (2002) in spreadsheet software format that embodies established principles of food safety risk assessment. Microbial predictive modeling techniques have been developed by many workers (George et al. 1996, Murphy et al. 1996, McClure et al. 1997, Xanthiakos et al. 2006, Membre and Lambent 2008). The validation of such models has been investigated (Murphy et al. 1996, Ross 1996, Baranyi et al. 1999, te Giffel and Zwietering 1999). Notermans et al. (1995) suggested the use of quantitative risk assessment for setting critical limits at the CCPs of a HACCP system for realistic levels of control.

Several computer programs have been launched for estimating bacterial growth and inactivation in different products such as the Pathogen Modeling Program, PMP (United States Department of Agriculture-Agricultural Research Service 2006), Combase Predictor (Combase 2010), the online resource for food safety risk analysis (Joint Institute for Food Safety and Applied Nutrition 2010), the @RISK spreadsheet software (Palisade 2012), the Crystal Ball Oracle Risk assessment software (Oracle 2012), Seafood Spoilage and Safety Predictor (National Institute of Aquatic Resources 2009), and Safe Foods (2010) for helping in development of HACCP-systems or in performing quantitative risk assessment (McMeekin et al. 2006, 2008, Hignette et al. 2008).

International agencies and all levels of government are increasingly relying on, or at least recognizing the need to rely on, risk assessments for decision-making in public health protection, international trade, and to support cost-effective resource allocation including prioritizing research directions (Council for Agricultural Science and Technology 1994, International Life Science Institute 1996, 2005, Food and Agriculture Orginisation/World Health Orginisation 1997) and several authors have highlighted the need for the application of risk assessment methods to food safety (Jaykus 1996, Kindred 1996, Lammerding 1997, Buchanan et al. 1998, Voysey and Brown 2000). For these reasons, a web-based quantitative risk assessment system was developed (Chen et al. 2013), which enables users to assess, compare and rank the risks posed by multiple food-hazard pairs at all stages of the food supply system, from primary production, through manufacturing and processing, to retail distribution and, ultimately, to the consumer.

Full quantitative risk assessments are provided by the three Joint Food and Agriculture Organisation/World Health Organisation Expert Bodies: the Joint Expert Committee on Food Additives (JECFA); the Joint Meeting on Pesticide Residues (JMPR); and the Joint Expert Meeting on Microbiological Risk Assessment (JEMRA). Additional risk assessments may be provided, on occasion, by ad hoc expert consultations, and by member governments that have conducted their own assessments

(FAO/WHO 2004). Methods for microbial food safety risk assessment are being developed by various organizations (Food and Agriculture Organisation/World Health Organisation 1995, PCCRARM 1997, Codex Alimentarius Commission 1999, FAO/WHO 1997, International Life Science Institute 1996, 2005, 2012) and, since the mid-1990s, a number of microbiological risk assessments have been presented (Schlundt 2000).

## Conclusions

The integration of HACCP plans with the development of fully quantitative risk assessments offer very useful tools for controlling the entire farm-to-table food chain. Moreover, the importance of dairy factory hygiene needs to be highlighted, as well as the need for efficient controls of the feed administered to production animals. For example, *Salmonella* spp. is a food pathogen that according to the Rapid Alert System for Food and Feed (RASFF) of the European Union (EU) is isolated in animal's feed. Finally, one must be aware and be able to follow changes regarding microbiological standards as these change both in EU and globally.

## References Cited

2005. ISO 22000:2005, Food safety management systems—Requirements for any organization in the food chain. Geneva, Switzerland.

2007. Global Food Safety Initiative Guidance Document, 5th edition. URL http://www.ciesnet.com/pfiles/programmes/foodsafety/GFSI_Guidance_Document_5th%20Edition%20_September%202007.pdf.

2009. ISO 22002-1:2009—Prerequisite programmes on food safety—Part 1: Food manufacturing. Geneva, Switzerland.

2010. A combined database for predictive microbiology. URL http://combase.arserrc.gov/BrowserHome/SearchOptions/Search.aspx.

2010. CIES—The Food Business Forum Global Food Safety Initiative URL http://http://www.ciesnet.com/4-press/4.2-press-release/index.asp.

2010. The Community Summary Report on Trends and Sources of Zoonoses and Zoonotic Agents and Foodborne Outbreaks in the European Union in 2008. The EFSA Journal 1496: 1–288.

2011. ISO 22002-3:2011—Prerequisite programmes on food safety—Part 3: Farming. Geneva, Switzerland.

2013. ISO 22002-2:2013—Prerequisite programmes on food safety—Part 2: Catering. Geneva, Switzerland.

Albert, I., R. Pouillot and J.B. Denis. 2005. Stochastically modeling *Listeria monocytogenes* growth in farm tank milk. Risk Analysis 25(5): 171–185.

Ali, A.A. and R.M. Fischer. 2002. Implementation of HACCP to bulk condensed milk production line. Food Reviews International 18(2-3): 177–190.

Ali, A.A. and R.M. Fischer. 2005. Implementation of HACCP to bulk cream and butter production Line. Food Reviews International 21189–210.

Arvanitoyannis, I.S. and A. Traikou. 2005. A comprehensive review of the implementation of hazard analysis critical control point (HACCP) to the production of flour and flour-based products. Critical Reviews in Food Science and Nutrition 45: 327–370.

Arvanitoyannis, I.S. and A.A. Mavropoulos. 2000. Implementation of the hazard analysis critical control point (HACCP) system to Kasseri/Kefalotiri and Anevato cheese production lines. Food Control 11(1) 31–40.

Arvanitoyiannis, I.S., T.H. Varzakas and M. Koukaliaroglou-van Houwelingen. 2009. Implementing HACCP and ISO 22000 for foods of animal origin—Dairy products. pp 91–180. *In*: I.S. Arvanitoyiannis [ ed.]. HACCP and ISO 22000—Application to Foods of Animal Origin. Wiley—Blackwell, Oxford, UK.

Aspán, A. and E. Eriksson. 2010. Verotoxigenic *Escherichia coli* O157:H7 from Swedish cattle; isolates from prevalence studies versus strains linked to human infections—A retrospective study. BMC Veterinary Research 6: 7–14.

Baranyi, J., C. Pin and T. Ross. 1999. Validating and comparing predictive models. International Journal of Food Microbiology 48: 159–166.

Barkema, H.M., Y.H. Schukken, T.G. Lam, M.L. Beiboer, H. Wilmine, G. Benediktus and A. Brand. 1998. Incidence of clinical mastitis in dairy herds grouped in three categories by bulk milk somatic cell counts. Journal of Dairy Science 81: 411–419.

Bemrah, N., M. Sanaa, M.H. Cassin, M.W. Griffiths and O. Cerf. 1999. Quantitative risk assessment of human listeriosis from consumption of soft cheese made from raw milk. Preventative Veterinary Medicine 37: 129–145.

Bernard, D. 1998. Developing and implementing HACCP in the USA. Food Control 9(2-3): 91–95.

British Retail Consortium Global Standard. 2005. British Retail Consortium Global Standard —Food Issue 4. URL http://www.brc.org.uk/standards/downloads/food_std_background.pdf.

British Standard Institute. 2008. British Standard Institute Group. URL http://www.bsigroup.com/en/Assessment-and-certification-services/management-systems/Standards-and-Schemes/PAS-220.

Brouillaud-Delattre, A., M. Maire, C. Collette, C. Mattei and C. Lahellec. 1997. Predictive microbiology of dairy products: influence of biological factors affecting growth of *Listeria monocytogenes*. Journal of AOAC International 80(4): 913–919.

Buchanan, R.L. 1995. The role of microbiological criteria and risk assessment in HACCP. Food Microbiology 12: 421–424.

Buchanan, R.L. and R.C. Whiting. 1998. Risk assessment: a means for linking HACCP plans and public health. Journal of Food Protection 61(11): 1531–1534.

Buchanan, R.L., A.M. Lammerding, M. van Schothorst and T.A. Roberts. 1998. International Commission on Microbiological Specifications for Foods working group on microbial risk assessment. Potential application of risk assessment techniques to microbiological issues related to international trade in food and food products. Journal of Food Protection 61: 1075–1086.

Burgess, K., C. Heggum, S. Walker and M. van Schothorst. 1994. Recommendations for the hygienic manufacture of milk and milk based products. Bulletin of the IDF 292: 12–19.

Center for Science in the Public Interest. 2008. Outbreak alert 2008. URL http://www.cspinet.org/new/pdf/outbreak_alert_2008_report_final.pdf.

Center for Science in the Public Interest. 2009. Outbreak alert Analyzing foodborne outbreaks 1998–2007. URL http://cspinet.org/new/pdf/outbreakalertreport09.pdf.

Center for Science in the Public Interest. 2010. Outbreaks Alert Database – Dairy. URL http://www.cspinet.org/foodsafety/outbreak/outbreaks.php?column=food&colval=Dairy.

Centers for Disease Control and Prevention. 2007. Centers for Disease Control and Prevention, MMWR Weekly Report, http://www.cdc.gov/mmwr/preview/mmwrhtml/mm5608a3.htm.

Chen, Y., S.B. Dennis, E. Hartnett, G. Paoli, R. Pouillot, T. Ruthman and M. Wilson. 2013. FDA-iRISK—A comparative risk assessment system for evaluating and ranking food-hazard pairs: case studies on microbiological hazards. Journal of Food Protection 3: 376–385.

Clough, H.E., D. Clancy and N.P. French. 2006. Vero-Cytotoxigenic *Escherichia coli* O157 in pasteurized milk containers at the point of retail: a qualitative approach to exposure assessment. Risk Analysis 26(5): 1291–1309.

Codex Alimentarius Commission. 1999. Principles and Guidelines for the Conduct of Microbiological Risk Assessment. URL *www.codexalimentarius.net/download/standards/.../CXG_030e.pdf.*

Codex Alimentarius Commission. 2003. Recommended international code of practice—General principles of food hygiene. CAC/RCP 1–1969, Rev. 4–2003. Rome: FAO.

Codex Alimentarius Commission. 2004. Code of Hygiene Practice for Milk and Milk Products. Codex Alimentarius Committee. CAC/RCP 57–2004. http://www.codexalimentarius.org/input/download/standards/10087/CXP_057e.pdf.

Codex Alimentarius Commission. 2009. Food Hygiene—Basic texts. 4th Edition. Codex Alimentarius Commission. Joint FAO/WHO Food Standards Programme. Rome.

Coffie, R., E. Cummins and S. Ward. 2009. Exposure assessment of mycotoxins in dairy milk. Food Control 20: 239–249.

Council for Agricultural Science and Technology. 1994. Foodborne pathogens: risk and consequences. Task Force Report No 122. Ames, Iowa, USA.

Cullor, J.S. 1997. HACCP: Is it coming to the dairy? Journal of Dairy Science 80: 3449–3452.

D'Amico, D.J., E. Groves and C.W. Donnelly. 2008. Low incidence of foodborne pathogens of concern in raw milk utilized for farmstead cheese production. Journal of Food Protection 71(8): 1580–1589.

Danielsson-Tham, M.L., E. Eriksson, S. Helmersson, M. Leffler, L. Lüdtke, M. Steen, S. Sørgjerd and W. Tham. 2004. Causes behind a human cheese-borne outbreak of gastrointestinal listeriosis. Foodborne Pathogens Diseases 1(3): 153–159.

Dijkers, J.H., T. Huurnink, P.P.L. Pennings and M.G. van den Berg. 1995. An example of HACCP application in an existing pasteurized milk plant, following the Codex Alimentarius model. Bulletin of the IDF 302: 11–34.

EC. 2002. Regulation (EC) No. 178/2002 of the European Parliament and of the Council of 28 January 2002 laying down the general principles and requirements of food safety law, establishing the European Food Standards Agency and laying down procedures in matters of food safety. Official Journal of the European Union L31: 1–24.

EC. 2004a. Regulation (EC) No. 852/2004 of the European Parliament and of the Council of 29 April 2004 on the hygiene of foodstuffs ( including HACCP principles). Official Journal of the European Union L139: 1–54.

EC. 2004b. Regulation (EC) No.853/2004/EC of the European Parliament and of the Council of 29 April 2004 laying down specific hygiene rules for food of animal origin. Official Journal of the European Union L139: 55–205.

EC. 2005. Commission Regulation (EC) No. 2073/2005 of 15 November 2005 laying down microbiological criteria for certain products. Official Journal of the European Union L338: 1–26.

Fahey, T., D. Morgan, C. Gunneburg, G.K. Adak, F. Majid and E. Kaczmarski. 1995. An outbreak of *Campylobacter jejuni* enteritis associated with failed milk pasteurisation. Journal of Infections 31: 137–143.

Food and Agriculture Organization/World Health Organisation. 1995. Application of Risk Analysis to Food Standards Issues. Report of the Joint FAO/WHO Expert Consultation, World Health Organisation, Geneva, Switzerland.

Food and Agriculture Organization/World Health Organisation. 1997. Risk Management and Food Safety. Report of a Joint FAO/WHO Consultation. Food and Agriculture Organization of the United Nations, Rome, Italy.

Food and Agriculture Organization/World Health Organisation. 2004. Risk assessment of Listeria monocytogenes in ready-to-eat foods. Microbiological risk assessment series 5. Food and Agriculture Organization of the United Nations, Rome, Italy. URL http://www.fao.org/es/esn/food/risk_mra_listeria_report_en.stm.

Food and Drug Adminsitration/Health Canada. 2012. Quantitative Assessment of the Risk of Listeriosis from Soft-Ripened Cheese Consumption in the United States and Canada: Draft Report. Food Directorate—Health Canada/Center for Food Safety and Applied Nutrition—Food and Drug Administration, U.S. Department of Health and Human Services.

Forsythe, S.J. and P.R. Hayes. 1998. Food Hygiene, Microbiology and HACCP, 3rd ed. Maryland: Aspen Publishers, pp. 276–326.

Fox, E., T. O'Mahony, M. Clancy, R. Dempsey, M. O'Brien and K. Jordan. 2009. *Listeria monocytogenes* in the Irish dairy farm environment. Journal of Food Protection 72(7): 1450–1456.

Frey, H.C. 1992. Quantitative Analysis of Uncertainty and Variability in Environmental Policy Making. http://www4.ncsu.edu/~frey/reports/frey_92.pdf.

Frost. 2006. Early adopters underline benefits of new ISO standard for safe food supply chains ISO Management Systems—March–April 2006: 21– 23.

Gardner, I.A. 1997. Testing to Fulfill HACCP (Hazard Analysis Critical Control Points) Requirements: Principles and Examples. Journal of Dairy Science 80: 3453–3457.

George, S.M., L.C.C. Richardson and M.W. Peck. 1996. Predictive models of the effect of temperature, pH and acetic and lactic acids on the growth of Listeria monocytogenes. International Journal of Food Microbiology 32:(1-2) 73–90.

[Globalgap]. 2007. General Regulations Integrated Farm Assurance, Version 3.0, GLOBALG.A.P., Cologne, Germany.

Hahn, G. 1996. Pathogenic bacteria in raw milk-situation and significance. Proceedings of Symposium on Bacteriological Quality of Raw Milk—Wolfpassing Austria, March 13–15 1996, IDF Special Issue 9601: 67–83.

Hancock, D., T. Besser, J. Lejeune, M. Davis and D. Rice. 2001. The control of VTEC in the animal reservoir. International Journal of Food Microbiology 66: 71–78.

Hathaway, S.C. 1997. Development of food safety risk assessment guidelines for foods of animal origin in international trade. Journal of Food Protection 60(11): 1432–1438.

Hauser, R., E. Breidenbach and K.D.C. Stärk. 2007. Swiss federal veterinary office risk assessments: advantages and limitations of the qualitative method. pp. 519–526 *In*: J-.L. Auget, N. Balakrishman, M. Mesbah and G. Molenbergh [eds.]. Advances in Statistical Methods for the Health Sciences. Birkhäuser, Boston.

Heeschen, W.H. and A.H. Bluthgen. 2004. Carry-over of environmental contaminants into milk and food hygiene assessment/management. Bulletin of the IDF 386: 28–39.

Heuvelink, A.E., C. van Heerwaarden, A. Zwartkruis-Nahuis, J.H.C. Tilburg, M.H. Bos, F.G.C. Heilmann, A. Hofhuis, T. Hoekstra and E. de Boer. 2009. Two outbreaks of campylobacteriosis associated with the consumption of raw cows' milk. International Journal of Food Microbiology 134(1-2): 70–74.

Hignette, G., P. Buche, O. Couvert, J. Dibie-Barthélemy, D. Doussot, O. Haemmerlé, E. Mettler and L. Soler. 2008. Semantic annotation of Web data applied to risk in food. International Journal of Food Microbiology 128: 174–180.

Honish, L., G. Predy, N. Hislop, L. Chui, K. Kowalewska-Grochowska, L. Trottier, C. Kreplin and I. Zazulak. 2005. An outbreak of *E. coli* O157:H7 hemorrhagic colitis associated with unpasteurized gouda cheese. Canadian Journal of Public Health 96: 182–184.

Hussein, H.S. and T. Sakuma. 2005. Shiga toxin-producing Escherichia coli: pre- and postharvest control measures to ensure safety of dairy cattle products. Journal of Food Protection 68(1): 199–207.

International Life Science Institute. 1996. A conceptual framework to assess the risks of human disease following exposure to pathogens. International Life Science Institute, North America—Risk Science Institute Pathogen Risk Assessment Working Group, Risk Analysis 16: 841–848.

International Life Science Institute. 2005. Achieving continuous improvement in reductions in foodborne listeriosis—a risk-based approach. ILSI Research Foundation Risk Science Institute Journal of Food Protection 68(9): 1932–94.

International Life Science Institute. 2012. Tools for Microbiological Risk Assessment. ILSI Europe Report Series, ILSI Europe, Brussels, Belgium.

Jayarao, B.M. and D.R. Henning. 2001. Prevalence of Foodborne Pathogens in Bulk Tank Milk. Journal of Dairy Science 84: 2157–2162.

Jayarao, B.M., S.C. Donaldson, B.A. Straley, A.A. Sawant, N.V. Hegde and J.L. Brown. 2006. A Survey of Foodborne Pathogens in Bulk Tank Milk and Raw Milk Consumption Among Farm Families in Pennsylvania. Journal of Dairy Science 89: 2451–2458.

Jaykus, L.A. 1996. The application of quantitative risk assessment to microbial food safety risks. Critical Review in Microbiology 22: 279–293.

Joint Institute for Food Safety and Applied Nutrition. 2010. FoodRisk.org—The online Resource for Food Safety Risk Analysis. Joint Institute for Food Safety and Applied Nutrition [Internet document] URL http://www.foodrisk.org/commodity/animal/dairy/index.cfm. Accessed 29/01/2010.

Julien, M.-C., P. Dion, C. Lafreniere, H. Antoun and P. Drouin. 2008. Sources of clostridia in raw milk on farms. Applied and Environmental Microbiology 74(20): 6348–6357.

Junedahl, P., E. Øgaard, H. Alnås and U. Nilsson. 2008. Arla Foods sees ISO 22000 becoming international benchmark for food safety. ISO Management Systems May–June, 23–28.

Karns, J.S., J.S. Van Kessel, B.J. McClusky and M.L. Perdue. 2007. Incidence of Escherichia coli O157:H7 and E. coli Virulence Factors in US Bulk Tank Milk as Determined by Polymerase Chain Reaction. Journal of Dairy Science 90: 3212–3219.

Katz, S.E. and M.S. Brady. 2000. Antibiotic residues in food and their significance. Food Biotechnology 14: 147–171.

Kindred, T.P. 1996. Risk assessment and its role in the safety of foods of animal origin. Journal of the American Veterinary Medical Association 209: 2055–2056.

Kousta, M., M. Mataragas, P. Skandamis and E.H. Drosinos. 2010. Prevalence and sources of cheese contamination with pathogens at farm and processing levels. Food Control 21: 805–815.

Lammerding, A.M. 1997. An overview of microbial food safety risk assessment. Journal of Food Protection 60: 1420–1425.

Lejeune, J.T. and P.J. Rajala-Schultz. 2009. Food safety: unpasteurized milk: a continued public health threat. Clinical Infectious Disease 48(1): 93–100.

Lind, L., J. Reeser, K. Stayman, M. Deasy, M. Moll, A. Weltman, V. Urdaneta and M.D. Ostroff. 2008. *Salmonella Typhimurium* infection associated with raw milk and cheese Consumption—Pennsylvania, 2007. Journal of American Medical Associaton 299(4): 402–404.

Maunsell, B. and D.J. Bolton. 2004. Guidelines for Food Safety Management on Farms, The Food Safety Department. Teagasc—The National Food Centre, Dublin.

Mauropoulos, A.A. and I.S. Arvanitoyannis. 1999. Implementation of hazard analysis critical control point to Feta and Manouri cheese production lines. Food Control 10: 213–219.

McClure, P.J., A.L. Beaumont, J.P. Sutherland and T.A. Roberts. 1997. Predictive modelling of growth of Listeria monocytogenes The effects on growth of NaCl, pH, storage temperature and $NaNO_2$. International Journal of Food Microbiology 34(3): 221–232.

McMeekin, T.A. and T. Ross. 2002. Predictive microbiology: providing a knowledge-based framework for change management. International Journal of Food Microbiology 78: 133–153.

McMeekin, T.A., J. Baranyi, J. Bowman, P. Dalgaard, M. Kirk, T. Ross, S. Schmid and M.H. Zwietering. 2006. Information systems in food safety management. International Journal of Food Microbiology 112: 181–194.

McMeekin, T.A., J. Bowman, O. McQuestin, L. Mellefont, T. Ross and M. Tamplin. 2008. The future of predictive microbiology: Strategic research, innovative applications and great expectations. International Journal of Food Microbiology 128: 2–9.

Membré, J.-M. and R.J.W. Lambent. 2008. Application of predictive modelling techniques in industry: From food design up to risk assessment. International Journal of Food Microbiology 128: 10–15.

Mortimore, S. 2001. How to make HACCP really work in practice. Food Control 12: 209–215.

Mortimore, S. and C. Wallace. 1998. HACCP—A practical approach. 2nd Edition. Chapman and Hall, London.

Murinda, S.E., L.T. Nguyen, S.J. Ivey, B.E. Gillespie, R.A. Almeida, F.A. Draughon and S.P. Oliver. 2002. Prevalence and molecular characterization of Escherichia coli O157:H7 in bulk tank milk and fecal samples from cull cows: a 12-month survey of dairy farms in East Tennessee. Journal of Food Protection 65: 752–759.

Murphy, P.M., M.C. Rea and D. Harrington. 1996. Development of a predictive model for growth of Listeria monocytogenes in a skim milk medium and validation studies in a range of dairy products. Journal of Applied Bacteriology 80(5): 557–564.

National Advisory Committee on Microbiological Criteria for Foods. 1997. Hazard Analysis and Critical Control Point Principles and Application Guidelines. National Advisory Committee on Microbiological Criteria for Foods. 14 August 1997. URL http://haccpalliance.org/alliance/microhaccp.pdf.

[National Institute of Aquatic Resources]. 2009. Seafood Spoilage and Safety Predictor (SSSP) software v. 3.1 URL http://sssp.dtuaqua.dk.

Noordhuizen, J.P.T.M. 2003. HACCP, total quality management and dairy herd health. pp. 1281–1289. *In*: H. Roginski, J.W. Fuquay and P.F. Fox [eds.]. Encyclopedia of Dairy Sciences, Vol. 3. Academic Press, London.

Noordhuizen, J.P.T.M. and J.H.M. Metz. 2005. Quality control on dairy farms with emphasis on public health, food safety, animal health and welfare. Livestock Production Science 94: 51–59.

Noordhuizen, J.P.T.M. and R. Jorritsma. 2006. The role of animal hygiene and animal health in dairy operations URL http://www.isah-soc.org/documents/2005/keynotespeakers/mon/nordhuisen%20.doc.

Notermans, S., A.W. Barendsz and F. Rombouts. 2002. The evolution of microbiological risk assessment in food production. pp. 5–43. *In*: M. Brown and M. Stringer [eds.]. Microbiological risk assessment in food processing. Woodhead Publishing Ltd., Cambridge, UK.

Notermans, S., G. Gallhoff, M.H. Zwietering and G.C. Mead. 1995. The HACCP concept: specification of criteria using quantitative risk assessment. Food Microbiology 12: 81–90.

Oliver, S.P., B.M. Jayarao and R.A. Almeida. 2005. Foodborne pathogens in milk and the dairy farm environment: food safety and public health implications. Foodborne Pathogen Disease 2(2): 115–129.

Oliver, S.P., K.J. Boor, S.C. Murphy and S.E. Murinda. 2009. Food safety hazards associated with consumption of raw milk. Foodborne Pathogen Disease 6(7): 793–806.

[Oracle]. 2012. The Crystal Ball Oracle Risk assessment software. http://www.oracle.com/us/products/applications/crystalball/overview/index.html.

[Palisade]. 2012. The @RISK software maker. http://www.palisade.com/risk/?gclid=CPmg gpv877UCFUbMtAodUnIAYg.

Papademas, P. and T. Bintsis. 2002. Microbiology of ice cream and related products. *In*: Dairy Microbiology Handbook—The Microbiology of Milk and Milk Products, pp. 213–260. Robinson R.K. (ed.). John Wiley and Sons, Inc., New York.

Papademas, P. and T. Bintsis. 2010. Food safety management systems (FSMS) in the dairy industry: A review. International Journal of Dairy Technology 63: 1–15.

Peeler, J.T. and V.K. Bunning. 1994. Hazard assessment of *Listeria monocytogenes* in the processing of bovine milk. Journal of Food Protection 57: 689–697.

Peterson, M.C. 2003. *Campylobacter jejuni* enteritis associated with consumption of raw milk. Journal of Environmental Health 65: 20–21.

Presidential Congressional Commission on Risk Assessment and Risk Management. 1997. Framework for environmental health risk management. Washington DC, USA: The Presidential/Congressional Commission on Risk Assessment and Risk Management.

Reybroeck, W. 1996. Modern methods for bacteriological quality control of raw milk. Proceedings of Symposium on Bacteriological Quality of Raw Milk, Wolfpassing Austria, March 13–15, 1996, IDF Special Issue 9601: 131–140.

Ross, T. 1996. Indices for performance evaluation of predictive models in food microbiology. Journal of Applied Bacteriology 81(5) 501–508.

Ross, T. and J. Sumner. 2002. A simple, spreadsheet-based, food safety risk assessment tool. International Journal of Food Microbiology 77(1-2): 39–53.

Ryan, M.J., P.G. Wall, G.K. Adak, H.S. Evans and J.M. Cowden. 1997. Outbreaks of infectious intestinal disease in residential institutions in England and Wales 1992–1994. Journal of Infection 34: 49–54.

[Safe Foods]. 2010. Promoting Food Safety through a New Integrated Risk Analysis Approach for Foods URL http://www.safefoods.nl/ default.aspx.

Sanaa, M., B. Pourel, J.L. Menard and F. Serieys. 1993. Risk factors associated with contamination of raw milk by *Listeria monocytogenes* in dairy farms. Journal of Dairy Science 76: 2891–2898.

Sanaa, M., L. Coroller and O. Cerf. 2004. Risk assessment of listeriosis linked to the consumption of two soft cheeses made from raw milk: Camembert of Normandy and Brie of Meaux. Risk Analysis 24(2): 389–399.

Sandrou, D.K. and I.S. Arvanitoyannis. 2000a. Implementation of Hazard Analysis Critical Control Point (HACCP) system to the dairy industry: current status and perspectives. Food Reviews International 16(1): 77–111.

Sandrou, D.K. and I.S. Arvanitoyannis. 2000b. Application of Hazard Analysis Critical Control Point (HACCP) system to the cheese-making industry: a review. Food Reviews International 16(3): 327–368.

Sansawat, S. and V. Muliyil. 2009. Understanding the FSSC 22000 Food Standard. URL http://www.sgs.com/foodsafety.

Sawant, A.A., N.V. Hegde, B.A. Straley, S.C. Donaldson, B.C. Love, S.J. Knabel and B. Jayarao. 2007. Antimicrobial-Resistant Enteric Bacteria from Dairy Cattle. Applied and Environmental Microbiology 73(1): 156–163.

Schlundt, J. 2000. Comparison of microbiological risk assessment studies published. International Journal of Food Microbiology 58: 197–202.

Srinivasan, V., H.M. Nam, L.T. Nguyen, B. Tamilselvam, S.E. Murinda and S.P. Oliver. 2005. Prevalence of antimicrobial resistance genes in *Listeria monocytogenes* isolated from dairy farms. Foodborne Pathogens Disease 2: 201–211.

Straley, B.A., S.C. Donaldson, N.V. Hedge, A.A. Sawant, V. Srinivasan, S.P. Oliver and B.M. Jayarao. 2006. Public health significance of antimicrobial-resistant Gram-negative bacteria in raw bulk tank milk. Foodborne Pathogens Disease 3: 222–233.

te Giffel, M.C. and M.H. Zwietering. 1999. Validation of predictive models describing the growth of Listeria monocytogenes. International Journal of Food Microbiology 46(2): 135–149.

The International Commission on Microbiological Specifications for Foods. 2006. A simplified guide to understanding and using Food Safety Objectives and Performance Objectives.

The International Dairy Federation/Food and Agriculture Organization of the United Nations. 2004. Guide to good dairy farming practice. A joint publication of the International Dairy Federation and the Food and Agriculture Organization of the United Nations. FAO, Rome. Troutt, H.F., J. Gillespie and B.I. Osburn 1995. Implementation of HACCP program on farms and ranches. pp 36–57. *In*: A.M. Pearson and T.R. Dutson [eds.]. HACCP in Meat, Poultry and Fish Processing—Advances in Meat Research Series, Vol. 10, Chapman & Hall, New York.

United States Department of Agriculture—Agricultural Research Service. 2006. Pathogens Modeling Program, United States Department of Agriculture—Agricultural Research Service, Microbial Food Safety Research Unit URL http://ars.usda.gov/services/docs.htm?Docid =6786.

Unnevehr, L.J. and H.H. Jensen. 1999. The economic implications of using HACCP as a food safety regulatory standard. Food Policy 4: 625–635.

Van Kessel, J.S., J.S. Karns, L. Gorski, B.J. McCluskey and M.L. Perdue. 2004. Prevalence of *Salmonellae, Listeria monocytogenes,* and Fecal Coliforms in Bulk Tank Milk on US Dairies. Journal of Dairy Science 87: 2822–2830.

Vissers, M.M.M. and F. Driehuis. 2009. On-farm Hygienic milk production. pp 1–22. *In*: A.Y. Tamime [ed.]. Milk Processing and Quality Management. Blackwell Publishing, Ltd.

Vissers, M.M.M., F. Driehuis, M.C. Te Giffel, P. De Jong and J.M.G. Lankveld. 2006. Improving farm management by modeling the contamination of farm tank milk with butyric acid bacteria. Journal of Dairy Science 89: 850–858.

Vissers, M.M.M., F. Driehuis, M.C. Te Giffel, P. De Jong and J.M.G. Lankveld. 2007. Concentrations of Butyric Acid Bacteria Spores in Silage and Relationships with Aerobic Deterioration. Journal of Dairy Science 90: 928–936.

Vissers, M.M.M., F. Driehuis, M.C. Te Giffel, P. De Jong and J.M.G. Lankveld. 2007a. Short Communication: quantification of the transmission of microorganisms to milk via dirt attached to the exterior of teats. Journal of Dairy Science 90: 3579–3582.

Voysey, P.A. and M. Brown. 2000. Microbiological risk assessment: a new approach to food safety control. International Journal of Food Microbiology 58(3): 173–179.

Wiedmann, M. 2006. ADSA Foundation Scholar Award—An Integrated Science Based Approach to Dairy Food Safety: *Listeria monocytogenes* as a Model System. Journal of Dairy Science 86: 1865–1875.

Xanthiakos, K., D. Simos, A.S. Angelidis, G.J.-E. Nychas and K. Koutsoumanis. 2006. Dynamic modeling of *Listeria monocytogenes* growth in pasteurized milk. Journal of Applied Microbiology 100: 1289–1298.

Young, I., S. Hendrick, S. Parker, A. Rajić, J.T. McClure, J. Sanchez and S.A. Mc Ewen. 2010. Attitudes towards the Canadian quality milk program and use of good production practices among Canadian dairy producers. Preventive Veterinary Medicine 94: 43–53.

# Index

**R**



**S**



**V**

# Color Plate Section

## Chapter 3

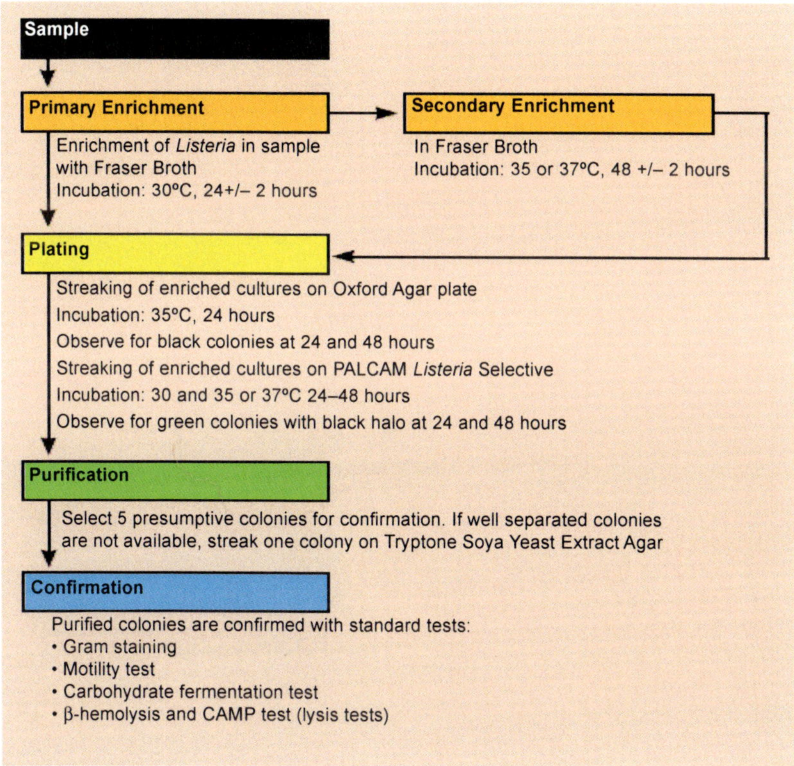


**Figure 2.** ISO 11290 for detection of *L. monocytogenes* in food samples (After: Scharlau 2007).

| Pre-enrichment medium | Enrichment media | Isolation media | Confirmation |
|---|---|---|---|
| **Buffered Peptone Water**<br>*Incubation:* 37°C for 18 ± 2 h | Rappaport-Vassiliadis Soya Broth **(RVS)**<br>*Incubation:* 41,5°C for 24 ± 3 h | Isolation in Xylose Lysine Deoxycholate Agar **(XLD)** and a second agar choosed by the laboratory<br>*Incubation:* 37°C for 24 ± 3 h | <u>Biochemical Confirmation</u><br>Triple Sugar iron Agar ( TSI)<br>Urea Agar<br>Lysine Descarboxilation<br>ß-galactosidase<br>Voges-Proskauer<br>Indole<br>*Incubation:* 37°C for 24 ± 3 h<br><u>Serological Confirmation</u><br>Antigens Confirmation O, H and Vi |
| | Tetrathionate Muller-Kauffmann Broth with Novobiocine **(MKTTn)**<br>*Incubation:* 37°C for 24 ± 3 h | Isolation in Xylose Lysine Deoxycholate Agar **(XLD)** and a second agar choosed by the laboratory<br>*Incubation:* 37°C for 24 ± 3 h | |

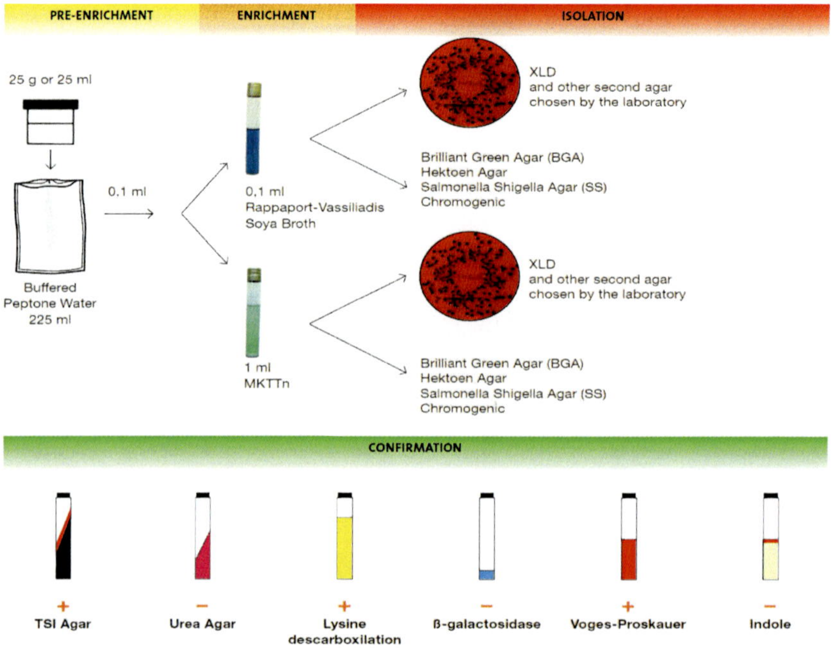


**Figure 3.** ISO 6579 Detection of Salmonella in Food (After: Scharlau 2007).

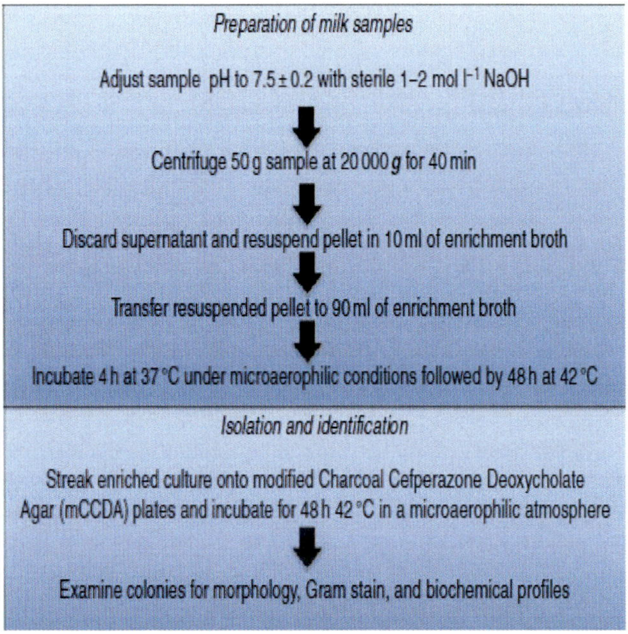


**Figure 4.** Procedure for Isolation and Identification of *Campylobacter* spp. from Milk (After: United States Food and Drug Administration).

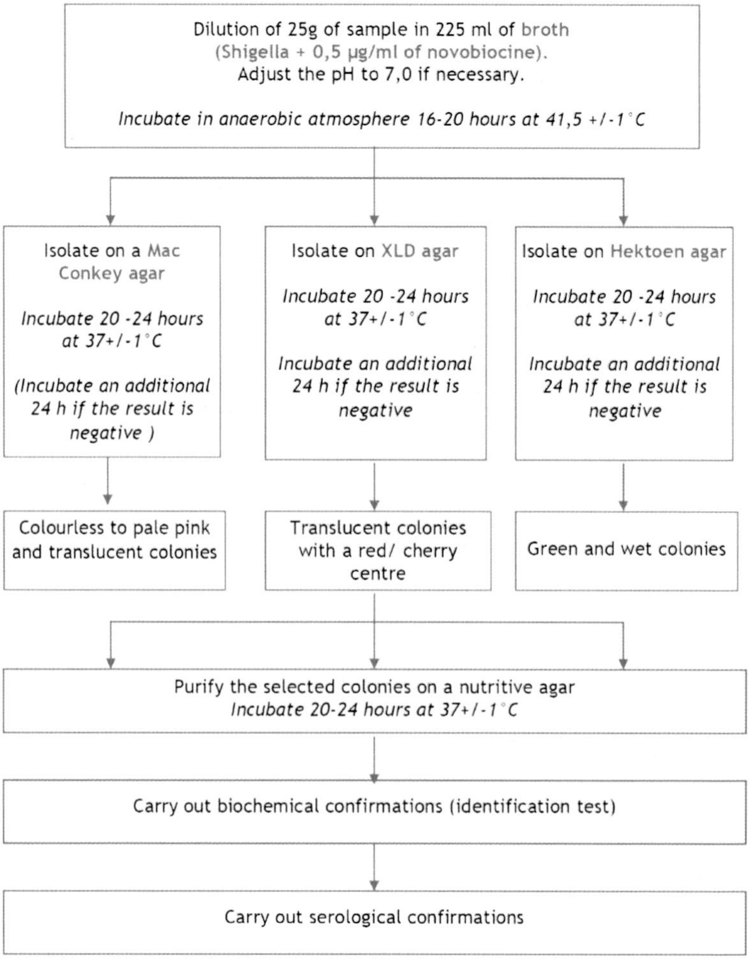


**Figure 5.** Detection of *Shigella* spp. in Food Samples (After: AES Chemunex 2008).

# Chapter 5

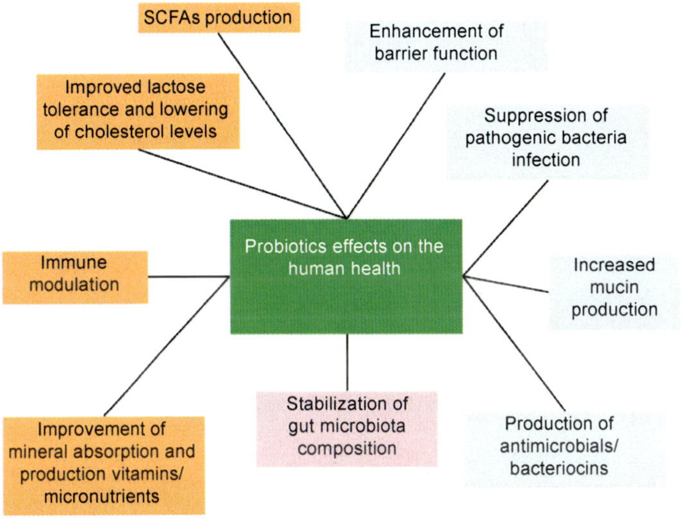


**Figure 2.** Effects of probiotics on the human health. Microbe-microbe interactions are represented in light blue, microbe-host interactions are represented in orange; interactions leading to effects directed either to host and microbes are represented in pink.